ANIMAL STRUCTURE AND FUNCTION

ANIMAL STRUCTURE AND FUNCTION

CECIE STARR

Belmont, California

RALPH TAGGART

Michigan State University

BIOLOGY:

THE UNITY AND DIVERSITY OF LIFE

SIXTH EDITION

Wadsworth Publishing Company

Belmont, California

A Division of Wadsworth, Inc.

BIOLOGY PUBLISHER: *Jack C. Carey*

EDITORIAL ASSISTANT: *Kathryn Shea*

ART DIRECTOR AND DESIGNER: *Stephen Rapley*

PRODUCTION EDITOR: *Mary Forkner Douglas*

COPY EDITOR: *Carolyn McGovern*

PRODUCTION COORDINATOR: *Jerry Holloway*

MANUFACTURING: *Randy Hurst*

MARKETING: *Todd Armstrong, Karen Culver*

EDITORIAL PRODUCTION: *Scott Alkire, John Douglas, Kathy Hart, Gloria Joyce, Ed Serdziak, Karen Stough*

PERMISSIONS: *Marion Hansen*

PHOTO RESEARCH: *Marion Hansen, Stuart Kenter*

ARTISTS: *Susan Breitbard, Lewis Calver, Joan Carol, Raychel Ciemma, Robert Demerest, Ron Erwin, Enid Hatton, Darwin Hennings, Vally Hennings, Joel Ito, Robin Jensen, Keith Kasnot, Julie Leech, Laszlo Mezoly, Leonard Morgan, Palay/Beaubois (Phoebe Gloeckner, Lynne Larson, Betsy Palay), Victor Royer, Jeanne Schreiber, Kevin Somerville, John Waller, Judy Waller, Jennifer Wardrip*

DESIGN CONSULTANT: *Gary Head*

COVER DESIGN: *Stephen Rapley*

COVER PHOTOGRAPH: *© Thomas D. Mangelsen*

COMPOSITOR: *G&S Typesetters, Inc.: Bill M. Grosskopf, Merry Finley, Pat Molenaar, Beverly Zigal, Maurine Zook*

COLOR SEPARATOR: *H&S Graphics, Inc.: Tom Andersen, Nancy Dean, Roger Tillander, Marty O'Dean, Dennis Schnell*

PRINTING: *R. R. Donnelley & Sons Company/Willard*

1 2 3 4 5 6 7 8 9 10   96 95 94 93

ISBN 0-534-30478-8

CONTENTS IN BRIEF

DETAILED TABLE OF CONTENTS

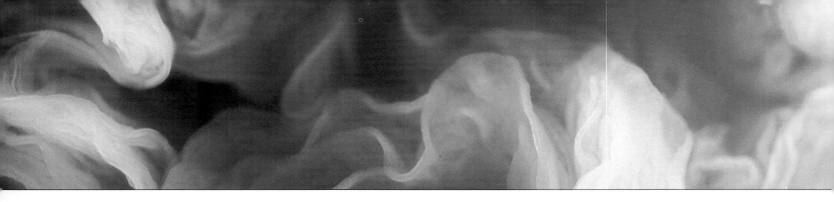

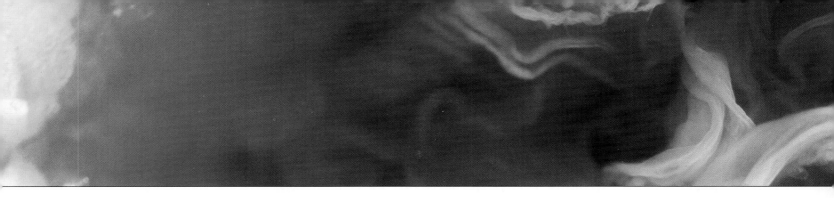

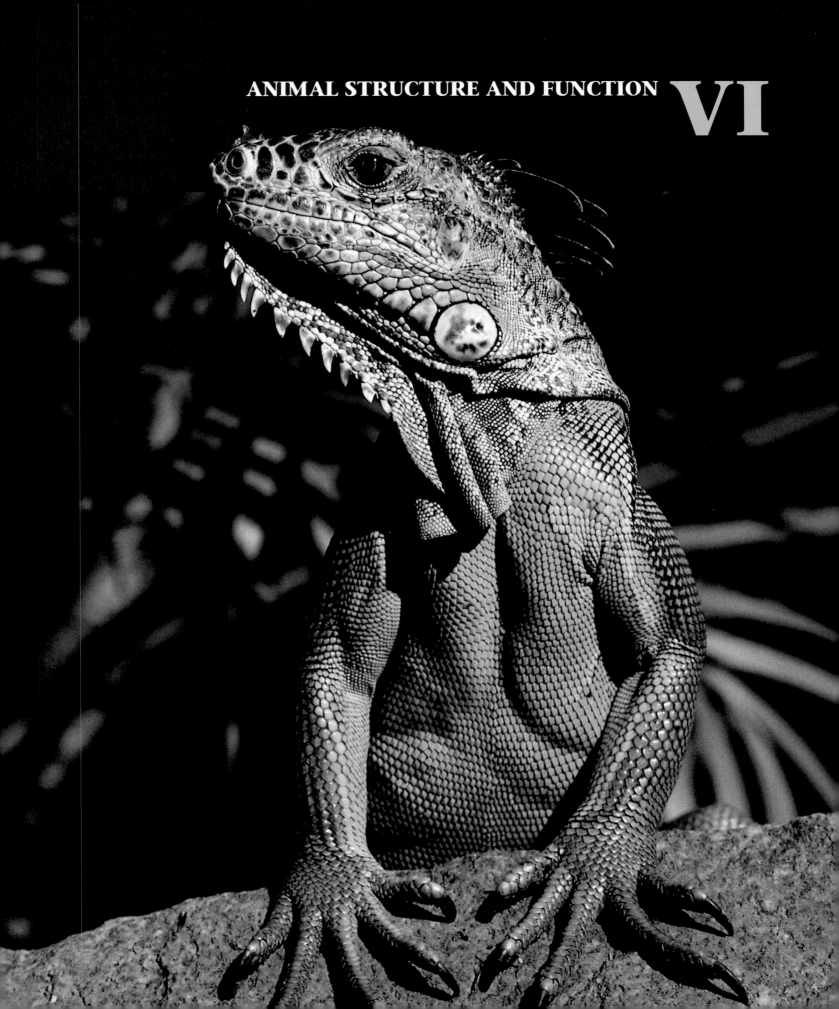

ANIMAL STRUCTURE AND FUNCTION VI

Meerkats, Humans, It's All the Same

After a bitingly cold night in the Kalahari Desert of Africa, animals small enough to fit in the pocket of an overcoat emerge from their underground burrows. These animals, called meerkats, stand on their hind legs and face eastward, exposing a large surface area of their chilled bodies to the warm rays of the morning sun (Figure 31.1).

The meerkats don't know it, but they are working to ensure good operating conditions for their enzymes. The animal body runs on enzymes, and if its temperature exceeds or falls below a tolerable range, the rate of enzyme activity drops sharply and metabolism suffers. Like many other animals, meerkats rely on behavioral adjustments to help maintain "inside" temperatures even though outside temperatures change.

After gathering warmth from the sun, the meerkats venture into the open and begin foraging. The farther they venture from their burrow, the more vulnerable they become to jackals and other predators—but they don't have much choice. They are hungry. The day before, nutritious little insects and the occasional lizard provided them with glucose and other required molecules. They used the glucose as a quick energy fix

and to keep their brain supplied with the stuff; glucose is *the* major nutrient that brain cells can actually use.

The meerkats that were lucky enough to take in many more food molecules than their cells could use stored away the excess, mostly by converting them to fats and to "animal starch" (glycogen). After returning to the safety of their burrow and hunkering in for the night, the animals relied on built-in controls in their body to trigger a shift in the type of molecules used to support their cells. Fats that had been stored away earlier in adipose tissue were broken down, transported by the bloodstream to the liver, and there converted to glucose. Glycogen that had been stored in liver cells was rapidly broken down to glucose, which was released into the bloodstream.

Each of those small furry bodies clearly depended on more than a digestive system. A circulatory system picked up nutrients that had been absorbed from the gut and transported them to cells throughout the body. A respiratory system helped cells use the nutrients by supplying them with oxygen (for aerobic respiration) and relieving them of carbon dioxide wastes. It did not matter that one meerkat had feasted on a tasty assort-

ment of insects, and another on a scrawny lizard. Even with variations in diet, their urinary system helped maintain the volume and composition of the vital fluids bathing their cells. As the meerkats slept, as they awakened and faced another day of hunger, thirst, and possibly heart-thumping flights from predators, they were never aware that their nervous system and endocrine system were working together in response to the challenges of the new day. The functioning of meerkats, humans, and all other complex animals depends on how well those two systems maintain operating conditions for all living cells in the body.

With no more than a cursory look at the meerkats, we have started thinking about the central topics of this unit. These topics are the structure of the animal body (its *anatomy*) and the mechanisms by which the body functions in the environment (its *physiology*). This chapter provides us with an overview of the kinds of tissues and organ systems we will be considering. It provides us also with an overview of the kinds of *homeostatic mechanisms* by which cells, tissues, organs, and organ systems work together in ways that maintain a stable environment *inside* the body.

Figure 31.1 In the vast Kalahari Desert of Africa, meerkats line up and face the warming rays of the morning sun, just as they do every morning.

This simple behavior helps meerkats maintain their internal body temperature even when the outside temperature changes. How animals function in their environment is the subject of this unit.

KEY CONCEPTS

1. The cells of most animals interact at three levels of organization—in *tissues*, many of which are combined in *organs*, which are components of *organ systems*.

2. Most animals are constructed of only four types of tissues: epithelial, connective, nervous, and muscle tissues.

3. Each cell engages in basic metabolic activities that assure its own survival. At the same time, cells of a tissue or organ perform activities that contribute to the survival of the animal as a whole.

4. The combined contributions of cells, tissues, organs, and organ systems help maintain a stable "internal environment" that is required for individual cell survival. This concept is central to understanding the functions of any organ system, regardless of its complexity.

ANIMAL STRUCTURE AND FUNCTION: AN OVERVIEW

Regardless of whether you are talking about a flatworm or salmon, a meerkat or human being, each animal is structurally and physiologically adapted to perform the following tasks:

1. Maintain internal "operating conditions" within some tolerable range even though external conditions change.

2. Locate and take in nutrients and other raw materials, distribute them through the body, and dispose of wastes.

3. Protect itself against injury or attack from viruses, bacteria, and other foreign agents.

4. Reproduce, and often help nourish and protect the new individuals during their early development.

Even the most complex animal is constructed of only four basic types of tissues, called epithelial, connective, muscle, and nervous tissues. A **tissue** is a group of cells and intercellular substances that function together in

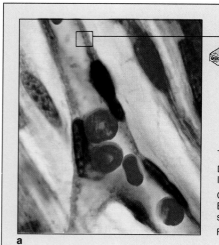
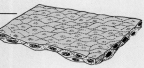

TYPE: Simple squamous

DESCRIPTION: Single layer flattened cells

COMMON LOCATION: Blood vessel walls, air sacs of lungs

FUNCTION: Diffusion

a

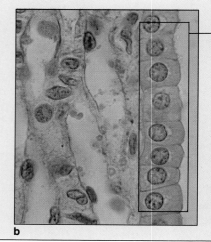
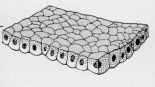

TYPE: Simple cuboidal

DESCRIPTION: Single layer cubelike cells; may have microvilli at its free surface

COMMON LOCATIONS: Part of gut lining, part of respiratory tract lining

FUNCTION: Secretion, absorption

b

one or more of the specialized tasks listed above. Complex animals consist of millions, even trillions of cells organized into tissues. Tissues split up the work, so to speak, in ways that contribute to the survival of the animal as a whole. This is sometimes called "a division of labor."

Different tissues become organized in specific proportions and patterns to form **organs**, such as a stomach. In **organ systems**, two or more organs interact chemically, physically, or both in performing a common task. For example, different organs of your digestive system ingest and prepare food for absorption by cells, then eliminate food residues.

Not all of the 2 million or so known species of animals have elaborate organ systems. Sponges and *Trichoplax adhaerens*, the organless, pancake-shaped marine animal shown on page 369, do not even have organs. The giant squid, one of the most complex invertebrates, has organ systems that are as sophisticated as yours. We mention these animals only to emphasize that there is no such thing as a "typical" animal. We humans tend to be interested in vertebrates (ourselves especially), and so they will be our focus in this unit. But comparative examples also will be drawn from the invertebrates to keep the structural and functional diversity of the animal kingdom in perspective.

ANIMAL TISSUES

Tissue Formation

Animal tissues are formed by cell divisions and cell differentiation that begin when an animal embryo embarks on its course of development. Chapter 42 sketches out the developmental mechanisms involved. Here it is enough to say that, at some point in the life cycle, meiotic cell division occurs in *germ cells*, which are immature reproductive cells that develop into gametes. In male animals, the gametes are sperm; in females, they are ova, or eggs. All other cells in the body are said to be *somatic* (from the Greek *soma*, meaning body). Fusion of a sperm and egg results in the formation of a zygote, which undergoes mitotic cell divisions that produce the early embryo.

In the case of vertebrates, cells in the early embryo soon become arranged into three "primary" tissues— ectoderm, mesoderm, and endoderm. These are the embryonic forerunners of all the specialized tissues of the body. *Ectoderm* gives rise to the outer layer of skin and to tissues of the nervous system. *Mesoderm* gives rise to muscle; the organs of circulation, reproduction, and excretion; most of the internal skeleton; and connective tissue layers of the gut and body covering. *Endoderm* gives rise to the lining of the gut and the major organs derived from it.

With this overview in mind, let's now consider some of the features that allow us to identify the four types of specialized tissues present in the adult body.

Epithelial Tissue

Figure 31.2 shows a few examples of epithelial tissues. An epithelial tissue also is called an **epithelium** (plural, epithelia), and different types serve different functions. The epidermis of skin, for example, is a protective covering for the body's exterior surface. Epithelia that line parts of the gut and other internal cavities function in secreting or absorbing substances.

General Features. The cells of epithelium adhere to one another closely, with little space or extracellular material between them, and they are organized as one or more layers. Epithelium always has one free surface,

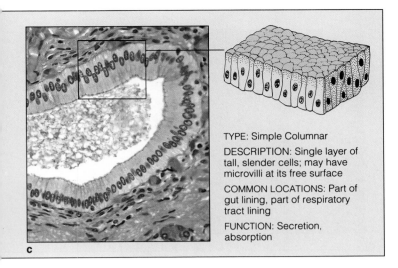

TYPE: Simple Columnar

DESCRIPTION: Single layer of tall, slender cells; may have microvilli at its free surface

COMMON LOCATIONS: Part of gut lining, part of respiratory tract lining

FUNCTION: Secretion, absorption

c

Figure 31.2 Examples of simple epithelium, showing the three basic cell shapes in this type of tissue.

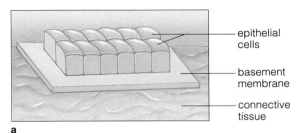

epithelial cells

basement membrane

connective tissue

a

which means that no other cells adhere to it. In many cases, cells at the free surface are crowned with cilia or microvilli (singular, microvillus), which are fingerlike protrusions of cytoplasm. These surface modifications have special roles, as you will see in later chapters. The opposite surface of epithelium adheres to a *basement membrane*. This is a noncellular layer, rich with proteins and polysaccharides, that lies between the epithelium and an underlying connective tissue (Figure 31.3a).

Simple epithelium consists of a single layer of cells, which typically have roles in the diffusion, secretion, absorption, or filtering of substances across the layer. For example, oxygen and carbon dioxide diffuse readily across the simple epithelium making up the wall of fine blood vessels, as shown in Figure 31.2a. *Stratified epithelium* consists of two or more layers of cells, as shown in Figure 31.3b, and it typically functions in protection.

Glandular Epithelium. It takes digestive enzymes, mucus, and many other glandular secretions to keep your body functioning properly. *Glands* are single cells or multicelled secretory structures that are derived from and composed of epithelia. Mucus-secreting, goblet-shaped cells, for instance, are embedded in epithelia that line some of the tubes leading to your lungs. The entire epithelium of your stomach is a sheet of glandular cells that secrete mucus and enzymes.

We classify glands according to how their products are distributed. **Exocrine glands** secrete products onto a free epithelial surface, usually through ducts or tubes. Exocrine products include cell-produced substances such as mucus, saliva, earwax, oil, milk, and digestive enzymes. **Endocrine glands** are ductless. As described in Chapter 34, their products (hormones) are secreted directly into the fluid bathing cells and are picked up for distribution by the bloodstream.

free surface

b

Figure 31.3 Characteristics of epithelium. (**a**) All epithelia have a free surface, and a basement membrane is interposed between the opposite surface and an underlying connective tissue. (**b**) In stratified epithelium, which consists of two or more layers, cells at the free surface typically are flattened as shown here.

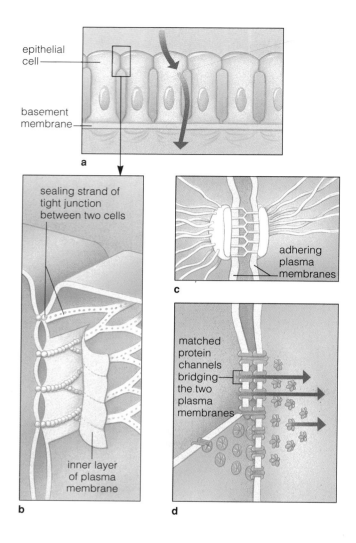

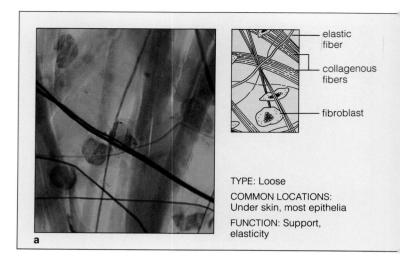

TYPE: Loose

COMMON LOCATIONS: Under skin, most epithelia

FUNCTION: Support, elasticity

Figure 31.4 Examples of cell-to-cell contacts in animal tissues.

(**a**) A complex of cell junctions serves as a barrier to the leakage of substances *between* the cells of some epithelia, such as the one lining the small intestine. Here, *tight junctions* ring each cell and seal it to its neighbors near their free surface. Protein strands are the sealing devices; they are embedded in the lipid bilayer of the plasma membranes (**b**). To move across the lining, any substance must be absorbed by the cells themselves.

(**c**) Below the tight junction are *desmosomes*, which are something like spot welds that mechanically link cells and permit them to function as a structural unit. Each has cytoskeletal filaments attached to dense, "cementing" material on the cytoplasmic side of the plasma membrane, and these extend into both of the adjoining cells. Desmosomes occur within all types of animal tissues. They are especially abundant in the surface layer of skin and other kinds of epithelia that are subjected to abrasion and other mechanical insults.

(**d**) *Gap junctions* also occur in epithelia. Here, clustered protein channels across the plasma membrane of one cell match up with clustered channels of an adjoining cell. Gap junctions provide a low-resistance path for the diffusion of ions and small molecules from cell to cell. They are abundant in certain tissues of the heart, liver, and other organs that depend on rapid coordination of chemical activities among their cells.

Cell-to-Cell Contacts in Epithelium. With few exceptions, epithelial cells adhere strongly to one another by means of specialized attachment sites. These sites of cell-to-cell contact are especially thick when substances must not leak from one compartment to another in the body.

If the potently acidic fluid in your stomach were to leak through its epithelial lining, for instance, it would start digesting your body's own proteins instead of the ones brought in with your meals. (Actually, this is an end result of a peptic ulcer, as described on page 646.) This epithelium and others have intricate cell-to-cell contacts that make up a leakproof barrier between cells near their free surface. As Figure 31.4 shows, epithelia also include junctions that serve as spot welds and as open channels between cells.

Connective Tissue

Connective tissues are diverse, and they serve very diverse functions. They range from the ones called connective tissue proper to the specialized types, which include cartilage, bone, adipose tissue, and blood (Table 31.1).

Spaces intervene between the cells of connective tissue. In this respect, connective tissue is notably different from epithelia. Except for blood, some of the cells produce fibers that serve as structural elements. The fibers contain collagen (which makes them strong) or elastin (which makes them elastic). The cells of connective tissue also secrete a *ground substance* which, together with the fibers, make up the extracellular matrix (page 70).

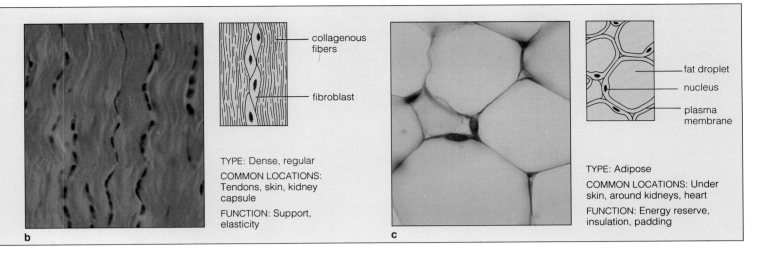

Figure 31.5 Three examples drawn from the diverse kinds of tissues collectively called connective tissues. (**a**) Loose connective tissue, (**b**) dense connective tissue, and (**c**) adipose tissue.

Connective Tissue Proper. The components of both types of connective tissue in this category (loose or dense) are irregularly arranged. **Loose connective tissue** supports epithelia and many organs, and it surrounds blood vessels and nerves. As the name implies, it contains loosely arranged collagenous and elastic fibers as well as fibroblasts, a type of cell that produces the fibers and ground substance (Figure 31.5a). It also contains macrophages and other cells that migrate through tissues or take up residence in them. These cells perform housekeeping tasks and help protect the body against disease. In fact, the ones in connective tissue underlying the skin serve as a first line of defense when bacteria and other agents enter the body through cuts, abrasions, and other wounds.

Compared to its loose counterpart, **dense, irregular connective tissue** has thicker fibers and more of them, but far fewer cells. Its fibers interweave with one another but not in a regular orientation. If this suggests to you that dense connective tissues occur in organs that are not subjected to continual stretching, you would be correct. For example, such tissues occur in the protective capsule around the testis, the primary reproductive organ in males.

Specialized Connective Tissue. Unlike the tissues just described, **dense, regular connective tissue** has its collagen fibers oriented in parallel (Figure 31.5b). With this organized orientation, the fibers strongly resist being pulled apart under tension when the tissue is stretched. Often, parallel bundles of fibers have rows of fibroblasts between them. This is the arrangement you see in ligaments (which attach bone to bone) and tendons (which attach muscle to bone).

Table 31.1 Types of Connective Tissue

Connective tissue proper:

Loose connective tissue
Dense, irregular connective tissue

Specialized connective tissue:

Dense, regular connective tissue (ligaments, tendons)
Cartilage
Bone
Adipose tissue
Blood

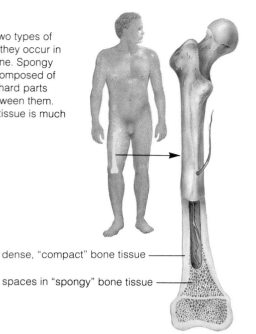

Figure 31.6 Two types of bone tissue, as they occur in a human leg bone. Spongy bone tissue is composed of tiny, needlelike hard parts with spaces between them. Compact bone tissue is much more dense.

dense, "compact" bone tissue ——

spaces in "spongy" bone tissue ——

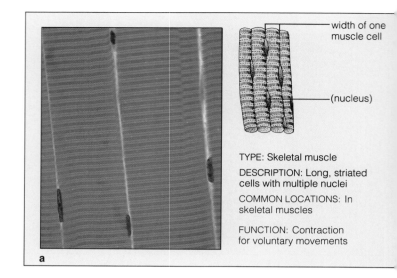

width of one muscle cell

(nucleus)

TYPE: Skeletal muscle
DESCRIPTION: Long, striated cells with multiple nuclei
COMMON LOCATIONS: In skeletal muscles

FUNCTION: Contraction for voluntary movements

a

Cartilage, another specialized connective tissue, cushions and helps maintain the shape of many body parts. It has a solid yet somewhat pliable matrix through which substances diffuse from nearby blood vessels. This pliability allows cartilage to resist compression and stay resilient, like a piece of solid rubber. As you will see, cartilage is an important tissue in growing bones. It also is present at the ends of many mature bones as well as parts of the nose, ear, and backbone.

Bone is unlike all other connective tissues in being mineralized and hardened; its collagen fibers and ground substance are loaded with calcium salts. Bone tissue is not completely solid, however. A variety of spaces occur within the ground substance, some of which harbor living bone cells (Figure 36.13). This specialized connective tissue becomes organized into bony structures that support and protect softer tissues and organs. Figure 31.6 shows such a bone in the human body. The body's arm and leg bones act with muscles; they form a leverlike system for movement. Some bones also function in red blood cell production.

The connective tissue called **adipose tissue** has large, densely clustered cells that are specialized for fat storage (Figure 31.5c). The animal body can store only so much carbohydrate and protein, and the excess is converted to storage fats that are tucked away in these cells. The tissue has a rich supply of blood, which serves as an immediately accessible "highway" for the movement of fats to and from individual adipose cells.

The specialized connective tissue called **blood** transports oxygen to cells and wastes away from them; it also transports hormones and enzymes. Some of its compo-

nents protect against blood loss (through clotting mechanisms), and others defend against disease-causing agents. Chapter 38 describes this tissue and its complex functions.

Muscle Tissue

All types of **muscle tissue** contain specialized cells that contract (shorten) in response to stimulation, then passively lengthen and so return to their resting state. Muscle tissue helps move the whole body as well as its individual parts. There are three categories, called skeletal, smooth, and cardiac muscle tissues.

Skeletal muscle tissue contains many long, cylindrical cells (Figure 31.7a). Typically, a number of skeletal muscle cells are bundled together, then several bundles are enclosed in a tough connective tissue sheath to form "a muscle," such as a biceps. Chapter 36 describes how this type of tissue functions.

Smooth muscle tissue consists of spindle-shaped cells, which are tapered at both ends (Figure 31.7b). Connective tissue holds the cells together. Smooth muscle tissue occurs in walls of blood vessels, the stomach, and other internal organs. In vertebrates, smooth muscle is said to be "involuntary," because the individual usually cannot directly control its contraction.

Cardiac muscle tissue is the contractile tissue of the heart (Figure 31.7c). The plasma membranes of adjacent cardiac muscle cells are fused together. Cell junctions at these fusion points allow the cells to contract as a unit. When one muscle cell receives a signal to contract, its neighbors are also stimulated into contracting.

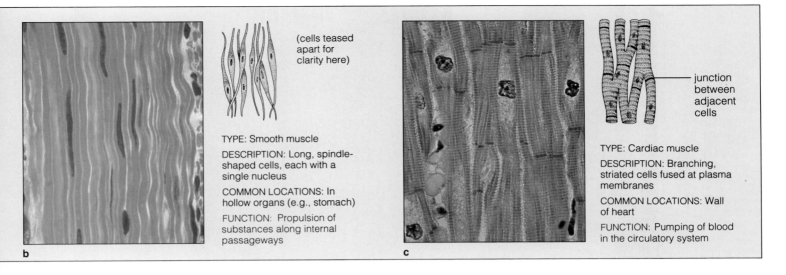

TYPE: Smooth muscle

DESCRIPTION: Long, spindle-shaped cells, each with a single nucleus

COMMON LOCATIONS: In hollow organs (e.g., stomach)

FUNCTION: Propulsion of substances along internal passageways

(cells teased apart for clarity here)

TYPE: Cardiac muscle

DESCRIPTION: Branching, striated cells fused at plasma membranes

COMMON LOCATIONS: Wall of heart

FUNCTION: Pumping of blood in the circulatory system

junction between adjacent cells

b
c

Figure 31.7 Examples of skeletal, smooth, and cardiac muscle tissues.

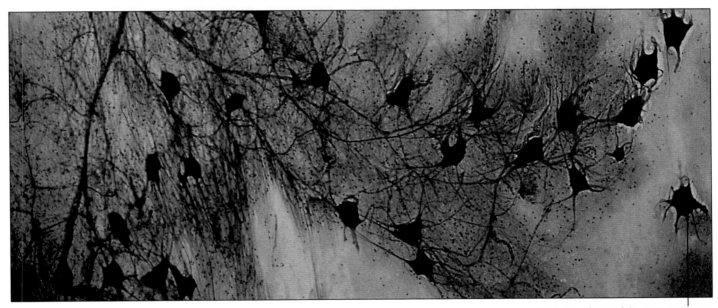

Cell body of one of the motor neurons in this nervous tissue sample

Nervous Tissue

In **nervous tissue**, cells called neurons are organized as lines of communication that extend throughout the body. Some types of neurons detect specific changes in environmental conditions. Others coordinate the body's immediate and long-term responses to change. Still others relay signals to muscles and glands that can carry out those responses. Examples of this last type are shown in Figure 31.8, and their functioning will be a key topic in Chapter 33.

Figure 31.8 A sampling of the millions of neurons that form communication lines within and between different regions of the human body. Shown here, motor neurons, which relay signals from the brain or spinal cord to muscles and glands. Collectively, these and other neurons sense environmental change, integrate a great number and variety of signals about those changes, and initiate appropriate responses.

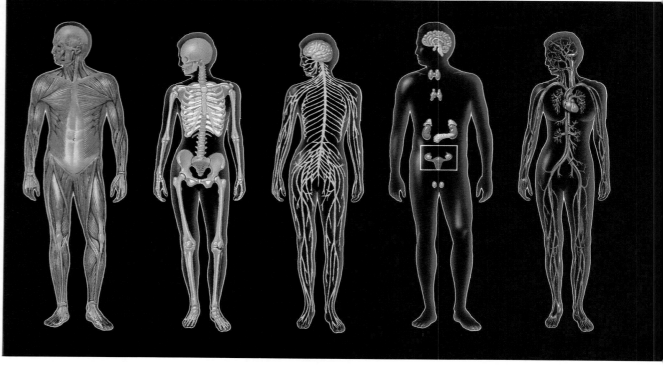

INTEGUMENTARY SYSTEM	MUSCULAR SYSTEM	SKELETAL SYSTEM	NERVOUS SYSTEM	ENDOCRINE SYSTEM	CIRCULATORY SYSTEM
Protection from injury and dehydration; body temperature control; excretion of some wastes; reception of external stimuli; defense against microbes.	Movement of internal body parts; movement of whole body; maintenance of posture; heat production.	Support, protection of body parts; sites for muscle attachment, blood cell production, and calcium and phosphate storage.	Detection of external and internal stimuli; control and coordination of responses to stimuli; integration of activities of all organ systems.	Hormonal control of body functioning; works with nervous system in integrative tasks.	Rapid internal transport of many materials to and from cells; helps stabilize internal temperature and pH.

Figure 31.9 Organ systems of the human body. All vertebrates have the same types of systems, serving similar functions.

MAJOR ORGAN SYSTEMS

During embryonic development, recall, epithelial tissues, connective tissues, nervous tissue, and muscle tissue begin their formation in the animal embryo. They eventually become organized into organs and then into organ systems that are much the same in all vertebrates. Figure 31.9 shows the general arrangement of these systems in humans. Each type of organ system will be described in subsequent chapters.

You might think we are stretching things a bit when we say that each one of those organ systems contributes to the survival of all living cells in the body. After all, what could the body's skeleton and musculature have to do with the life of a tiny cell? Yet interactions between

the skeletal and muscular systems provide a means of moving about—toward sources of nutrients and water, for example. Some of their components help circulate blood through the body, as when contractions of certain leg muscles help move blood in veins back to the heart. Through blood circulation, nutrients and other substances are transported to individual cells, and wastes are carried away from them.

Throughout this unit, we will be using some standard terms for describing the location of organs and organ systems in the vertebrate body. Take a moment to study Figure 31.10a, which shows the location of some major body cavities in which organs occur. Also study Figure 31.10b, which defines some anatomical terms that apply to most animals.

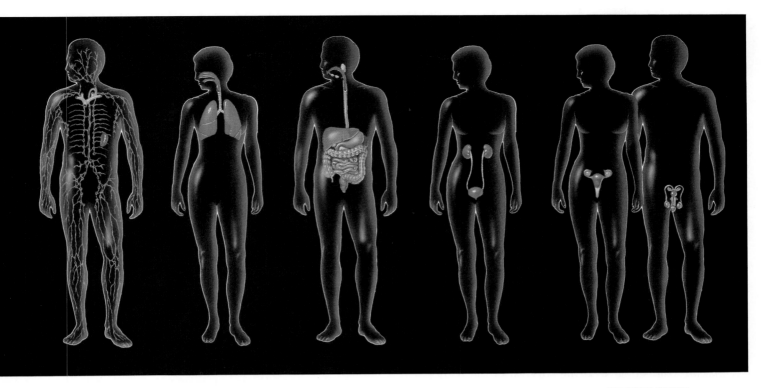

LYMPHATIC SYSTEM

Return of some tissue fluid to blood; roles in immunity (defense against specific invaders of the body).

RESPIRATORY SYSTEM

Provisioning of cells with oxygen; removal of carbon dioxide wastes produced by cells; pH regulation.

DIGESTIVE SYSTEM

Ingestion of food, water; preparation of food molecules for absorption; elimination of food residues from the body.

URINARY SYSTEM

Maintenance of the volume and composition of extracellular fluid. Excretion of blood-borne wastes.

REPRODUCTIVE SYSTEM

Male: production and transfer of sperm to the female. Female: production of eggs; provision of a protected, nutritive environment for developing embryo and fetus. Both systems have hormonal influences on other organ systems.

SUPERIOR
(of two body parts, the one closer to head)

distal (farthest from trunk or from point of origin of a body part)

frontal plane (aqua)

midsagittal plane (green)

proximal (closest to trunk or to point of origin of a body part)

ANTERIOR (at or near front of body)

POSTERIOR (at or near back of body)

transverse plane (yellow)

INFERIOR
(of two body parts, the one farthest from head)

a

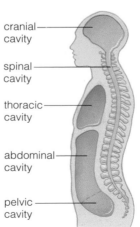

cranial cavity

spinal cavity

thoracic cavity

abdominal cavity

pelvic cavity

b

Figure 31.10 (**a**) Directional terms and planes of symmetry for the human body. The midsagittal plane divides the body into right and left halves. The frontal plane divides it into anterior (front) and posterior (back) parts. For humans, the main body axis is perpendicular to the earth. For rabbits and other animals that move with the main body axis parallel to the earth, *ventral* corresponds to anterior, and *dorsal* corresponds to posterior. (Compare Figure 25.2.)

(**b**) Major cavities in the human body.

HOMEOSTASIS AND SYSTEMS CONTROL

The Internal Environment

To stay alive, your body's cells must be continually bathed in fluid that supplies them with nutrients and carries away metabolic wastes. In this they are no different from an amoeba or any other free-living, single-celled organism. However, many *trillions* of cells are crowded together in your body—and they all must draw nutrients from and dump wastes into the same 15 liters of fluid. That is less than 16 quarts.

The fluid *not* inside cells is called **extracellular fluid**. Much of it is *interstitial*, meaning it occupies the spaces between cells and tissues. The rest is *plasma*, the fluid portion of blood. Interstitial fluid exchanges substances with blood and with the cells it bathes.

In functional terms, the extracellular fluid is continuous with the fluid inside cells. That is why drastic changes in its composition and volume have drastic effects on cell activities. Its concentrations of hydrogen, potassium, calcium, and other ions are especially important in this regard. Those concentrations must be maintained at levels that are compatible with the survival of the body's individual cells. Otherwise, the animal itself cannot survive.

It makes no difference whether the animal is simple or complex. *The component parts of any animal work together to maintain the stable fluid environment required by its living cells.* This concept is absolutely central to our understanding of the structure and function of animals, and it may be summarized this way:

1. Each cell of the animal body engages in basic metabolic activities that ensure its own survival.

2. Concurrently, the cells of a given tissue typically perform one or more activities that contribute to the survival of the whole organism.

3. The combined contributions of individual cells, organs, and organ systems help maintain the stable internal environment—that is, the extracellular fluid—required for individual cell survival.

Mechanisms of Homeostasis

The word **homeostasis** refers to stable operating conditions in the internal environment. This state is maintained through homeostatic controls, which operate in coordinated ways to keep physical and chemical aspects of the body within tolerable ranges. Many of the controls work through feedback mechanisms.

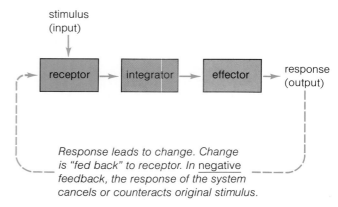

Response leads to change. Change is "fed back" to receptor. In negative feedback, the response of the system cancels or counteracts original stimulus.

Figure 31.11 Components necessary for negative feedback at the organ level.

In a **negative feedback mechanism**, an activity changes some condition in the internal environment, and this in turn triggers a response that reverses the changed condition. Think about a furnace with a thermostat. The thermostat senses the air temperature and "compares" it to a preset point on a thermometer built into the furnace control system. When the temperature falls below the preset point, the thermostat signals a switching mechanism that turns on the heating unit. When the air becomes heated enough to match the prescribed level, the thermostat signals the switching mechanism, which shuts off the heating unit.

Similarly, meerkats and many other animals rely on feedback mechanisms to raise or lower body temperature so that it is maintained near 37°C (98.6°F) even during extremely hot or cold weather. When the body senses that its skin is getting too hot outside in the summer sun, for example, mechanisms are set in motion that slow down metabolic activity and overall body activity. Thus internal controls work to counteract the possibility of an intolerably high body temperature by slowing down the body's heat-generating activities.

Under some circumstances, **positive feedback mechanisms** set in motion a chain of events that *intensify* the original condition. Positive feedback is associated with instability in a system. For example, sexual arousal leads to increased stimulation, which leads to more stimulation, and so on until an explosive, climax level is reached (page 766). As another example, during childbirth, pressure of the fetus on the uterine walls stimulates production and secretion of the hormone oxytocin. Oxytocin causes muscles in the walls to contract, this increases pressure on the fetus, and so on until the fetus is expelled from the mother's body.

Homeostatic control mechanisms require three components: sensory receptors, integrators, and effectors (Figures 31.11 and 31.12). *Sensory receptors* are cells or parts of cells that can detect a specific change in the

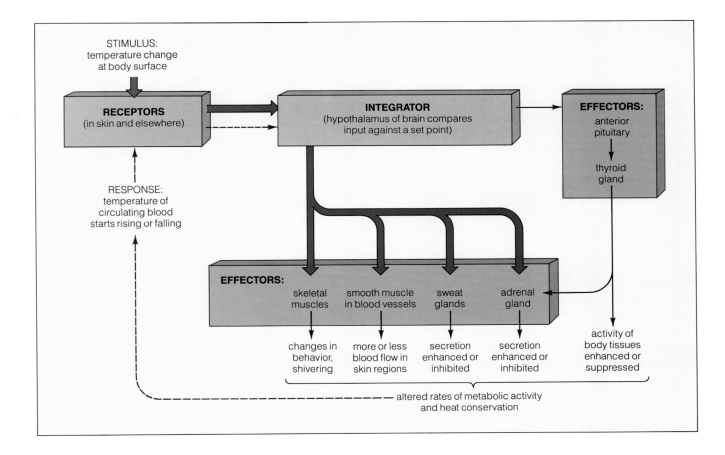

STIMULUS:
temperature change
at body surface

RECEPTORS
(in skin and elsewhere)

INTEGRATOR
(hypothalamus of brain compares
input against a set point)

EFFECTORS:
anterior
pituitary

thyroid
gland

RESPONSE:
temperature of
circulating blood
starts rising or falling

EFFECTORS:

skeletal
muscles

smooth muscle
in blood vessels

sweat
glands

adrenal
gland

changes in
behavior,
shivering

more or less
blood flow in
skin regions

secretion
enhanced or
inhibited

secretion
enhanced or
inhibited

activity of
body tissues
enhanced or
suppressed

altered rates of metabolic activity
and heat conservation

environment. For example, when someone kisses you, there is a change in pressure on your lips. Receptors in the skin of your lips translate the stimulus energy into a signal, which can be sent to the brain. Your brain is an *integrator*, a control point where different bits of information are pulled together in the selection of a response. It can send signals to your muscles or glands (or both), which are *effectors* that carry out the response. In this case, the response might include flushing with pleasure and kissing the person back. Or it might include flushing with rage and shoving the person away from your face.

Thus your brain continually receives information about how things *are* operating—that is, the information from receptors. It also receives information about how things *should be* operating—that is, information from "set points." When conditions deviate significantly from a set point, the brain functions to bring them back to the most effective operating range. It does this by way of signals that cause specific muscles and glands to increase or decrease their activity.

Homeostatic control mechanisms maintain physical and chemical aspects of the internal environment within ranges that are most favorable for cell activities.

Figure 31.12 Homeostatic controls over the internal temperature of the human body. The dashed line shows how the feedback loop is completed. The blue arrows indicate the main control pathways.

What we have been describing here is a general pattern of monitoring and responding to a constant flow of information about the external and internal environments. During this activity, organ systems operate together in coordinated fashion. Throughout this unit, we will be asking the following questions about their operation:

1. What physical or chemical aspect of the internal environment are organ systems working to maintain as conditions change?

2. By what means are organ systems kept informed of change?

3. By what means do they process incoming information?

4. What mechanisms are set in motion in response?

As you will see in the chapters that follow, the operation of all organ systems is under neural and endocrine control.

SUMMARY

1. A tissue is an aggregation of cells and intercellular substances united in the performance of a specialized activity. An organ is a structural unit in which tissues are combined in definite proportions and patterns that allow them to perform a common task. An organ system has two or more organs interacting chemically, physically, or both in ways that contribute to the survival of the animal as a whole.

2. Epithelial tissues are one or more layers of adhering cells with one free surface and the opposite surface resting on a basement membrane that intervenes between it and an underlying connective tissue. Epithelia cover external body surfaces and line internal cavities and tubes.

3. Connective tissues bind together other tissues or offer them mechanical or metabolic support. They include connective tissue proper and specialized connective tissues (such as cartilage, bone, and blood).

4. Muscle tissue, which is specialized for contraction, functions in the movements of body parts and move-

ment through the environment. Nervous tissue detects and coordinates information about change in the internal and external environments, and it controls responses to those changes.

5. Organ systems work largely through homeostatic control mechanisms that help maintain a stable internal environment. In negative feedback (the most common mechanism), the response to a disturbance in a system decreases the original disturbance. In positive feedback, a response intensifies the original disturbance.

6. Tissues, organs, and organ systems work in ways that help maintain the stable internal environment (the extracellular fluid) required for individual cell survival.

7. Control of the internal environment depends on the body's receptors, integrators, and effectors. Receptors detect stimuli, which are specific changes in the environment. Integrating centers (such as the brain) receive information from receptors about how some aspect of the body's receptors, integrators, and effectors. Receptors detect stimuli, which are specific changes in the environment. Integrating centers (such as the brain) receive both), which carry out the appropriate response.

Review Questions

1. Label the following tissues appropriately: *534–539*

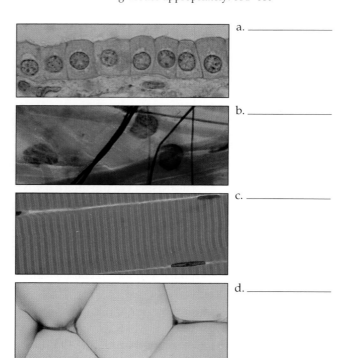

a. _____

b. _____

c. _____

d. _____

2. What is an animal tissue? An organ? An organ system? List the major organ systems of the human body, along with their functions. *533–534, 540*

3. Define extracellular fluid and interstitial fluid. *542*

4. Epithelial tissue and connective tissue differ from each other in overall structure and function. Describe how. *534–538*

5. State the overall functions of (a) muscle tissue and (b) nervous tissue. *538–539*

6. A major important concept in animal physiology relates the functioning of cells, organs, and organ systems to the internal environment. Can you state the three main points of this concept? *542*

7. Define homeostasis. What are the three components necessary for homeostatic control over the internal environment? *542–543*

8. What are the differences between negative feedback and positive feedback mechanisms? *542*

Self-Quiz *(Answers in Appendix IV)*

1. The four main types of tissues in most animals are _____, _____, _____, and _____.

2. Animals are structurally and functionally adapted for these tasks:
   a. maintenance of the internal environment
   b. nutrient acquisition, processing, distribution, disposal
   c. self-protection against injury or attack
   d. reproduction
   e. all of the above

3. _____ tissues cover external body surfaces, line internal cavities and tubes, and some form the secretory portions of glands.
   a. Muscle
   b. Nervous
   c. Connective
   d. Epithelial

4. Most _____ tissues bind or mechanically support other tissues; but one type functions in physiological support of other tissues.
   a. muscle
   b. nervous
   c. connective
   d. epithelial

5. _____ tissues detect and coordinate information about environmental changes and control responses to those changes.
   a. Muscle
   b. Nervous
   c. Connective
   d. Epithelial

6. _____ tissues contract and make possible internal body movements as well as movements through the external environment.
   a. Muscle
   b. Nervous
   c. Connective
   d. Epithelial

7. Cells in the animal body _____ .
   a. engage in metabolic activities that ensure their survival
   b. perform activities that contribute to the survival of the animal
   c. contribute to maintaining the extracellular fluid
   d. all of the above

8. In a state of _____ , physical and chemical aspects of the body are being kept within tolerable ranges by controlling mechanisms.
   a. positive feedback
   b. negative feedback
   c. homeostatis
   d. metastasis

9. In negative feedback mechanisms, _____ .
   a. a detected change brings about a response that tends to return internal operating conditions to the original state
   b. a detected change suppresses internal operating conditions to levels below the set point
   c. a detected change raises internal operating conditions to levels above the set point
   d. fewer solutes are fed back to the affected cells

10. _____ detect specific environmental changes, an _____ pulls different bits of information together in the selection of a response, and _____ carry out the response.

11. Match the concepts.
    _____ muscles and glands
    _____ positive feedback
    _____ body receptors
    _____ negative feedback
    _____ brain

    a. integrating center
    b. the most common homeostatic mechanism
    c. eyes and ears
    d. effectors
    e. chain of events intensifies the original condition

Selected Key Terms

adipose tissue 538
blood 538
bone 538
cardiac muscle 538
cartilage 538
dense, irregular connective tissue 537
dense, regular connective tissue 537
endocrine gland 535
effector 543
epithelium 534
exocrine gland 535
extracellular fluid 542
germ cell 534
homeostasis 542

integrator 543
loose connective tissue 537
muscle tissue 538
negative feedback mechanism 542
nervous tissue 539
organ 534
organ system 534
positive feedback mechanism 542
sensory receptor 542
skeletal muscle 538
smooth muscle 538
somatic cell 534
tissue 533

Readings

Bloom, W., and D. W. Fawcett. 1986. *A Textbook of Histology*. Eleventh edition. Philadelphia: Saunders. Outstanding reference text.

Leeson, C. R., T. Leeson, and A. Paparo. 1985. *Textbook of Histology*. Philadelphia: Saunders.

Ross, M., and E. Reith. 1985. *Histology: A Text and Atlas*. New York: Harper & Row.

Vander, A., J. Sherman, and D. Luciano. 1990. "Homeostatic Mechanisms and Cellular Communication" in *Human Physiology*. Fifth edition. New York: McGraw-Hill.

32 INFORMATION FLOW AND THE NEURON

Tornado!

It is spring in the American Midwest, and you are fully engrossed in photographing the magnificent wildflowers all around you in an expanse of shortgrass prairie. So intent are you on capturing all the different species on film that you fail to notice the rapidly darkening sky. By the time you finally look up, the sky is darkening ominously. What's that rumbling you hear in the distance? It sounds like a freight train. Yet how can that be, when there is no track, anywhere, in this part of the prairie? You turn to identify the source of the sound. And you see it—but you don't want to believe your eyes. A funnel cloud is advancing across the prairie and heading right for you! *TORNADO!* The terrifying image rivets your attention as nothing else has ever done before.

You sense instantly that you cannot remain where you are and survive. Suddenly you remember that hiding in a low area is better than standing out in the open. Commands flash from your brain to your limbs: *GET MOVING OUT OF HERE!* With heart thumping, you start to run along a path, looking frantically for safety. You're in luck! Just ahead is a steep-banked creek. With a tremendous burst of speed you reach the creek in less than a minute, then scramble down the muddy bank. There you find a small ledge hanging over the water. You wedge yourself under it, hoping wildly to be inconspicuous, to be overlooked by a force of nature on the rampage.

It takes a few minutes before you realize that the tornado has roared past the creek. You remain motionless, heart still thumping, fingers still clutching mud. Finally, cautiously, you sit up and look around. Some distance away, you see a swath of twisted prairie grass that marks the tornado's path. Your heart no longer feels like it is slamming against your chest wall, although your legs feel like rubber when you stand up.

You can thank your nervous system for every perception, every memory, and nearly every action that helped you escape the tornado. You can thank the coordinated signals that traveled rapidly through its communication lines, which are composed of cells called neurons.

Were all of those lines silent until they received signals from the outside, much as telephone lines wait to carry calls from all over the country? Not even

Figure 32.1 A truly terrifying view across the Kansas prairie— a tornado about to touch down. Imagine yourself alone in the prairie when you first see the tornado, knowing you have but minutes to remove yourself from harm's way. What thoughts flash through your mind? How rapidly do you accept or reject possible plans of action?

remotely. Even before you were born, your newly developing neurons became organized in vast gridworks and started chattering among themselves. All through your life, in moments of danger or reflection, excitement or sleep, their chatter has never ceased. Constant communication among neurons, such as the ones controlling the muscles concerned with breathing, keep you alive. It keeps your body primed for rapid response to change—even to the totally unexpected appearance of a tornado.

This chapter begins with the structure and function of individual neurons. Then it starts us thinking about how neurons interact. The next two chapters deal with the nervous system and endocrine system, which are inseparable in their integrative and control functions. Chapter 35 provides a closer look at sensory structures and sensory organs, including eyes, by which you detect tornadoes and other events that may have bearing on whether you survive from one day to the next.

KEY CONCEPTS

1. A nervous system senses, interprets, and issues commands for responses to specific aspects of the environment. Its communication lines are highly organized gridworks of cells called neurons.

2. A steady difference in electric charge exists across a neuron's plasma membrane. Sudden, brief reversals in that difference are the basis of messages sent through the nervous system. The reversals, called action potentials, occur when a neuron is adequately stimulated. They occur from the point of stimulation to the neuron's signal output zone, where it forms a junction with another cell.

3. At junctions called chemical synapses, action potentials trigger the release of a chemical substance that stimulates or inhibits the activities of the next cell in line.

CELLS OF THE NERVOUS SYSTEM

In all nervous systems, the basic unit of communication is the nerve cell, or **neuron**. Neurons do not act alone. They *collectively* sense changing conditions, integrate sensory inputs, then activate different body parts that can carry out responses. These tasks involve signals among different classes of nerve cells, called sensory neurons, interneurons, and motor neurons.

We can define each class in terms of its role in a control scheme, described in Chapter 31, by which the nervous system monitors and responds to change:

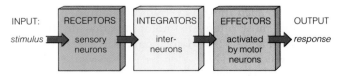

INPUT: RECEPTORS → INTEGRATORS → EFFECTORS → OUTPUT
stimulus → sensory neurons → inter-neurons → activated by motor neurons → *response*

Sensory neurons have receptor regions that can detect specific stimuli, such as light energy. These neurons relay signals about the stimulus *to* the brain and spinal cord (the integrators in our control scheme). In the brain and spinal cord are **interneurons**, which integrate infor-

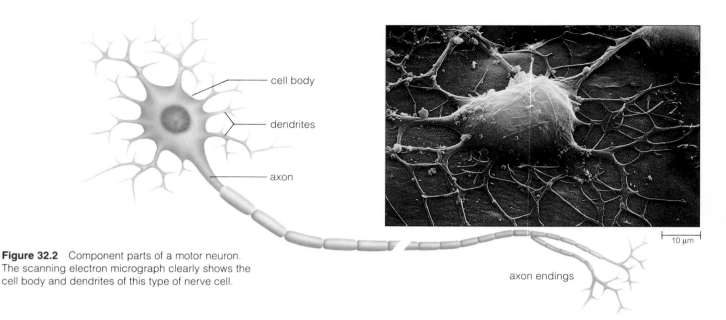

Figure 32.2 Component parts of a motor neuron. The scanning electron micrograph clearly shows the cell body and dendrites of this type of nerve cell.

10 µm

cell body

dendrites

axon

axon endings

mation arriving on sensory lines and then influence other neurons in turn. **Motor neurons** relay information *away* from the integrator to muscle cells or gland cells. Muscles and glands are the body's effectors, which carry out responses.

Keep in mind that the three classes of neurons just defined make up only about half the volume of vertebrate nervous systems. A variety of **neuroglial cells** also are present in great numbers. These specialized cells physically support and protect the neurons and help them carry out their tasks. Some impart structure to the brain, much as connective tissues do for other body regions. Some segregate groups of neurons. Others wrap like a series of jellyrolls around parts of sensory and motor neurons, and they affect how fast a signal travels along the neuron.

FUNCTIONAL ZONES OF THE NEURON

All neurons have a cell body, which contains the nucleus and the metabolic machinery for protein synthesis. Most neurons have slender cytoplasmic extensions of the cell body, although these differ enormously in number and length. So great are the structural differences that there really is no such thing as a "typical" neuron. The ones described most often are motor neurons of the sort shown in Figure 32.2. Motor neurons have many **dendrites**, which are short, slender extensions of the cell body. They also have one long, cylindrical extension called an **axon**.

An axon of motor neurons has finely branched endings that terminate on muscle cells. As Figure 32.3 suggests, dendrites and axons also occur on many sensory neurons and interneurons.

Generally speaking, we can think of dendrites and the cell body as "input zones," where signals about changing conditions are received. Axon endings are "output zones," where messages are sent to other cells.

Each neuron has input zones for receiving signals as well as output zones, where signals are sent on to other cells.

NEURAL MESSAGES

Membrane Excitability

Different kinds of signals, some electrical, some chemical, constitute the messages that travel through a nervous system. Let's start with the signals that travel along the plasma membrane of individual neurons.

A neuron "at rest," not doing anything special, shows a steady difference in electric charge—a *voltage difference*—across its plasma membrane. In this case, the fluid just inside the membrane is more negatively charged than the fluid outside. That steady voltage difference is the **resting membrane potential**. Think of it as a "potential for activity" across the membrane.

When a neuron receives signals at its input zone, the resting membrane potential there may change temporarily. For example, imagine you accidentally trip over a cat that is snoozing on the kitchen floor. In response to the unfortunate stimulation, a veritable tidal wave of signals washes swiftly through the cat's nervous system, and those that reach, say, motor neurons with axons in the cat's tail will no doubt disturb the resting membrane potential of those neurons.

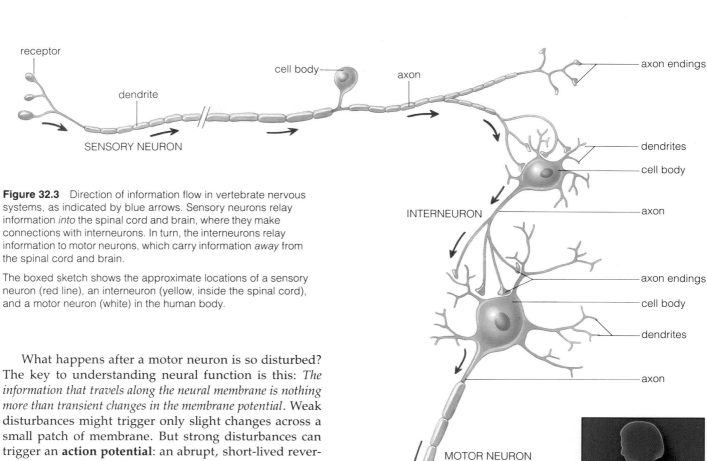

receptor

cell body

axon

axon endings

dendrite

SENSORY NEURON

dendrites

cell body

axon

INTERNEURON

axon endings

cell body

dendrites

axon

MOTOR NEURON

Figure 32.3 Direction of information flow in vertebrate nervous systems, as indicated by blue arrows. Sensory neurons relay information *into* the spinal cord and brain, where they make connections with interneurons. In turn, the interneurons relay information to motor neurons, which carry information *away* from the spinal cord and brain.

The boxed sketch shows the approximate locations of a sensory neuron (red line), an interneuron (yellow, inside the spinal cord), and a motor neuron (white) in the human body.

What happens after a motor neuron is so disturbed? The key to understanding neural function is this: *The information that travels along the neural membrane is nothing more than transient changes in the membrane potential.* Weak disturbances might trigger only slight changes across a small patch of membrane. But strong disturbances can trigger an **action potential**: an abrupt, short-lived reversal in the polarity of charge across the plasma membrane. For a fraction of a second, the cytoplasmic side of a patch of membrane becomes positive with respect to the outside. This transient change in membrane potential moves rapidly along the entire length of the axon.

Any cell that can respond to stimulation by producing action potentials is said to show *membrane excitability*.

Keep in mind that action potentials usually begin and end on the same neuron; they are not transferred to neighboring cells. When action potentials reach the axon endings of the cat's motor neurons, for example, they trigger the release of molecules that serve as chemical signals to adjacent muscle cells. Muscles contract in response to the signals and the cat's tail shoots straight up. The cat simultaneously makes other, violent responses to the stimulation, but these need not concern us here.

A neuron at rest shows a steady voltage difference across its plasma membrane, with the inside more negative than the outside. That difference is the resting membrane potential.

A neuron shows membrane excitability. It can produce action potentials in response to strong stimulation.

An action potential is an abrupt, transient reversal in the voltage difference across the membrane: The inside becomes more positive with respect to the outside.

Neurons "At Rest"

For information to flow along a neuron, the neuron must first be in the "resting" state. Only then can it undergo the changes that result in an action potential. The question becomes this: What establishes the resting membrane potential and maintains it between action potentials? The answer starts with three factors.

First, the concentrations of potassium ions (K^+), sodium ions (Na^+), and other charged substances are not the same on the two sides of the plasma membrane. *Second*, channel proteins that span the membrane control the diffusion of specific types of ions across it. *Third*, transport proteins that span the membrane actively pump sodium and potassium ions across it.

The unequal concentration of ions across the membrane can be depicted this way:

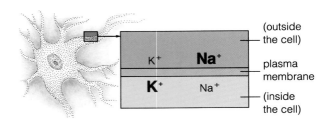

In the above sketch, the large letters denote which side of the membrane has the greater concentration of the two ions. How great is "greater"? Think about a motor neuron in the tail of that cat on the kitchen floor. For every 150 potassium ions on the cytoplasmic side of the plasma membrane, there are only 5 in the same volume of fluid outside. For every 15 sodium ions inside, there are about 150 on the outside.

If K^+, Na^+, and other charged substances could move freely across the neural membrane, they would each dif-

Figure 32.4 (*Below*) Pathways for ions across the plasma membrane of a neuron. These pathways are provided by proteins embedded in the lipid bilayer. Compare Figure 32.6 to this model of membrane structure.

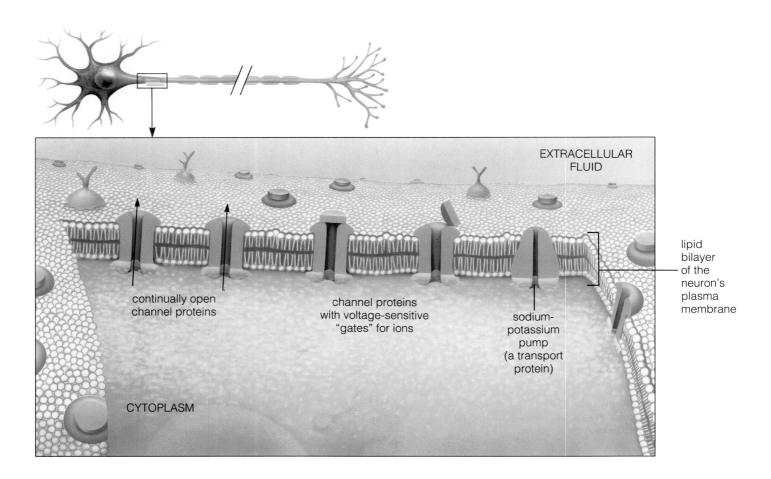

EXTRACELLULAR FLUID

lipid bilayer of the neuron's plasma membrane

continually open channel proteins

channel proteins with voltage-sensitive "gates" for ions

sodium-potassium pump (a transport protein)

CYTOPLASM

fuse down their respective concentration gradients—and those gradients eventually would disappear. However, the ions can move only through the interior of proteins that span the membrane (Figure 32.4). Their passage through the type called channel proteins is called facilitated diffusion. (As described on page 80, a channel protein passively allows the ions to move through their interior in the direction that a concentration gradient would take them.) Some channel proteins permit ions to "leak" (diffuse) through them all the time. Others have "gates" that open only when stimulated.

Imagine now that you are small enough to stand on an input zone of a motor neuron in the cat's tail. A profusion of surface bumps—the tops of membrane proteins—spread out before you. The sleeping cat has not yet been tripped over, so the neuron is at rest. More precisely, most of its channels for sodium are shut. And some channels for potassium are open, so potassium is leaking out through channel proteins, following its concentration gradient. This makes the interior of the neuron more negative than the extracellular fluid, and some potassium ions (which are positively charged) are attracted back inside.

When the inward pull of opposite charge balances the outward force of diffusion, there is no more *net* movement of potassium across the membrane. There is a steady voltage difference across the membrane, which is the amount of energy inherent in the concentration and electric gradients between the two differently charged regions. For many neurons, this amount—the resting membrane potential—is about −70 millivolts.

Neurons cannot maintain a resting potential indefinitely without expending energy. It happens that a small amount of sodium does leak into the neuron through a few open sodium channels. Unless the inward leakage of positive ions is countered, the resting membrane potential eventually will disappear. Transport proteins called **sodium-potassium pumps** do the countering. Using energy from ATP, they actively transport potassium into the neuron, and they pump sodium ions out at the same time. This is an example of the active transport mechanism shown earlier in Figure 5.10.

Figure 32.5 summarizes the balancing effect of the pumping and leaking mechanisms that maintain membrane conditions between action potentials.

Passive transport mechanisms establish the concentration and electric gradients across the plasma membrane of a neuron.

In a resting neuron, an active transport mechanism maintains those gradients by pumping potassium ions into the neuron and sodium ions out of it.

Local Disturbances in Membrane Potential

In all neurons, not just motor neurons, stimulation at an input zone produces localized signals that do not spread very far. When you tripped over the cat, for example, you disturbed many sensory neurons that had their receptors (input zones) embedded in the connective tissue beneath the cat's skin. At each receptor site, the stimulus—in this case, mechanical pressure—affected ion movements across a small patch of plasma membrane. The voltage difference changed slightly at this patch, producing a type of graded, local signal.

"Graded" means the signals can vary in magnitude—they can be small or large—depending on the intensity and duration of the stimulus. In the cat's case, the stronger the pressure on its skin, the greater the disturbance to the sensory receptors.

"Local" means the signal does not spread far—half a millimeter or less, most often. Input zones simply do not have the type of ion channels needed to propagate a signal farther than this. However, when stimulation is intense or prolonged, graded signals can spread into an adjacent **trigger zone** of the membrane—the site where action potentials can be initiated.

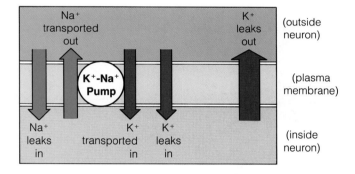

Figure 32.5 Balance between pumping and leaking processes that maintain the distribution of sodium and potassium ions across the plasma membrane of a neuron at rest. The relative widths of the arrows indicate the magnitude of the movements. The total inward movement counteracts the total outward movement for each kind of ion; hence the ion distributions are maintained.

A CLOSER LOOK AT ACTION POTENTIALS

Mechanism of Excitation

The action potential is analogous to a pulse of electrical activity (hence its original name, nerve impulse). Measurements of the voltage difference across a membrane before, during, and after an action potential reveal this pattern:

1. The inside of a neuron at rest is more negative with respect to the outside (its membrane is *polarized*).

2. During an action potential, the inside is more positive than the outside (the membrane is *depolarized*).

3. Following an action potential, resting conditions are restored (the membrane is *repolarized*).

An action potential is triggered when a disturbance causes the voltage difference to change by a certain minimum amount, a *threshold* level. The change occurs when voltage-sensitive gated channels for sodium ions open in an accelerating way (Figure 32.6). The inward flow of these positively charged sodium ions makes the fluid on the cytoplasmic side of the membrane less negative. This causes more gates to open, more sodium to enter, and so on until the charge difference reverses. The accelerating flow of sodium is an example of positive feed-

back, whereby an event intensifies as a result of its own occurrence:

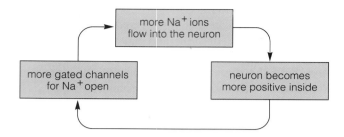

Once threshold is reached, the opening of more sodium gates no longer depends on the strength of the stimulus. It proceeds automatically because the positive-feedback cycle has started. That is why all action potentials in a given neuron "spike" to the same level above threshold as an *all-or-nothing event*. In other words, if threshold is reached, nothing can stop the full spiking. If threshold is not reached, the membrane disturbance will subside when the stimulus is removed.

The squid *Loligo* helped provide researchers with evidence of the spiking that occurs during an action potential. Certain squid axons have such large diameters that a fine electrode can be inserted easily into one of them, then another electrode can be positioned outside the axon membrane. Electrodes are devices used to measure voltage changes. When the two electrodes are connected to a voltage source and an oscilloscope, voltage changes show up as deflections of a beam traveling across the scope's fluorescent screen. Figure 32.7 shows examples of this.

Figure 32.6 Propagation of action potentials along the axon of a neuron. The plasma membrane is shown in yellow.

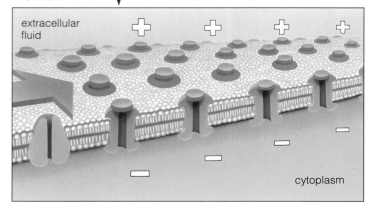

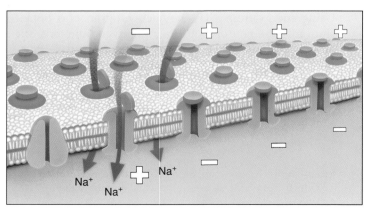

a Membrane at rest (inside negative with respect to the outside). An electrical disturbance (red arrow) spreads from an input zone to an adjacent trigger region of the membrane, which has many gated sodium channels.

b A strong disturbance initiates an action potential. Sodium gates open, the inflow decreases the negativity inside; this causes more gates to open, and so on, until threshold is reached and the voltage difference across the membrane reverses.

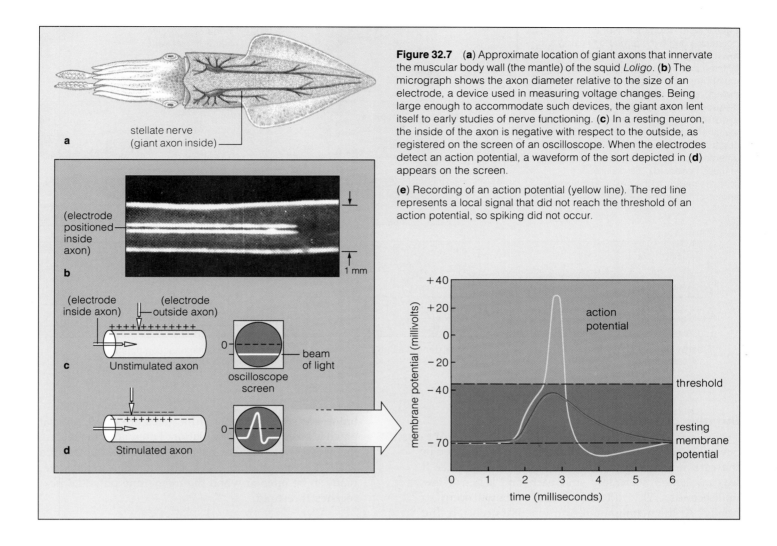

Figure 32.7 (**a**) Approximate location of giant axons that innervate the muscular body wall (the mantle) of the squid *Loligo*. (**b**) The micrograph shows the axon diameter relative to the size of an electrode, a device used in measuring voltage changes. Being large enough to accommodate such devices, the giant axon lent itself to early studies of nerve functioning. (**c**) In a resting neuron, the inside of the axon is negative with respect to the outside, as registered on the screen of an oscilloscope. When the electrodes detect an action potential, a waveform of the sort depicted in (**d**) appears on the screen.

(**e**) Recording of an action potential (yellow line). The red line represents a local signal that did not reach the threshold of an action potential, so spiking did not occur.

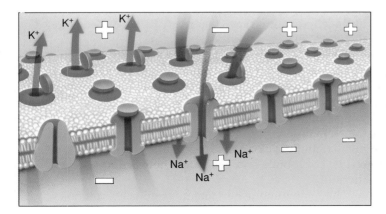

c The reversal causes sodium gates to shut and potassium gates to open (at purple arrows). Potassium follows its gradient (out of the neuron). Voltage is restored. The disturbance produced by the action potential triggers another action potential at the adjacent membrane site, and so on, away from the point of stimulation.

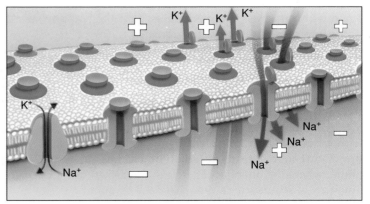

d The inside of the membrane becomes negative again following each action potential, but the sodium and potassium concentration gradients are not yet fully restored. Active transport at sodium-potassium pumps restores the gradients.

Figure 32.8 Propagation of an action potential along a motor neuron having a myelin sheath. (**a**) An action potential is initiated at a trigger zone in the axon membrane. (**b**) The sheath hinders ion movements across the membrane, so the disturbance spreads rapidly down the axon. (**c**) The small nodes are not sheathed, and they have very dense arrays of gated sodium channels. The voltage difference across the membrane reverses at these nodes. (**d**) The disturbance spreads rapidly to the next node in line, and so on down the axon (**e**).

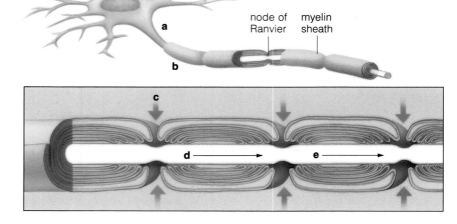

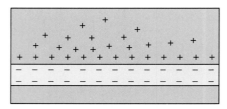

Charge density along an unsheathed axon

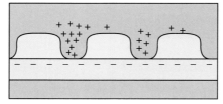

Charge density along a sheathed axon

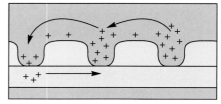

Propagation of action potential along a sheathed axon

Duration of Action Potentials

Several hundred new action potentials were triggered in the cat's nervous system in the single second after it was awakened so rudely, and each one lasted a few milliseconds. Why did each action potential occur so briefly? At the membrane region where it occurred, the gates of some channel proteins closed and shut off the flow of sodium ions across the membrane. Then, about halfway through the action potential, the gates of other channel proteins opened and potassium ions were free to diffuse out of the neuron. The increased outward flow of many more potassium ions restored the original voltage difference across the membrane. And the sodium-potassium pumps restored the gradients.

Propagation of Action Potentials

Refractory Period. After an action potential occurs in a trigger zone, it is propagated along the membrane without becoming diminished in magnitude. In brief, the disturbance spreads to adjacent patches of membrane, where the opening of gated channels is repeated. The new disturbance causes channels to open in the next patch of membrane, and so on away from the stimulation site. Notice, in Figure 32.6, how action potentials travel *away* from the stimulation site. A *refractory period* following each one helps prevent backflow. This is the period when the sodium gates at a given patch of mem-

brane are shut and potassium gates are open, so that patch is insensitive to stimulation. Later, after the resting membrane potential has been restored, most potassium gates close and sodium gates return to their initial state, ready to be opened when the membrane potential next reaches threshold.

Sheathed Axons. A *myelin sheath* wraps around the axons of many sensory and motor neurons that serve as cord. The sheath consists of the plasma membranes of specialized neuroglial cells called *Schwann cells*. of specialized neuroglial cells called Schwann cells. These cells grow around and around the axon, rather like a jellyroll, and they form many layers of insulation (Figure 32.8).

Each Schwann cell is separated from adjacent ones by a small, exposed gap, or node, where the axon membrane is loaded with voltage-sensitive gated sodium channels. In a manner of speaking, the action potentials jump from node to node. The sheathed regions hinder the flow of ions across the membrane, and this forces the ions to flow along the length of the axon until they can exit at a node and generate a new action potential there.

The node-to-node hopping in myelinated neurons is called *saltatory conduction*, after the Latin word meaning "to jump." Saltatory conduction affords the most rapid signal propagation with the least metabolic effort by the cell. In the largest myelinated axon, signals travel 120 meters per second. That's 270 miles per hour.

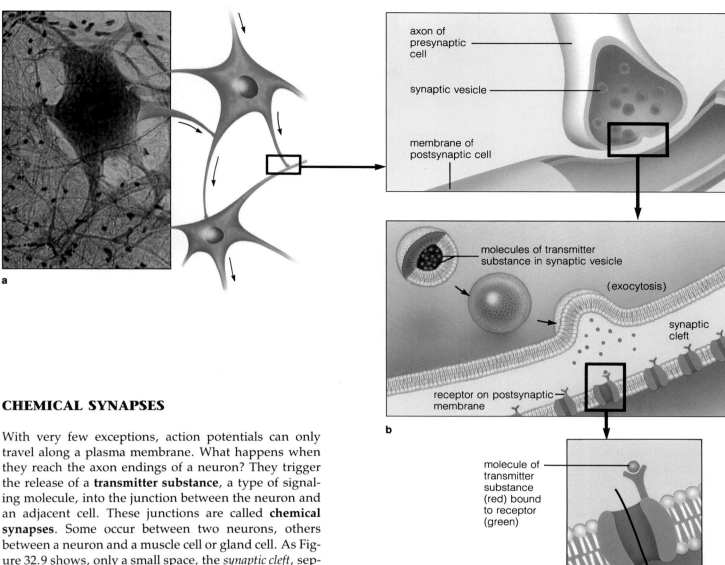

a

b

c

Figure 32.9 Chemical synapses. Typically, action potentials spread along axons, away from the neuron cell body (**a**). In (**b**), the axon terminates next to another neuron, this being an example of a chemical synapse. Information flows from the presynaptic cell to the postsynaptic cell by way of a transmitter substance (**c**).

Labels in figure:
- axon of presynaptic cell
- synaptic vesicle
- membrane of postsynaptic cell
- molecules of transmitter substance in synaptic vesicle
- (exocytosis)
- synaptic cleft
- receptor on postsynaptic membrane
- molecule of transmitter substance (red) bound to receptor (green)
- Na⁺

CHEMICAL SYNAPSES

With very few exceptions, action potentials can only travel along a plasma membrane. What happens when they reach the axon endings of a neuron? They trigger the release of a **transmitter substance**, a type of signaling molecule, into the junction between the neuron and an adjacent cell. These junctions are called **chemical synapses**. Some occur between two neurons, others between a neuron and a muscle cell or gland cell. As Figure 32.9 shows, only a small space, the *synaptic cleft*, separates the two cells.

A neuron that releases a transmitter substance into the cleft is called the "*pre*synaptic cell." The one whose behavior is affected by the transmitter substance is the "*post*synaptic cell." The presynaptic cell contains numerous vesicles filled with transmitter substances (Figure 32.9).

When an action potential arrives at the presynaptic cell membrane facing the cleft, it causes voltage-sensitive gated channels for *calcium ions* to open. Calcium ions are more concentrated outside the cell, and when they move inside (down their gradient), they cause synaptic vesicles to fuse with the plasma membrane. Thus the contents of the vesicles are released into the cleft, and diffusion carries them to the postsynaptic cell. There the molecules of transmitter substance bind briefly to membrane receptor molecules, and after they exert their effects, they are picked up again by the presynaptic cell or inactivated by enzymes.

Figure 32.10 (a) Neuromuscular junction, a region of chemical synapse between a motor neuron and a muscle cell. (b) At this junction, axon terminals act on troughs in the muscle cell membrane (the motor end plate). The myelin sheath of the motor axon terminates before the junction, so that the membranes of the two cells are exposed to each other. (c) Scanning electron micrograph of a portion of a neuromuscular junction.

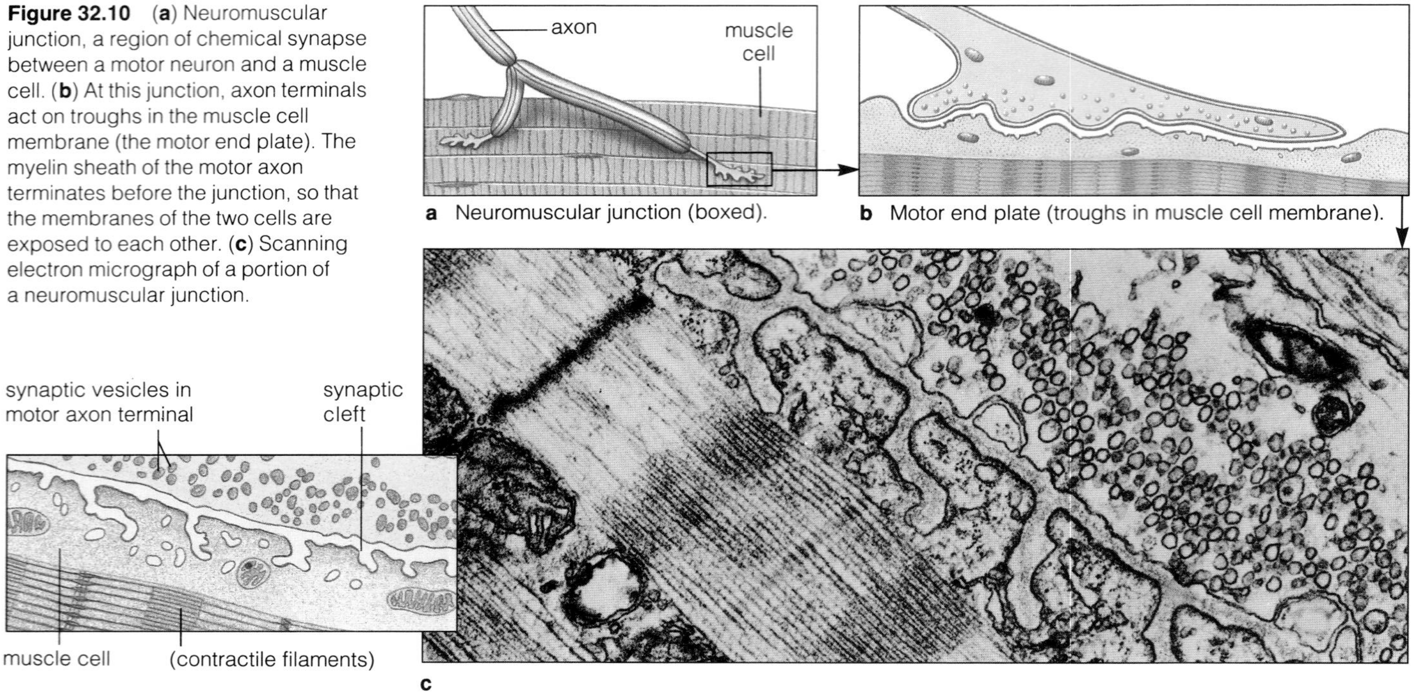

a Neuromuscular junction (boxed).

b Motor end plate (troughs in muscle cell membrane).

synaptic vesicles in motor axon terminal

synaptic cleft

muscle cell   (contractile filaments)

c

Effects of Transmitter Substances

A transmitter substance affects a postsynaptic cell in one or two ways. If it has an *excitatory* effect, it helps drive the cell's membrane toward the threshold of an action potential. If it has an *inhibitory* effect, it helps drive the membrane away from threshold. A given transmitter substance can have either excitatory or inhibitory effects, depending on which type protein channels it opens up in the postsynaptic membrane.

Acetylcholine (ACh) is a transmitter substance with excitatory and inhibitory effects on the cells of muscles and glands throughout the body. It also acts on certain cells in the brain and spinal cord. Think about what ACh does at **neuromuscular junctions**, which are synapses between a motor neuron and muscle cells. At this type of junction, the branched axon endings of the motor neuron are positioned on the muscle cell membranes (Figure 32.10). An action potential traveling down the motor neuron spreads through all the endings and causes the release of ACh into each synaptic cleft. When ACh binds to receptors on the muscle cell membrane, it has an excitatory effect. It may trigger action potentials, which in turn initiate events in the muscle cells that lead to contraction. Those events are a topic of Chapter 35.

Serotonin, another transmitter substance, acts on brain cells that govern sleeping, sensory perception, temperature regulation, and emotional states. *Norepinephrine* affects brain regions that apparently are concerned with emotional states as well as dreaming and awaking. Other transmitter substances that act on different parts of the brain are *dopamine* and *GABA* (gamma aminobutyric acid).

Neuromodulators

The effects of transmitter substances are often influenced by signaling molecules known generally as **neuromodulators**. These molecules enhance or reduce membrane responses in target neurons. Among them are the *endorphins*, which function in inhibiting perceptions of pain. These naturally occurring peptide molecules are much more potent than morphine, a painkiller derived from opium poppies. Endorphins also may have roles in memory and learning, temperature regulation, and sexual behavior as well as emotional depression and other mental disorders.

Synaptic Integration

How will a given postsynaptic cell respond to incoming signals? That depends on which of its membrane receptors are called into play and on the nature of the signals reaching those receptors. *At any given moment, excitatory and inhibitory signals are competing for control of the membrane.*

In a process called **synaptic integration**, the competing signals at an input zone of a neuron are summed up. Those signals are two kinds of graded potentials that often are abbreviated EPSP and IPSP. An *excitatory post-*

Commentary

Deadly Imbalances at Chemical Synapses

An anaerobic bacterium, *Clostridium tetani*, lives in the gut of horses and other grazing animals and its endospores can survive in soil, especially soil enriched with manure. This bacterium can enter a human body through a deep puncture or cut, and it can multiply if tissues around the wound die off and become anaerobic. A product of this bacterium's metabolic activities is a neurotoxin. It interferes with the effect of acetylcholine (ACh) on motor neurons. The result is a severe disorder called *tetanus*, the symptoms of which are prolonged, spastic paralysis that can lead to death.

During normal muscle contraction, action potentials from the brain travel through the spinal cord. They excite motor neurons that trigger the release of acetylcholine at neuromuscular junctions, and the muscle cells are stimulated into contracting. Many of the body's muscles occur as paired sets, such as those shown in Figure *a*. When one set contracts, an opposing set is stretched. Bend your arm at the elbow and you can feel two such sets (biceps and triceps) in your upper arm. When the biceps contracts, inhibitory signals are sent to the triceps and it

relaxes. The tetanus toxin blocks the release of inhibitory signals—so *both* sets of muscles contract! Within four to ten days, paired muscles attempt to work in opposition

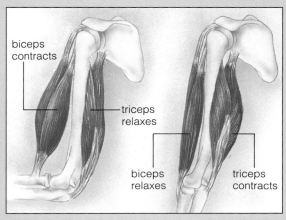

biceps contracts

triceps relaxes

biceps relaxes

triceps contracts

a Antagonistic muscle pair.

---

synaptic potential (EPSP) brings the membrane closer to threshold; it has a *de*polarizing effect. An *inhibitory postsynaptic potential* (IPSP) drives the membrane away from threshold—it has a *hyper*polarizing effect—or maintains the membrane at its resting level.

Take a look at Figure 32.11. The yellow line in this diagram shows how an EPSP of a given magnitude would register on an oscilloscope screen *if it were occurring alone*. The purple line shows the same thing for an IPSP. The red line shows what happens when the two occur simultaneously. In this case, synaptic integration pulls the membrane potential away from threshold.

The balancing acts that go on at many chemical synapses are essential for survival. We know this because foreign substances that interfere with synaptic integration can have deadly consequences, as the *Commentary* suggests.

Synaptic integration is the moment-by-moment combining of excitatory and inhibitory signals acting on adjacent membrane regions of a neuron.

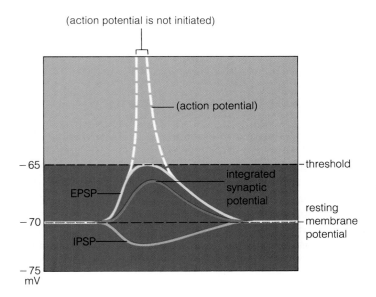

(action potential is not initiated)

(action potential)

−65

threshold

EPSP

integrated synaptic potential

−70

resting membrane potential

IPSP

−75 mV

Figure 32.11 Synaptic integration. In this example, an excitatory synapse and an inhibitory synapse nearby are activated at the same time. The IPSP reduces the magnitude of the EPSP from what it could have been, pulling it away from threshold. The red line represents the integration of these two synaptic potentials. Threshold is not reached in this case; hence an action potential cannot be initiated.

to each other. This is the start of spastic paralysis—the muscles simply cannot be released from contraction. The increase in muscle tension (spasms) can become violent enough to break bones in the body. Fists and jaws may undergo prolonged clenching (hence the name lockjaw, which is sometimes used for the disorder). The back may become paralyzed in a permanent arch. Muscles of the respiratory system and heart also may undergo spastic paralysis, in which case the affected individual nearly always dies.

Since the development of effective vaccines, tetanus occurs only rarely in the United States. But the disease was terrifying to the soldiers of early wars, when battlefields were littered with manure from calvary horses and with corpses of the horses themselves. At that time, *C. tetani* endospores were like profuse biological landmines. Battle wounds commonly became contaminated with those endospores. We sense the agony of one young victim of such contamination; the dramatic painting in Figure *b* was made as he lay dying in a military hospital.

C. botulinum, a relative of the bacterial agent of tetanus, has a different effect on synaptic integration. It produces a toxin that can block the release of ACh from motor neurons. In this case, muscle contraction cannot occur, so the body shows the symptoms of the disease *botulism*. It shows flaccid paralysis, meaning its muscles remain relaxed. Without prompt treatment, victims simply will stop breathing.

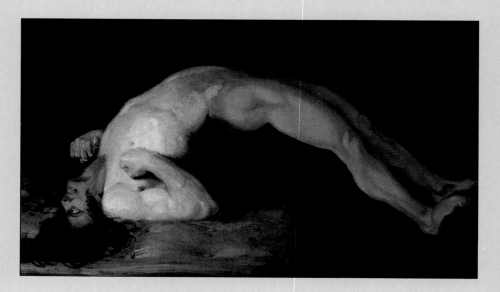

b A soldier in 1809 dying of tetanus.

PATHS OF INFORMATION FLOW

Through synaptic integration, signals arriving at any given neuron in the body can be reinforced or dampened, sent on or suppressed. What determines the direction in which a given signal will travel? That depends on the organization of neurons into circuits or pathways.

The brain has many "local" circuits in which the chattering of neurons is confined to a single region. In contrast, signals between the brain or spinal cord and other body regions travel by cordlike communication lines called **nerves** (Figure 32.12). Axons of sensory neurons, motor neurons, or both are bundled together in a nerve. Within the brain and spinal cord, such bundles are called nerve "tracts."

The sensory and motor neurons of many nerves take part in reflexes, which are simple, stereotyped movements made in response to sensory stimulation. In the simplest **reflex arc**, sensory neurons directly synapse on motor neurons. The *stretch reflex* is an example; it works to contract a muscle when that muscle has been stretched.

Think about how you can hold out a large glass and keep it stationary when someone pours lemonade into it. As the lemonade adds weight to the glass and your hand starts to drop, a muscle in your arm (the biceps) is stretched. The stretching activates certain receptors in the muscle. The stretch-sensitive receptors are part of **muscle spindles**—sensory organs in which small, specialized cells are enclosed in a sheath that runs parallel with the muscle itself. The receptors are the input zone of sensory neurons that synapse with motor neurons in the spinal cord (Figure 32.13). Axons of the motor neurons lead right back to the stretched muscle, and action potentials that reach the axon endings trigger the release of ACh, which initiates contraction. Continued receptor activity excites the motor neurons further, allowing them to maintain your hand's position.

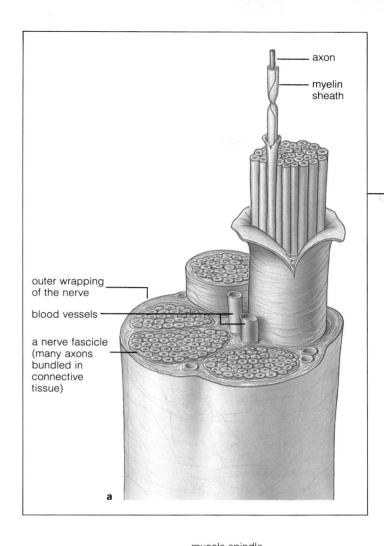

axon

myelin sheath

outer wrapping of the nerve

blood vessels

a nerve fascicle (many axons bundled in connective tissue)

a

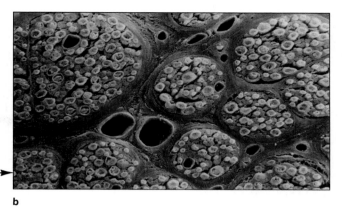

b

Figure 32.12 Structure of a nerve. The sketch (**a**) and the scanning electron micrograph (**b**) show bundles of axons in cylindrical wrappings of connective tissue inside the nerve.

Figure 32.13 Simple reflex arc governing the stretch reflex. A sensory axon is shown in purple, a motor axon in red. Stretch-sensitive receptors of the sensory neuron are located in muscle spindles within a skeletal muscle. Stretching the muscle disturbs the receptors, and action potentials are generated in the sensory neuron. They travel to the axon endings that synapse with motor neurons—which have axons leading right back to the stretched muscle. Signals from the motor neuron can stimulate the muscle cell membrane and initiate contraction (page 634).

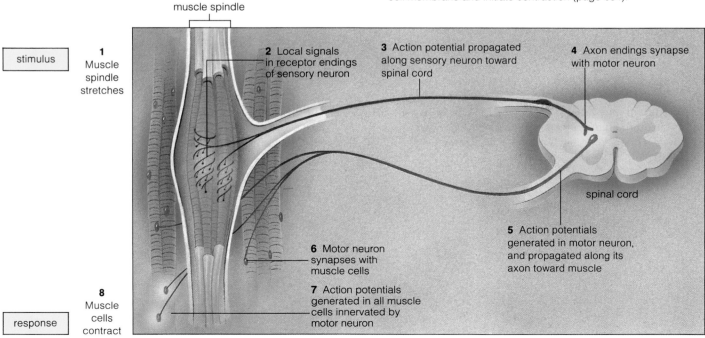

muscle spindle

stimulus

1 Muscle spindle stretches

2 Local signals in receptor endings of sensory neuron

3 Action potential propagated along sensory neuron toward spinal cord

4 Axon endings synapse with motor neuron

spinal cord

5 Action potentials generated in motor neuron, and propagated along its axon toward muscle

6 Motor neuron synapses with muscle cells

7 Action potentials generated in all muscle cells innervated by motor neuron

8 Muscle cells contract

response

In the vast majority of reflexes, sensory neurons make connections with a number of interneurons, which then activate or suppress the motor neurons necessary for a coordinated response. An example is the *withdrawal reflex*, a rapid pulling away from an unpleasant or harmful stimulus. If you have ever accidentally touched a hot stove, you know this reflex action can be completed even before you are conscious it has occurred.

SUMMARY

1. The neuron, or nerve cell, is the basic unit of communication in all nervous systems. Neurons *collectively* sense environmental change, swiftly integrate sensory inputs, and then activate muscle cells and gland cells (effectors) that can carry out responses.

2. Vertebrate nervous systems contain sensory neurons, interneurons, and motor neurons. They also contain neuroglial cells, which support and protect the neurons.

3. Like all cells, the neuron at rest shows a steady voltage difference across its plasma membrane, with the inside more negative than the outside. This results from differences in the concentrations of potassium ions, sodium ions, and other charged substances present in cytoplasm and extracellular fluid. The amount of energy inherent in the concentration and electric gradients across the membrane is the resting membrane potential.

4. Most neurons and some other cells (including muscle cells) show membrane excitability. In response to stimulation, the voltage difference across the membrane can undergo brief but sudden reversals, called action potentials—the inside becomes positive with respect to the outside.

5. Excitability depends on these membrane features:
   a. The lipid bilayer is impermeable to ions.
   b. Certain ions can diffuse passively across the membrane through channel proteins. Some channel proteins are always open, others have gates that open only under stimulation.

   c. Transport proteins (sodium-potassium pumps) in the membrane restore and maintain concentration gradients between action potentials.

6. The neuron receives and integrates signals at *input* zones, usually dendrites and the cell body. The disturbance may produce local, graded potentials that may spread to a *trigger* zone, where action potentials can be generated. Action potentials travel to *output* zones (axon terminals), where other kinds of signals are sent to target cells.

7. Action potentials occur when the voltage difference across the membrane changes dramatically, past a certain minimum amount called the threshold level. Then, gates on channel proteins open in an accelerating way and suddenly reverse the voltage difference, which registers as a spike on recording devices.

8. The overall pattern before, during, and after an action potential is this: polarization (inside negative with respect to outside), depolarization (inside becomes positive), and repolarization (inside becomes negative again).

9. Chemical synapses are junctions between two neurons, or between a neuron and a muscle cell or gland cell. Here, the *pre*synaptic cell releases a transmitter substance into a cleft between it and the *post*synaptic cell. A transmitter substance may have an excitatory or inhibitory effect, depending on which type of ion channels it opens up in the postsynaptic cell membrane. Integration is the moment-by-moment combining of all signals—excitatory and inhibitory—acting at all the different synapses on a neuron.

10. The direction of information flow through the body depends on the organization of neurons into circuits and pathways. Local circuits are sets of interacting neurons confined to a single region in the brain or spinal cord. Nerve pathways extend from neurons in one body region to neurons in different regions.

11. Reflexes (simple, stereotyped movements made in response to sensory stimuli) are examples of how signals are sent through nervous systems. In simple reflexes, sensory neurons directly signal motor neurons that act on muscle cells. In more complex reflexes, interneurons coordinate and refine the responses.

Review Questions

1. Define sensory neuron, interneuron, and motor neuron. *547–548*

2. What is the difference between a neuron and a nerve? *547, 558*

3. Two major concentration gradients exist across a neural membrane. What substances are involved, and how are the gradients maintained? *550–551*

4. An electric gradient also exists across a neural membrane. Explain what the electric and concentration gradients together represent. *548–549, 550–551*

5. Label the functional zones of a motor neuron on the following diagram: *548*

6. Distinguish between an action potential and a graded potential. What is meant by "all-or-nothing" messages? *549, 551, 552*

7. What is a synapse? Explain the difference between an excitatory and an inhibitory synapse. Define neural integration. *555–557*

8. What is a reflex? Describe the sequence of events in a stretch reflex. *558–560*

Self-Quiz *(Answers in Appendix IV)*

1. The communication lines of vertebrate nervous systems are organized networks of cells called _____.

2. In vertebrate nervous systems, _____ are receptors of specific stimuli, _____ integrate information and send commands for responses to muscles and glands by way of _____.

3. In a neuron at rest, there is a steady voltage difference across the plasma membrane, with the _____ being more _____ than the _____.
   a. inside; negative; outside
   b. outside; negative; inside
   c. inside; positive; outside
   d. outside; positive; inside

4. The "resting membrane potential" is _____ across the neuron's plasma membrane.
   a. an action potential
   b. a graded potential
   c. a steady voltage difference
   d. both a and c are correct

5. Action potentials occur _____.
   a. when a neuron is adequately stimulated
   b. when potassium gates open in an accelerating way
   c. when sodium-potassium pumps kick into action
   d. both a and b are correct

6. An action potential lasts only briefly because _____ at the membrane region where it occurred.
   a. gates for sodium open, gates for potassium close
   b. gates for sodium close, gates for potassium open
   c. sodium-potassium pumps restore gradients
   d. both b and c are correct

7. All transmitter substances diffuse across a _____.
   a. neuromuscular junction
   b. synaptic cleft
   c. myelin sheath
   d. both a and b are correct

8. _____ is an example of a transmitter substance; _____ is an example of a neuromodulator.
   a. Serotonin; an endorphin
   b. Serotonin; GABA
   c. Ach; an endorphin
   d. both a and c are correct

9. A nerve may consist of bundled-together axons of _____.
   a. sensory neurons
   b. motor neurons
   c. sensory and motor neurons
   d. all of the above are correct

10. Match the following concepts and descriptions.
   _____ simple reflex
   _____ local circuit
   _____ transmitter substances
   _____ neuron
   _____ complex reflex

   a. have excitatory or inhibitory effects on postsynaptic cells
   b. basic unit of communication in all nervous systems
   c. sensory neuron directly signals motor neuron
   d. set of interacting neurons in region of brain or spinal cord
   e. interneurons coordinate and refine responses

Selected Key Terms

acetylcholine (ACh) *556*
action potential *549*
axon *548*
chemical synapse *555*
dendrite *548*
excitatory postsynaptic potential (EPSP) *557*
graded potential *551*
inhibitory postsynaptic potential (IPSP) *557*
interneuron *547*
membrane excitability *549*
motor neuron *548*
muscle spindle *558*
myelin sheath *554*
nerve *558*
neuroglial cell *548*
neuromodulator *556*
neuromuscular junction *556*
neuron *547*
reflex arc *558*
refractory period *554*
resting membrane potential *548*
saltatory conduction *554*
Schwann cell *554*
sensory neuron *547*
sodium-potassium pump *551*
stretch reflex *558*
synaptic integration *556*
threshold *552*
transmitter substance *555*
trigger zone *551*

Readings

Berne, R., and M. Levy (editors). 1988. *Physiology*. Second edition. St. Louis: Mosby. Section II is an authoritative introduction to neural functioning.

Dunant, Y., and M. Israel. April 1985. "The Release of Acetylcholine." *Scientific American* 252(4):58–83. Experiments showing how this major neurotransmitter functions.

Lent, C., and M. Dickenson. June 1988. "The Neurobiology of Feeding Behavior in Leeches." *Scientific American* 258(6):98–103. Describes the relationship between serotonin (a transmitter substance) and feeding behavior in an invertebrate.

Why Crack the System?

James Kalat, a professor at North Carolina State University, sometimes asks students to volunteer for an experiment. He tells them he would like to implant a device inside their brain that will make them feel really good. There are risks. The device compromises health, reduces life expectancy by a decade or so, and possibly causes permanent brain damage. The behavior of the volunteers will change for the worse, so they might have trouble completing their education, getting a job, keeping a job, or holding a family together. Volunteers can quit the experiment at any time but the longer the device is in their brain, the harder it will be to get out. They will not be paid. Rather, they must pay the experimenter—at bargain rates at first, then a little more each week. The device is illegal, so if volunteers get caught using it, they as well as the experimenter will probably go to jail.

Very, very few students volunteer for the experiment (which of course is hypothetical). Yet when Kalat changes "brain device" to *drug* and "experimenter" to *drug dealer*, an amazing number come forward. Like 30 million people in the United States alone, those students appear more than willing to engage in the self-destructive use of the so-called psychoactive drugs, which alter emotional and behavioral states.

The consequences show up in unexpected places. Each year, about 300,000 women addicted to crack give birth—and their newborns are already addicts. *Crack* is a cheap, potent form of cocaine, and it disrupts basic functions of the nervous system. Remember those chemical synapses described in the preceding chapter? Crack disrupts synapses between the neurons of a "pleasure center" within the brain. It stimulates the release of transmitter substances from presynaptic cells into the synaptic cleft—then blocks their reabsorption. The transmitter substances accumulate next to the postsynaptic cells and relentlessly stimulate them. Normal impulses to eat and sleep are suppressed. Blood

pressure rises. Feelings of euphoria and sexual desire become intense. All the while, the brain demands constant stimulation—but the molecules of transmitter substances in the synaptic cleft gradually break down, and the presynaptic cells cannot keep up with the incessant demand to provide more. Crack users become frantic, then profoundly depressed. Only crack makes them "feel good" again.

Addicted newborns cannot know all of this, of course. They can only quiver with "the shakes." Overstimulation of their brain neurons makes them chronically irritable. Their body is abnormally small. As they were developing inside their mother, their body tissues were not provided with enough oxygen and nutrients—crack causes blood vessels to constrict. Paradoxically, even though crack babies are abnormally fussy, they do not respond to rocking and other kinds of normal stimulation. It may be a year or more before they recognize their mother. Without treatment they are likely to grow up as emotionally unstable children, prone to aggressive outbursts and stony silences.

Each of us possesses a body of great complexity. Its architecture, its functioning are legacies of millions of years of evolution. It is unique in the living world because of its highly developed nervous system—a system that is capable of processing far more than the experience of the individual. One of its most astonishing products is language, the encoding of shared experiences of groups of individuals in time and space. Through the evolution of our nervous system, the sense of history was born, and the sense of destiny. Through this system we can ask how we have come to be what we are, and where we are headed from here. Perhaps the sorriest consequence of drug abuse is its implicit denial of this legacy—the denial of self when we cease to ask, and cease to care.

Figure 33.1 Owners of an evolutionary treasure—a complex brain that is the foundation for our memory and reasoning, and our future.

KEY CONCEPTS

1. Nervous systems provide a means of sensing specific information about external and internal conditions, integrating that information, and issuing commands for response from the body's effectors (muscles and glands).

2. The simplest nervous systems are the nerve nets of sea anemones and other cnidarians. The vertebrate nervous system shows pronounced cephalization and bilateral symmetry. It includes a brain, spinal cord, and many paired nerves. The brain is complex, with centers for receiving, integrating, processing, and responding to information.

3. The oldest parts of the vertebrate brain provide reflex control over breathing, blood circulation, and other essential functions. During the evolution of certain vertebrates, the brain expanded in complexity and its newer regions appropriated more and more control over the ancient reflex functions.

Neurologically speaking, sponges are just about the simplest members of the animal kingdom. Prick one with a pin and it will contract slowly, in a diffuse sort of way, and its response will never extend more than a few millimeters beyond the point of stimulation. Even a Venus flytrap makes a showier response to stimulation, as when its spiny trap slams shut around insect prey. However, animals as a group are unexcelled in their means of detecting specific stimuli and responding swiftly to them.

Almost all animals reach out or lunge after food; they pull back, crawl, swim, run, or fly when they are about to become food themselves. And think about what they have to do to find a mate and slow it down or hold its attention. (Think about all the things you have to do.) *The more complex the life-style, the more elaborate are the animal modes of receiving, integrating, and responding to information about the external and internal world.* Nervous systems provide for those three functions. As we saw in the preceding chapter, the communication lines of such systems are composed of nerve cells, or neurons. Messages traveling along those lines are coordinated with one another to produce complex patterns of behavior.

There are more than a million known species of animals, each with special features in its neural wiring, so the examples used in this chapter are necessarily limited. Even so, the following list is a useful starting point for understanding the organization of nearly all nervous systems:

---

1. Reflexes provide the basic operating machinery of nervous systems. *Reflexes* are simple, stereotyped movements made in response to specific types of stimulation. In the simplest reflexes, a sensory neuron signals a motor neuron, which acts on muscle cells that help carry out responses to the stimulation (page 558).

2. Nervous systems evolved as more nervous tissue became layered over ancient reflex pathways. In bilateral animals, the layerings became most pronounced at the head end of the body, with many neurons becoming concentrated into a brain. This evolutionary process is called *cephalization*.

3. In existing vertebrates, the oldest parts of the brain still deal with reflex coordination of vital functions, such as blood circulation and breathing. Other parts deal with storing and comparing information about experiences— and using that information to initiate novel actions. The neural connections within the most recent layerings are the basis of memory, learning, and reasoning.

4. Nervous systems coevolved with complex sensory organs, such as eyes, and motor structures, such as legs and wings. The coevolution of nervous, sensory, and motor systems was central to the development of more intricate behavior.

INVERTEBRATE NERVOUS SYSTEMS

Nerve Nets

Sea anemones, hydras, jellyfishes, and other cnidarians are aquatic animals, and they have the simplest nervous systems. As described in Chapter 25, these animals show radial symmetry. Their body parts are arranged about a central axis, much like the spokes of a bike wheel. Their nervous system extends through all the spokes, so to speak, and allows the animal to respond to food and danger coming from any direction.

The cnidarian nervous system, a **nerve net**, has sensory cells, nerve cells, and contractile epithelial cells (Figure 33.2). The three types of cells interact in simple reflex pathways. In the pathway concerned with feeding behavior, for example, nerve cells extend from sensory receptors in tentacles to contractile cells around the mouth. In jellyfishes, reflex pathways permit slow swimming movements and keep the body right-side up.

Bilateral, Cephalized Nervous Systems

Flatworms are the simplest animals with bilateral symmetry, meaning their body has equivalent parts on the left and right side of its midsagittal plane. (Imagine yourself sliding down a staircase banister that turns into a razor and you probably never will forget where the midsagittal plane is.) Muscles that move the body forward are arranged the same way on both sides of the body, not just one. Nerves controlling the muscles are arranged the same way on both sides, and so on.

Bilateral nervous systems may have evolved from arrangements as simple as nerve nets. Some cnidarians pass through a self-feeding larval stage during the life cycle. Like flatworms, the larva (a planula) has a somewhat flattened body, and it uses cilia to swim or crawl about before it develops into an adult:

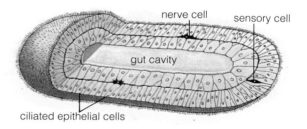

Imagine a few such planulas crawling about on the seafloor in Cambrian times. Suppose mutations of regulatory genes blocked their transformation into the adult form but allowed their reproductive organs to mature. As indicated on page 446, this actually happens among certain animals. The planulas would keep on crawling, they would reproduce—and so pass on the mutated genes.

The forward end of a crawling, aquatic animal is the first to encounter the presence or odor of food in the water. We can speculate that selective agents favored planulalike animals in which sensory cells became concentrated at the forward end, for that arrangement would permit more rapid and effective responses to important stimuli.

All flatworms have a ladderlike nervous system (Figure 33.3). Intriguingly, some also have **ganglia** (singular, ganglion). These are small, regional clusterings of the cell bodies of neurons. They also have **nerves** and two **nerve cords**, these being bundled-together axons of neu-

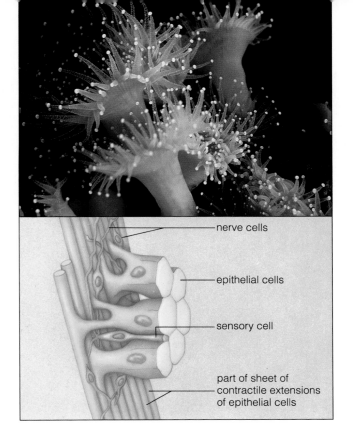

nerve cells

epithelial cells

sensory cell

part of sheet of
contractile extensions
of epithelial cells

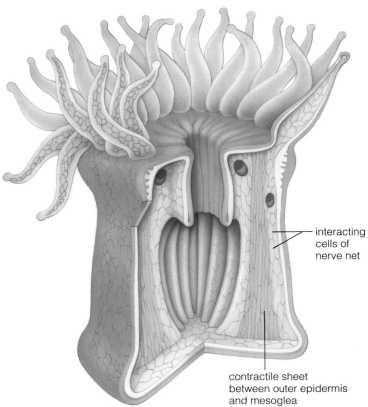

interacting
cells of
nerve net

contractile sheet
between outer epidermis
and mesoglea

Figure 33.2 (*Above*) Nerve net of a sea
anemone, one of the cnidarians.

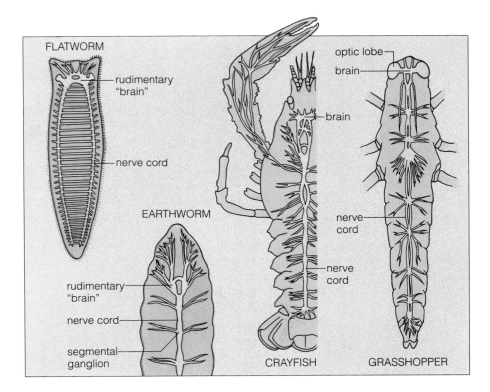

FLATWORM

rudimentary
"brain"

nerve cord

EARTHWORM

rudimentary
"brain"

nerve cord

segmental
ganglion

CRAYFISH

brain

nerve
cord

nerve
cord

optic lobe

brain

nerve
cord

GRASSHOPPER

Figure 33.3 Bilateral symmetry evident
in the nervous systems of a few
invertebrates. The sketches are not to
scale relative to one another.

rons. Some flatworm ganglia are arranged as a two-part
brainlike structure at the head end. They coordinate signals from paired sensory organs, including two eyespots, and provide some control over the nerve cords.

As you will now see, *the patterns of bilateral symmetry
and cephalization have echoes in the paired nerves and muscles,
paired sensory structures, and paired brain center of yourself
and all other vertebrates.*

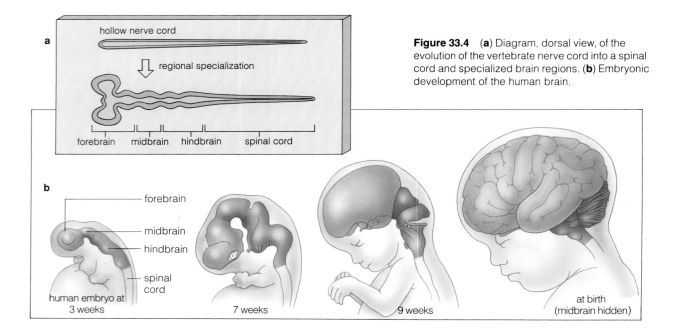

Figure 33.4 (**a**) Diagram, dorsal view, of the evolution of the vertebrate nerve cord into a spinal cord and specialized brain regions. (**b**) Embryonic development of the human brain.

VERTEBRATE NERVOUS SYSTEMS

How the System Developed

The vertebrate nervous system shows more than cephalization and bilateral symmetry. It shows features that were shaped by a shift from reliance on a notochord to reliance on a vertebral column and nerve cord.

The *notochord*, a long rod of stiffened tissue, helps support the body. All vertebrate embryos have one, but it is greatly reduced or absent in adults. Most often, a backbone takes over the function of the notochord. The backbone contains hard, bony segments (vertebrae) arranged one after the other in a *vertebral column*. As described in Chapter 26, the vertebral column evolved many millions of years ago, and it was pivotal in the evolution of fast-moving, jawed predators.

The first vertebrates also had a *nerve cord*, a hollow, tubular structure running dorsally above the notochord. As Figure 33.4 suggests, this nerve cord was the forerunner of the spinal cord and brain. In the changing world of fast-moving vertebrates, predators and prey that were better equipped to sense and respond to one another's presence had the competitive edge. Their senses of smell, hearing, and balance became keener. The brain itself became variably thickened with nervous tissue that could integrate the rich sensory information and issue commands for complex, coordinated responses. In time, the thickening brain tissues became divided into three functionally specialized parts, called the forebrain, midbrain, and hindbrain.

Today, a nerve cord still develops in all vertebrate embryos. We call it the "neural tube." The neural tube undergoes expansion and regional modification into the brain and spinal cord, and it becomes enclosed within the vertebral column. Adjacent tissues in the embryo give rise to nerves that thread through all body regions and connect with the spinal cord and brain. Figure 33.5 will give you a sense of how intricate the communication lines become in the human nervous system.

Functional Divisions of the System

For descriptive purposes, we can divide the vertebrate nervous system into central and peripheral regions. The **central nervous system** includes the brain and spinal cord. The **peripheral nervous system** includes all the nerves carrying signals to and from the brain and spinal cord. Both divisions also have neuroglial cells, which protect or assist neurons. The Schwann cells described in the preceding chapter are an example. They wrap around the axons of neurons that are bundled together in nerves of the peripheral nervous system (see, for example, page 554).

PERIPHERAL NERVOUS SYSTEM

The peripheral nervous system of humans has thirty-one pairs of spinal nerves, which connect with the spinal cord. It also has twelve pairs of cranial nerves, which connect directly with the brain. Some nerves of the peripheral system carry only sensory information. The optic nerves, which carry visual signals from the eyes, are like this. Other nerves contain both sensory and motor axons. For example, the vagus nerves have sensory axons leading into the brain as well as motor axons leading out to the lungs, gut, and heart.

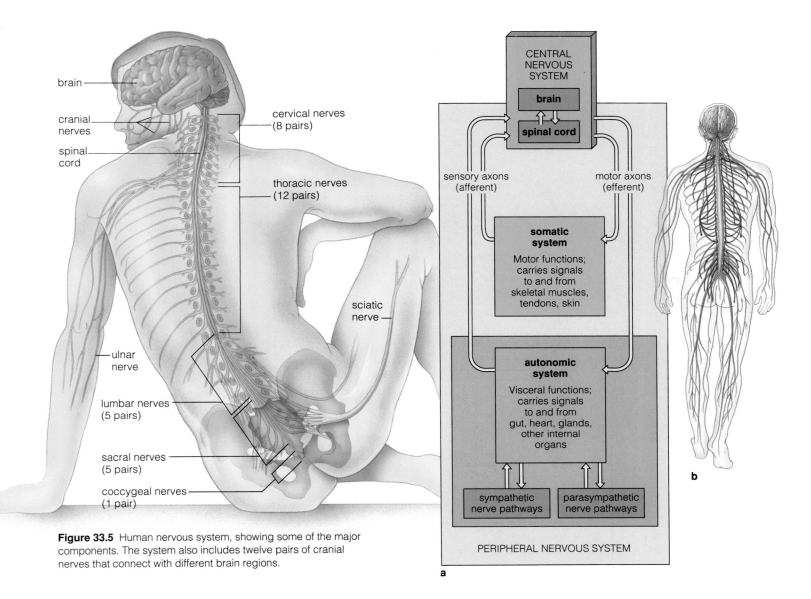

Figure 33.5 Human nervous system, showing some of the major components. The system also includes twelve pairs of cranial nerves that connect with different brain regions.

Labels (left figure): brain; cranial nerves; spinal cord; cervical nerves (8 pairs); thoracic nerves (12 pairs); sciatic nerve; ulnar nerve; lumbar nerves (5 pairs); sacral nerves (5 pairs); coccygeal nerves (1 pair)

Labels (right figure a): CENTRAL NERVOUS SYSTEM; brain; spinal cord; sensory axons (afferent); motor axons (efferent); somatic system — Motor functions; carries signals to and from skeletal muscles, tendons, skin; autonomic system — Visceral functions; carries signals to and from gut, heart, glands, other internal organs; sympathetic nerve pathways; parasympathetic nerve pathways; PERIPHERAL NERVOUS SYSTEM

Figure 33.6 (**a**) Functional divisions of the vertebrate nervous system. In (**b**), the central nervous system is color-coded blue, the somatic nerves green, and the autonomic nerves, red. Sometimes the nerves carrying sensory input to the central nervous system are said to be *afferent* (a word meaning "to bring to"). The ones carrying motor output away from the central nervous system to muscles and glands are *efferent* ("to carry outward").

Somatic and Autonomic Subdivisions

The peripheral nervous system has two subdivisions, called somatic and autonomic (Figure 33.6). The **somatic system** deals with movements of the body's head, trunk, and limbs. Its sensory axons carry signals inward from receptors in the skin, skeletal muscles, and tendons; and its motor axons carry signals out to the body's skeletal muscles. The **autonomic system** deals with the "visceral" portion of the body—that is, the internal organs and structures. Its sensory and motor axons carry signals from and to smooth muscle, cardiac (heart) muscle, and glands in different regions inside the body.

The reflex pathway shown earlier in Figure 32.14 is a simple example of how somatic nerves work. Autonomic nerves are more intricate than this, for they play off one another during the body's overall functioning in ways that will now be described.

The Sympathetic and Parasympathetic Nerves

There are two subdivisions of autonomic nerves, called parasympathetic and sympathetic. Excitatory and inhibitory signals from **parasympathetic nerves** tend to slow down the body overall and divert energy to basic "housekeeping" tasks, such as digestion. This nerve action dominates when the body is not receiving much outside stimulation. Signals from its **sympathetic nerves** tend to slow down housekeeping tasks and increase overall body

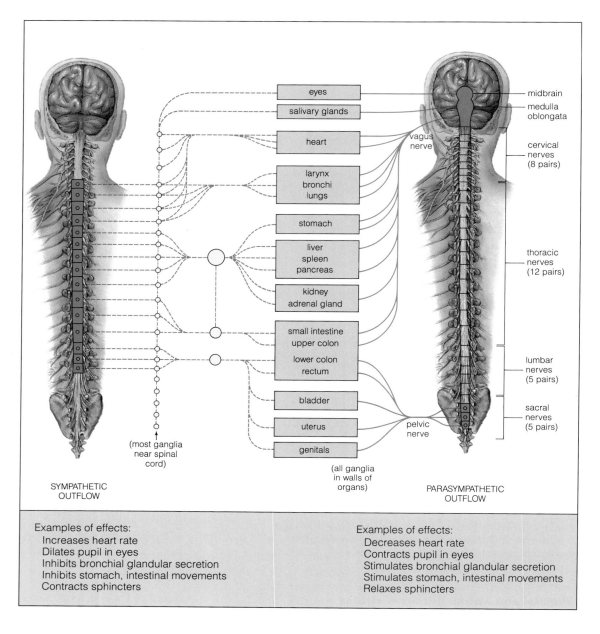

Figure 33.7 Autonomic nervous system. Shown are the main sympathetic and parasympathetic pathways leading out from the central nervous system to some major organs. Keep in mind that both systems have *paired* nerves leading out from the central nervous system. Ganglia (singular, ganglion) are simply clusters of cell bodies of the neurons that are bundled together in nerves.

eyes
salivary glands
heart
larynx
bronchi
lungs
stomach
liver
spleen
pancreas
kidney
adrenal gland
small intestine
upper colon
lower colon
rectum
bladder
uterus
genitals

vagus nerve
pelvic nerve

midbrain
medulla oblongata
cervical nerves (8 pairs)
thoracic nerves (12 pairs)
lumbar nerves (5 pairs)
sacral nerves (5 pairs)

(most ganglia near spinal cord)

(all ganglia in walls of organs)

SYMPATHETIC OUTFLOW

PARASYMPATHETIC OUTFLOW

Examples of effects:
 Increases heart rate
 Dilates pupil in eyes
 Inhibits bronchial glandular secretion
 Inhibits stomach, intestinal movements
 Contracts sphincters

Examples of effects:
 Decreases heart rate
 Contracts pupil in eyes
 Stimulates bronchial glandular secretion
 Stimulates stomach, intestinal movements
 Relaxes sphincters

activities during times of heightened awareness, excitement, or danger. Sympathetic nerves prepare the animal to fight or flee when threatened or to frolic intensely, as in play and sexual behavior.

Both kinds of autonomic nerves are *continually* carrying signals to and from the central nervous system, and so help bring about minor adjustments in the activity of internal organs. Even while low levels of sympathetic signals are causing your heart to beat a little faster, low levels of parasympathetic signals are opposing this effect. At any moment, your heart rate is the net outcome of opposing signals (Figure 33.7). However, the parasympathetic input is reduced and the sympathetic system dominates in times of emergency or intense

excitement. Then, it calls a hormone (epinephrine) into action, the heart rate and breathing rate increase, and if the individual is capable of sweating, it sweats. In this state of intense arousal, the individual is primed to fight (or play) hard or get away fast. Hence the name, the *fight-flight response.*

Once the stimulus that triggered a fight-flight response is removed, sympathetic activity may decrease abruptly and parasympathetic activity may rise suddenly. Evidence of this "rebound effect" might be observed after a person has become instantly mobilized to rush onto a highway to save a child from an oncoming car. The person may well faint as soon as the child has been swept out of danger.

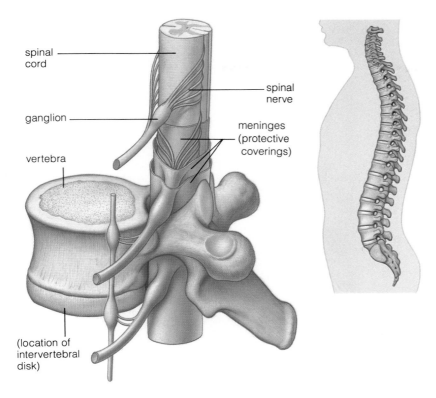

Figure 33.8 Organization of the spinal cord and its relation to the vertebral column.

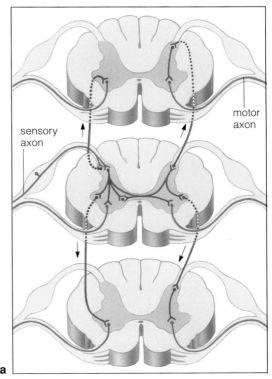

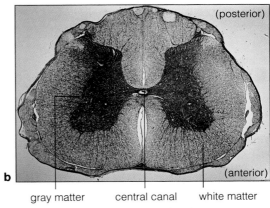

Figure 33.9 (**a**) Examples of vertical and lateral connections among interneurons (color-coded green) in the spinal cord. (**b**) Photograph showing the arrangement of gray matter and white matter of the spinal cord.

CENTRAL NERVOUS SYSTEM

The Spinal Cord

The **spinal cord** is a vital expressway for signals between the peripheral nervous system and the brain. It also is a center for controlling some reflex activities. Here, in the cord, the sensory and motor neurons governing the movements of the body's limbs make direct reflex connections. The stretch reflex, described in the preceding chapter, arcs through the spinal cord in this manner.

The spinal cord threads through a canal formed by the stacked bones of the vertebral column (Figure 33.8). The bones and ligaments attached to them protect the cord. So do the *meninges*—three tough, tubelike coverings around the spinal cord and brain.

Propagation of signals up and down the spinal cord occurs in major **nerve tracts**, which are bundles of sheathed axons. The glistening myelin sheaths of these axons give the tracts the name *white matter*. The *gray matter* of the spinal cord consists of dendrites, cell bodies of neurons, as well as neuroglial cells. In cross-section, the gray matter of the cord looks vaguely like a butterfly (Figure 33.9). This part of the spinal cord deals mainly with reflex connections for limb movements (such as walking) and internal organ activity (such as bladder emptying).

Maybe you have been on a farm when a chicken is destined for the stewpot. Even though the chicken has its head cut off, it still runs around for awhile. Chicken legs, you might correctly deduce, are governed to a great extent by stereotyped reflex pathways in the spinal cord. Experiments with frogs provide evidence of the importance of such pathways. Ascending nerve tracts between the frog's spinal cord and brain contain neu-

Divisions	Main Components	Some Functions
FOREBRAIN	Cerebrum	Two cerebral hemispheres. Centers for coordinating sensory and motor functions, for memory, and for abstract thought. Most complex coordinating center; intersensory association, memory circuits
	Olfactory lobes	Relaying of sensory input from the nose to olfactory structures of cerebrum
	Limbic system	Scattered brain centers. With hypothalamus, coordination of skeletal muscle and internal organ activity underlying emotional expression
	Thalamus	Major coordinating center for sensory signals; relay station for sensory impulses to cerebrum
	Hypothalamus	Neural-endocrine coordination of visceral activities (e.g., solute-water balance, temperature control, carbohydrate metabolism)
	Pituitary gland	"Master" endocrine gland (controlled by hypothalamus). Control of growth, metabolism, etc.
	Pineal gland	Control of some circadian rhythms; role in mammalian reproductive physiology
MIDBRAIN	Tectum	Largely reflex coordination of visual, tactile, auditory input; contains nerve tracts ascending to thalamus, descending from cerebrum
HINDBRAIN	Pons	"Bridge" of transverse nerve tracts from cerebrum to both sides of cerebellum. Also contains longitudinal tracts connecting forebrain and spinal cord
	Cerebellum	Coordination of motor activity underlying limb movements, maintaining posture, spatial orientation
	Medulla oblongata	Contains tracts extending between pons and spinal cord; reflex centers involved in respiration, cardiovascular function, gastric secretion, etc.

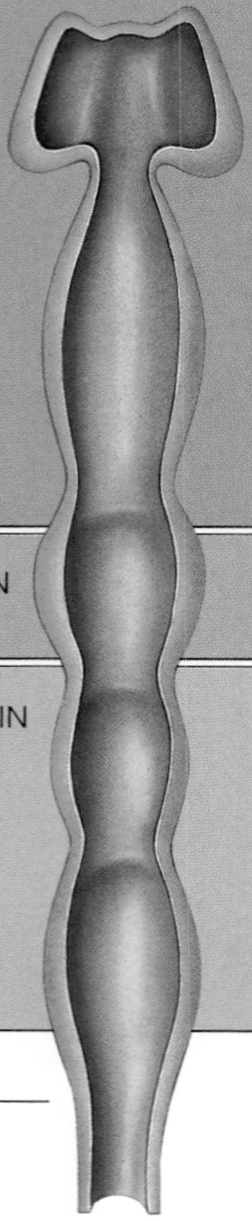

anterior
end of
spinal cord

Figure 33.10 Summary of the three parts of the vertebrate brain and their main subdivisions. The drawing is highly simplified and flattened. While the vertebrate embryo is developing, the brain bends forward, and complex folds form in its wall regions, as shown in Figure 33.4. The midbrain, pons, and medulla oblongata also are called the *brain stem*. A network of interneurons, the reticular formation, extends the length of the brain stem and helps govern the activity of the nervous system as a whole.

rons that deal with straightening the legs after they have been bent. If those neurons are severed at the base of a frog brain, the legs become paralyzed—but only for about a minute. The so-called extensor reflex pathways in its spinal cord recover quickly and have the frog hopping around in no time. By comparison, it takes a few days for such pathways to recover in cats, days or weeks in monkeys, and many months in humans (who show the greatest cephalization).

Divisions of the Brain

The **brain** is the body's master control panel. It receives, integrates, stores, and retrieves information, and it coordinates appropriate responses by intricately stimulating and inhibiting the activities of different body parts. The brain starts out as a continuation of the anterior end of the spinal cord. Like the spinal cord, it is protected by bones (of the cranial cavity) and meninges.

Figure 33.10 summarizes the functions of the three divisions of the brain—the hindbrain, midbrain, and forebrain. As we have seen, these regions develop from a hollow tube of nervous tissue in the embryo.

Hindbrain. The **hindbrain** consists of the medulla oblongata, cerebellum, and pons. The *medulla oblongata* has reflex centers for respiration, blood circulation, and other vital tasks. Here also, motor responses and complex reflexes (such as coughing) are coordinated. Its centers influence other brain centers that help you sleep or wake up.

The *cerebellum* has reflex centers for maintaining posture and refining limb movements. It integrates signals from the eyes, muscle spindles, skin, and elsewhere. It keeps other parts of your brain informed about how your trunk and limbs are positioned, how much different muscles are contracted or relaxed, and in which direction the body or limbs happen to be moving.

The *pons* (meaning bridge) is a major traffic center for nerve tracts passing between brain centers. The name refers to prominent bands of axons that extend into each side of the cerebellum.

Midbrain. The **midbrain** evolved as a coordinating center for reflex responses to visual and auditory input. Its roof of gray matter, the *tectum*, is important in fishes, amphibians, reptiles, and birds for integrating signals from the eyes and ears. (You can surgically remove a frog's cerebrum, its highest integrative center, and the frog can still do just about everything it normally does.) In mammals, sensory input still converges on the tectum, but it is rapidly sent on to higher centers.

The midbrain, pons, and medulla oblongata together represent the brain's "stem." Within the core of the brain stem and extending through its entire length is a major network of interneurons, the **reticular formation**. The reticular formation has extensive connections with the forebrain and helps govern the activity of the nervous system as a whole.

Forebrain. The **forebrain** has the most recently evolved layers of nerve tissues. A pair of olfactory lobes, which deal with the sense of smell, dominated early vertebrate forebrains. A brain center, the *cerebrum*, integrated input about odors and selected motor responses to it. Sensory signals were relayed and coordinated at the *thalamus*, a center below the cerebrum (some motor pathways also converged here). Another brain center, the *hypothalamus*, monitored internal organs and influenced forms of behavior related to their activities, such as thirst, hunger, and sex. In time, a thin layer of gray matter developed over each half of the cerebrum. In mammals, this *cerebral cortex* expanded into information-encoding and information-processing centers. It has become most highly developed in the human brain.

Cerebrospinal Fluid

To get an idea of how soft nervous tissue is, try holding a jiggling blob of Jell-O in your hands. The brain and spinal cord are fragile! Besides being protected by bones and meninges, both actually float in **cerebrospinal fluid**. This clear extracellular fluid cushions the brain and spinal cord from abrupt, jarring movements. The fluid also fills four ventricles, which are interconnected cavities

within the brain that connect with one another and with the central canal of the spinal cord (Figure 33.11).

The bloodstream exchanges substances with the cerebrospinal fluid, which in turn exchanges substances with neurons. Mechanisms called the **blood-brain barrier** help control *which* blood-borne substances are allowed to reach the neurons. The mechanisms are built into more than 99 percent of the blood capillaries servicing the brain. Endothelial cells making up the walls of those special capillaries are fused together by continuous tight junctions (Figure 31.0). This means that substances cannot reach the brain without passing *through* the cells. Transport proteins embedded in the plasma membrane of those cells selectively transport glucose and other water-soluble substances across the barrier. Lipid-soluble substances quickly diffuse through the lipid bilayer of the plasma membrane. This is why caffeine, nicotine, alcohol, barbiturates, heroin, and other lipid-soluble drugs have such rapid effects on brain function.

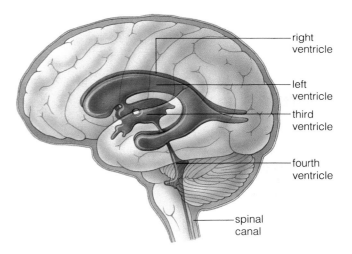

Figure 33.11 Location of the cerebrospinal fluid in the human brain. This extracellular fluid surrounds and cushions the brain and spinal cord. It also fills the four interconnected cavities (cerebral ventricles) within the brain and the central canal of the spinal cord.

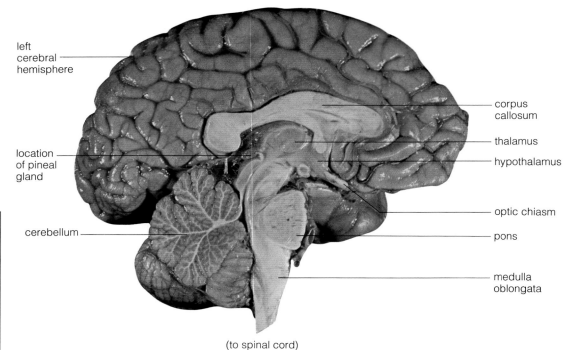

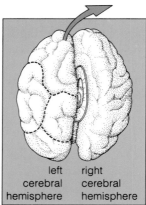

| left cerebral hemisphere | right cerebral hemisphere |

Figure 33.12 Human brain, sagittal section. The corpus callosum is a major nerve tract that runs transversely, connecting the two cerebral hemispheres. The boxed inset shows the two hemispheres pulled slightly apart; normally they are pressed together, with only a longitudinal fissure separating them.

THE HUMAN BRAIN

Although there may be occasional argument about how wisely we use it, we humans have an impressively large, intricate brain. On the average, the human brain weighs about 1,300 grams (3 pounds). It contains about a hundred billion neurons.

The Cerebrum

The human cerebrum vaguely resembles the much-folded nut inside a walnut shell. The folding suggests that expansion of the mammalian cerebrum outpaced the enlargement of the hard skullbones housing it. A deep fissure divides the human cerebrum into two parts, the left and right *cerebral hemispheres*. Other fissures and folds in each hemisphere follow certain patterns and divide it into the lobes named in Figure 33.12.

Much of the gray matter of the cerebral hemispheres is arranged as a thin surface layer, the **cerebral cortex**. The cerebral cortex weighs about a pound, and if you were to stretch it flat, it would cover a surface area of 2-1/2 square feet. The white matter consists of major nerve tracts that keep the hemispheres in communication with each other and with the rest of the body. Some

of the tracts originate in the brainstem, where they are rather jammed together, then fan out extensively in the cerebral hemispheres. Each hemisphere has its own set of tracts that serve as communication lines among its different regions. A prominent band of 200 million axons, the *corpus callosum*, keeps the two hemispheres in communication with each other. We know this through experiments of the sort described in the *Doing Science* essay.

The functioning of the cerebral hemispheres has been the focus of many experiments. Taken together, the results have revealed the following information:

1. Each cerebral hemisphere can function separately. However, the left cerebral hemisphere responds primarily to signals from the right side of the body. The opposite is true for the right cerebral hemisphere. Signals that travel by way of the corpus callosum coordinate the functioning of both hemispheres.

2. The main regions responsible for spoken language skills generally reside in the left hemisphere.

3. The main regions responsible for nonverbal skills (music, mathematics, and other abstract abilities) generally reside in the right hemisphere.

Doing Science

Sperry's Split-Brain Experiments

Experiments performed by Roger Sperry and his coworkers demonstrated some intriguing differences in perception between the two halves of the cerebrum. The subjects of the experiments were epileptics. Persons with severe epilepsy are wracked with seizures, sometimes as often as every half hour of their lives. The seizures have a neurological basis, analogous to an electrical storm in the brain.

What would happen if the corpus callosum of epileptics were cut? Would the electrical storm be confined to one cerebral hemisphere, leaving at least the other to function normally? Earlier studies of animals and of humans whose corpus callosum had been damaged suggested this might be so.

The surgery was performed. The electrical storms subsided in frequency and intensity. Cutting the neural bridge between the two hemispheres put an end to what must have been positive feedback of ever intensified electrical disturbances between them. Beyond this, the "split-brain" individuals were able to lead what seemed, on the surface, entirely normal lives.

But then Sperry devised some elegant experiments to find out whether the conscious experience of those individuals was indeed "normal." After all, the corpus callosum contains no less than 200 million axons; surely something was different. Something was. "The surgery," Sperry later reported, "left these people with two separate minds, that is, two spheres of consciousness. What is experienced in the right hemisphere seems to be entirely outside the realm of awareness of the left."

In Sperry's experiments, the left and right hemispheres of split-brain individuals were presented with different stimuli. It was known at the time that visual connections to and from one hemisphere are mainly concerned with the opposite visual field (Figure *a*).

Sperry projected words—say, COWBOY—onto a screen. He did this in such a way that COW fell only on the left visual field, and BOY fell on the right (Figure *b*). The subject reported seeing the word BOY. The left hemisphere, which controls language, received only the letters BOY. However, when asked to write the perceived word with the left hand—a hand that was deliberately blocked from the subject's view—the subject wrote COW, as shown in Figure *b*.

The right hemisphere, which "knew" the other half of the word (COW) had directed the left hand's motor response. (That hemisphere controls muscles on the left side of the body, and vice versa.) But it couldn't tell the

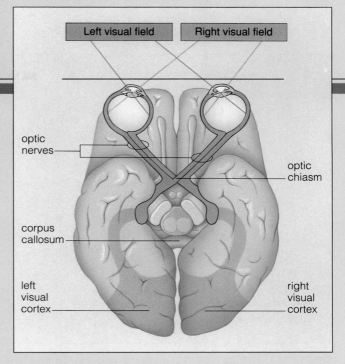

a In the human eye, visual information is gathered at the retina, a layer of densely packed light receptors. Light from the *left* half of the viewing field strikes receptors on the right side of both retinas. Parts of the two optic nerves carry signals from those receptors to the right cerebral hemisphere. Light from the *right* half of the viewing field strikes receptors on the left side of both retinas. Parts of the two optic nerves carry signals from those receptors to the left hemisphere.

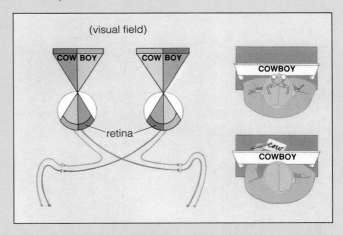

b Responses of split-brain individual to different visual stimuli.

left hemisphere what was going on because of the severed corpus callosum. The subject knew that a word was being written, but could not say what it was!

Thus, the two cerebral hemispheres were functioning separately, and each was responding to visual signals from the opposite side of the body. Sperry's work showed that the corpus callosum is necessary to coordinate their functioning.

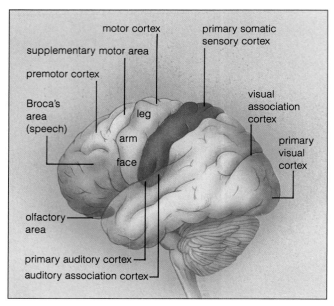

Figure 33.13 Primary receiving and association areas for the human cerebral cortex. Signals from receptors on the body's periphery enter primary cortical areas. Sensory input from different receptors is coordinated and processed in association areas. The text describes the main cortical regions. Also shown here are the *premotor area*, involved in intricate motor activity (as typified by a concert pianist performing); the *supplementary motor area*, which helps coordinate sequential voluntary movements; and *Broca's area*, which coordinates muscles required for speech.

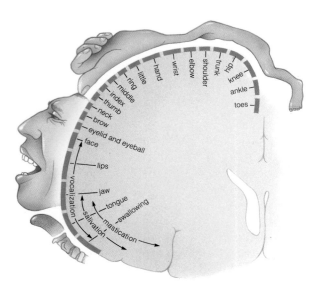

Figure 33.14 Section through the motor cortex of the right hemisphere of someone facing you. Notice the distorted human body draped over the diagram. The distortions show which body regions are controlled by different parts of the motor cortex.

Functional Regions of the Cerebral Cortex

Several regions of the cortex have been identified as having different functions. Some regions receive sensory input from the skin, muscles, and tendons. Others receive sensory input from the nose, eyes, and ears, and so on. Still other regions are concerned with coordinating and processing information, and with coordinating appropriate responses (Figure 33.13).

The motor centers, for example, coordinate instructions for motor responses. By experimentally stimulating different points on the motor cortex, researchers discovered that they could stimulate the contraction of different muscles. Much of the motor cortex deals with thumb and tongue muscles, indicating the extent of control required for hand movements and verbal expression (Figure 33.14). A primary center for receiving input from the skin and joints lies just behind the motor cortex. In regions called "association centers," information from memory stores is added to the primary sensory information to give it fuller meaning.

Memory

Have you heard the one about the elderly grandparents who got a craving for ice cream? Gramps went to the store and brought back bacon, and Granny yelled, "You darned fool, you forgot the eggs!"

Memory is the storage and retrieval of information about previous experiences. We take it for granted—that is, until it starts to falter as we age. Our memory begins to develop before birth. It is a treasurehouse of experience, and it underlies the capacity for learning.

Today we know that information becomes stored in stages. *Memory traces*, the chemical and structural changes necessary for storage, occur in many different brain regions (Figure 33.15). Short-term memory lasts a few seconds to a few hours and seems to be limited to only a few bits of information—words of a sentence, numbers, and so on. Long-term memory lasts more or less permanently, and it seems to be limitless in its capacity.

Observations of people suffering from *retrograde amnesia* tell us something about memory. These people can't remember anything that happened during the half hour or so before losing consciousness from a severe head blow. Yet memories of events before that time remain intact! Similarly, lesions in the hippocampus are known to cause the loss of short-term memory. (The hippocampus is part of the limbic system, as shown in Figure 33.15.) Affected individuals live entirely in the present.

These observations suggest that whereas short-term memory is a fleeting stage of neural excitation, long-term memory depends on chemical or structural changes in the brain.

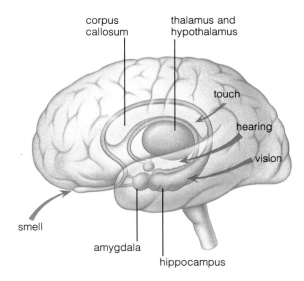

Figure 33.15 Some brain regions that play key roles in memory. Sensory input is processed by the cerebral cortex and sent into parts of the limbic system and parts of the forebrain. The limbic system, our "emotional brain," includes the regions called the thalamus, hypothalamus, amygdala, and hippocampus.

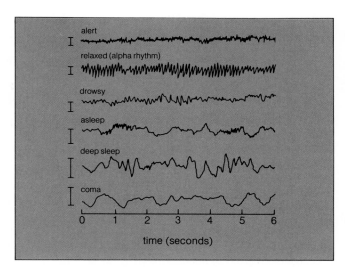

Figure 33.16 EEG patterns. Vertical bars indicate a range of 50 microvolts of electrical activity. The irregular horizontal graph lines indicate the electrical response with time.

The prominent wave pattern for someone who is relaxed, with eyes closed, is an *alpha rhythm*. The EEG waves are recorded in "trains" of one after the other, about ten per second. Alpha waves predominate during the state of meditation. With a transition to sleep, wave trains gradually become larger, slower, and more erratic. This *slow-wave sleep* pattern shows up about 80 percent of the total sleeping time for adults. It occurs when sensory input is low and the mind is more or less idling. Subjects awakened from slow-wave sleep usually report that they were not dreaming. Often they seemed to be mulling over recent, ordinary events.

Slow-wave sleep is punctuated by brief spells of *REM sleep. Rapid Eye Movements* accompany this pattern (the eyes jerk beneath closed lids). So do irregular breathing, faster heartbeat, and twitching fingers. Most people awakened from REM sleep report they were experiencing vivid dreams. The transition from sleep or deep relaxation into alert wakefulness is marked by a shift to low-amplitude, higher frequency wave trains. Associated with this accelerated brain activity are increased blood flow and oxygen uptake in the cortex. The transition, *EEG arousal*, occurs when conscious effort is made to focus on external stimuli or even on one's own thoughts.

Also, information seemingly forgotten can be recalled after being unused for decades, so individual memory traces must be encoded in a form somewhat resistant to degradation. Most molecules and cells in your body are used up, wear out, or age and are constantly being replaced—yet memories can be retrieved in exquisite detail after many years of such wholesale turnovers. Gramps may forget the ice cream, but he may also dredge up excruciatingly detailed examples of how well behaved *he* was when *he* was growing up.

Nerve cells are among the few kinds that are *not* replaced. You are born with billions, and as you grow older some 50,000 die off steadily each day. The nerve cells formed during embryonic development are the same ones present, whether damaged or otherwise modified, at the time of death.

The part about being "otherwise modified" is tantalizing. *There is evidence that neuron structure is not static, but rather can be modified in several ways.* Most likely, such modifications depend on electrical and chemical interactions with neighboring neurons. Electron micrographs show that some synapses "wither" as a result of disuse. Such regression weakens or breaks connections between neurons. The visual cortex of mice raised without visual stimulation showed such effects of disuse. Similarly, there is some evidence that intensively stimulated synapses may form stronger connections, grow in size, or sprout buds or spines to form more connections! The chemical and physical transformations that underlie changes in synaptic connections may correspond to memory storage.

States of Consciousness

States of consciousness include sleeping, dozing, daydreaming, and full alertness. The central nervous system governs these states, and psychoactive drugs can alter them, as described in the *Commentary* on the next page.

Throughout the spectrum of consciousness, neurons in the brain are constantly chattering among themselves. This neural chatter shows up as wavelike patterns in an *electroencephalogram* (EEG). An EEG is an electrical recording of the frequency and strength of potentials from the brain's surface. Figure 33.16 shows examples of EEG patterns. More recently, PET scans are providing information on the precise location of brain activity while it is occurring. Such scans are described on page 22.

Drug Action on Integration and Control

Broadly speaking, a drug is any substance introduced into the body to provoke a specific physiological response. Some drugs help a person cope with discomforts of an illness or stress. Others act on the *pleasure center* in the hypothalamus and artificially fan the sense of pleasure that we associate with eating, sexual activity, and other self-gratifying behaviors.

Many drugs are habit-forming. Even if the body functions well enough without them, they continue to be used for the real or imagined relief they afford. Often the body develops *tolerance* of such drugs, meaning it takes larger or more frequent doses to produce the same effect. Habituation and tolerance are signs of **addiction**, or chemical dependence, on a drug. *The drug has taken on an "essential" biochemical role in the body*. Abruptly deprive an addict of the drug, and he or she will suffer agonizing physical pain as well as mental anguish. Such withdrawal symptoms are manifestations of major biochemical upheavals throughout the body.

Here we consider four classes of psychoactive drugs, which act on brain regions governing states of consciousness and behavior. They are the depressants and hypnotics, stimulants, narcotic analgesics, and hallucinogens and psychedelics.

Stimulants

Stimulants include caffeine, nicotine, amphetamines, and cocaine. First they increase alertness and body activity, then they lead to depression.

Coffee, tea, chocolate, and many soft drinks contain caffeine, one of the most widely used stimulants. Low doses of caffeine stimulate the cerebral cortex first, and cause increased alertness and restlessness. Higher doses act at the medulla oblongata to disrupt motor coordination and intellectual coherence.

Nicotine, a component of tobacco, has powerful effects on the central and peripheral nervous systems. It mimics acetylcholine and can directly stimulate a number of sensory receptors. Its short-term effects include water retention, irritability, increased heart rate and blood pressure, and gastric upsets. Its long-term effects can be devastating.

Like dopamine and norepinephrine (which they resemble), the amphetamines (including "speed") stimulate the pleasure center. In time, the brain produces less and less of its own signaling molecules and comes to depend on artificial stimulation.

Cocaine stimulates the pleasure center in a different way. It produces a rush of pleasure by blocking the reabsorption of dopamine, norepinephrine, serotonin, and other signaling molecules that are normally released at synapses. Receptor cells are incessantly stimulated over an extended period. Heart rate and blood pressure rise; sexual appetite increases. But then the effects change. The signaling molecules that have accumulated in synaptic clefts diffuse away—but the cells that produce them cannot make up for the extraordinary loss. The sense of pleasure evaporates as the receptor cells (which are now hypersensitive to stimulation) demand stimulation. The cocaine user becomes anxious and depressed. After prolonged, heavy use of cocaine, "pleasure" is impossible to experience. The addict loses weight and cannot sleep properly. The immune system becomes compromised, and heart abnormalities set in.

Granular cocaine, which is inhaled (snorted), has been around for some time. Crack cocaine, a cheaper but more potent form, is burned and the smoke inhaled. As suggested at the start of this chapter, crack is incredibly addictive; its highs are higher, but the crashes are more devastating.

In the brain stem, a branch of the reticular formation controls the changing levels of consciousness. It sends signals to the spinal cord, cerebellum, and cerebrum as well as back to itself. The flow of signals along these circuits—and the inhibitory or excitatory chemical changes accompanying them—affects whether you stay awake or fall asleep. Damage to parts of the circuits can lead to unconsciousness and coma.

Interneurons of one of the "sleep centers" of the reticular formation release serotonin. This transmitter substance inhibits other neurons that arouse the brain and maintain wakefulness. Thus, high serotonin levels are linked to drowsiness and sleep. Substances released from another brain center counteract serotonin's effects and bring about wakefulness.

Emotional States

Our emotions are governed by the cerebral cortex and by different brain regions collectively called the **limbic**

Classes of Psychoactive Drugs	
Class	Examples
Depressants, hypnotics	Barbiturates (e.g., Nembutal, Quaalude) Antianxiety drugs (e.g., Valium, alcohol)
Stimulants	Caffeine Nicotine Amphetamines (e.g., Dexedrine) Cocaine
Narcotic analgesics	Codeine Opium Heroin
Psychedelics, hallucinogens	Lysergic acid diethylamide (LSD) *Cannabis* (marijuana)

Depressants, Hypnotics

These drugs lower the activity in nerves and parts of the brain, so they reduce activity throughout the body. Some act at synapses in the reticular formation system and in the thalamus.

Depending on the dosage, most of these drugs can produce responses ranging from emotional relief, sedation, sleep, anesthesia and coma, to death. At low doses, inhibitory synapses are often suppressed slightly more than excitatory synapses, so the person feels excited or euphoric at first. Increased doses also suppress excitatory synapses, leading to depression. Depressants and hypnotics have additive effects; one amplifies another. For example, combining alcohol with barbiturates amplifies behavioral depression.

Alcohol (ethyl alcohol) differs from the drugs just described because it acts directly on the plasma membrane to alter cell function. Some persons mistakenly think of it as a harmless stimulant (it produces an initial "high"). But alcohol is one of the most powerful psychoactive drugs and a major cause of death. Small doses even over the short term can produce disorientation, uncoordinated motor functions, and diminished judgment. Long-term addiction destroys nerve cells and causes permanent brain damage; it can permanently damage the liver (cirrhosis).

Analgesics

When stress leads to physical or emotional pain, the brain produces its own analgesics, or natural pain relievers. Endorphins and enkephalins are examples. These substances seem to inhibit activity in many parts of the nervous system, including brain centers concerned with emotions and perception of pain.

The narcotic analgesics, including codeine and heroin, sedate the body and relieve pain. They are extremely addictive. Deprivation following massive doses of heroin leads to fever, chills, hyperactivity and anxiety, violent vomiting, cramping, and diarrhea.

Psychedelics, Hallucinogens

These drugs, which alter sensory perception, have been described as "mind-expanding." Some skew acetylcholine or norepinephrine activity. Others, such as LSD (lysergic acid diethylamide), affect serotonin activity. Even in small doses, LSD dramatically warps perceptions.

Marijuana is another hallucinogen. The name refers to the drug made from crushed leaves, flowers, and stems of the plant *Cannabis*. In low doses marijuana is like a depressant. It slows down but does not impair motor activity; it relaxes the body and elicits mild euphoria. However, it can produce disorientation, increased anxiety bordering on panic, delusions (including paranoia), and hallucinations.

Like alcohol, marijuana can affect an individual's ability to perform complex tasks, such as driving a car. In one study, commercial pilots showed a marked deterioration in instrument-flying ability for more than two hours after smoking marijuana. Recent studies point to a link between marijuana smoking and suppression of the immune system.

system (Figure 33.15). The limbic system of humans is only distantly related to the sense of smell that figured so prominently in vertebrate evolution. Even so, interconnections have been maintained that play some role in memory function and cognition.

The connections between the sense of smell and the limbic system are the reason why you may "smell" a cologne all over again when you have a pleasant memory of the person who wore it; or why you smell a bad odor when you remember a confrontation with a skunk.

The hypothalamus is the gatekeeper of the limbic system. Many connections from the cerebral cortex and lower brain centers pass through it. Through these connections, the reasoning possible in the cerebral cortex can dampen rage, hatred, and other so-called "gut reactions."

The hypothalamus also monitors internal organs in addition to emotional states. This is what keeps your heart and stomach on fire when you are sick with passion (or indigestion).

SUMMARY

1. Nervous systems provide specialized means of detecting stimuli and responding to them swiftly.

2. The simplest nervous systems are nerve nets, such as those of sea anemones, hydra, and jellyfishes. They are based on reflex connections between nerve cells and contractile cells of the epithelium.

3. The nervous system of vertebrates shows pronounced cephalization and bilateral symmetry, as evident in its complex brain centers and its many paired nerves.

4. Parts of the vertebrate brain deal with reflex coordination of sensory inputs and motor outputs beyond that afforded by the spinal cord alone. Its most recent layerings also deal with storing, comparing, and using experiences to initiate novel, nonstereotyped action. These regions are the basis of memory, learning, and reasoning.

5. The brain and spinal cord represent the central nervous system. Many pairs of nerves carry signals between the central nervous system and the body's various organs and structures. These nerves are the basis of the peripheral nervous system.

6. The peripheral nervous system has a somatic subdivision, which deals with skeletal muscles concerned with voluntary body movements. It also has an autonomic subdivision, which deals with the functions of the heart, lungs, glands, and other internal organs.

7. The spinal cord has nerve tracts that carry signals between the brain and the peripheral nervous system. It also is a center for some direct reflex connections that underlie limb movements and internal organ activity.

8. The brain has three regional divisions (hindbrain, midbrain, and forebrain).

    a. The hindbrain includes the medulla oblongata, pons, and cerebellum and contains reflex centers for vital functions and muscle coordination.

    b. The midbrain functions in coordinating and relaying visual and auditory information.

    c. The medulla oblongata, pons, and midbrain constitute the brain stem. The reticular formation, an extensive network of interneurons, extends the length of the brain stem and helps govern activities of the nervous system as a whole.

    d. The forebrain includes the cerebrum, thalamus, hypothalamus, and limbic system. The thalamus relays sensory information and helps coordinate motor responses. The hypothalamus monitors internal organs and influences thirst, hunger, sexual activity, and other behaviors related to their functioning. It is gatekeeper to the limbic system, which has roles in learning, memory, and emotional behavior.

9. The cerebral cortex has regions devoted to specific functions, such as receiving information from the various sense organs, integrating this information with memories of past events, and coordinating motor responses.

10. Memory apparently occurs in two stages: a short-term formative period and long-term storage, which depends on chemical or structural changes in the brain.

11. States of consciousness vary between total alertness and deep coma. The levels are governed by the reticular activating system. They are subject to the influence of psychoactive drugs.

Review Questions

1. What are some of the organizational features that nearly all nervous systems have in common? *564*

2. What constitutes the central nervous system? The peripheral nervous system? *566*

3. Can you distinguish among the following:
    a. ganglia and nerves *564*
    b. spinal nerves and cranial nerves *566*
    c. somatic system and autonomic system *567*
    d. parasympathetic and sympathetic nerves *567–568*

4. Review Figure 33.10. Then, on your own, describe the components of the three main subdivisions of the vertebrate brain. *570*

5. What is a psychoactive drug? Can you describe the effects of one such drug on the central nervous system? *576–577*

6. Label the parts of the human brain: *572*

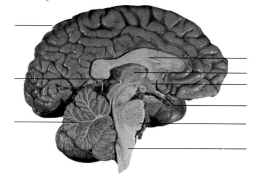

Self-Quiz *(Answers in Appendix IV)*

1. Sea anemones, hydras, and jellyfishes have simple nervous systems called _____.

2. Structurally, the vertebrate nervous system shows pronounced _____ and _____ symmetry.

3. The oldest parts of the vertebrate brain provide _____.
   a. reflex control of breathing, blood circulation, and other basic activities
   b. coordinating and relaying visual and auditory information
   c. storing, comparing, and using experiences to initiate novel, nonstereotyped action
   d. both a and c are correct

4. The central nervous system includes _____. The peripheral nervous system includes _____.
   a. nerves and ganglia; brain and spinal cord
   b. brain and spinal cord; nerves and ganglia
   c. spinal cord; brain
   d. nerves and interneurons; brain and spinal cord

5. Overall, _____ nerves slow down the body and divert energy to digestion and other housekeeping tasks; _____ nerves slow down housekeeping tasks and increase overall activity during times of heightened awareness, excitement, or danger.
   a. autonomic; somatic
   b. sympathetic; parasympathetic
   c. parasympathetic; sympathetic
   d. peripheral; central

6. Parasympathetic and sympathetic nerves carry signals to and from the nervous system _____ to bring about minor adjustments in internal organ activity.
   a. continually
   b. in separate, alternating fashion
   c. only during a fight-flight response
   d. none of the above is correct

7. The _____ of the spinal cord consists of nerve tracts; the _____ consists of dendrites, neuron cell bodies, and neuroglial cells.
   a. gray matter; white matter
   b. white matter; gray matter

8. The hindbrain (medulla oblongata, cerebellum, and pons) contains _____.
   a. the reticular formation
   b. major nerve tracts between brain centers
   c. reflex centers for limb movements, respiration, breathing, and other vital tasks
   d. both b and c are correct

9. Extending through the entire length of the brain stem (midbrain, pons, and medulla oblongata) is the _____.
   a. reticular formation
   b. blood-brain barrier
   c. olfactory lobe
   d. tectum

10. The most highly developed part of the human brain (the forebrain) includes the _____.
    a. medulla oblongata, pons, and cerebellum
    b. cerebrum, thalamus, hypothalamus, and limbic system
    c. medulla oblongata, pons, and cerebral cortex
    d. cerebellum, medulla oblongata, pons, and limbic system
    e. hypothalamus, limbic system, pons, and cerebral cortex

11. Match the central nervous system region with some of its functions.
    _____ spinal cord
    _____ medulla oblongata
    _____ hypothalamus
    _____ limbic system
    _____ cerebral cortex

    a. receives sensory input, integrates it with stored information
    b. monitors internal organs and related behavior (e.g., hunger)
    c. with cerebral cortex, governs emotions
    d. reflex control of respiration, blood circulation, other basic activities
    e. expressway for signals between brain and peripheral nervous system

Selected Key Terms

addiction *576*
autonomic system *567*
blood-brain barrier *571*
brain *570*
brain stem *570*
central nervous system *566*
cephalization *564*
cerebellum *571*
cerebral cortex *571*
cerebrospinal fluid *571*
cerebrum *571*
corpus callosum *572*
forebrain *571*
ganglion *564*
gray matter *569*
hindbrain *570*
hippocampus *574*
hypothalamus *571*
limbic system *576*
medulla oblongata *570*

memory *574*
meninges *569*
midbrain *571*
nerve *564*
nerve cord *564*
nerve net *564*
nerve tract *569*
notochord *566*
parasympathetic nerve *567*
peripheral nervous system *566*
pons *571*
reflex *564*
reticular formation *571*
somatic system *567*
spinal cord *569*
sympathetic nerve *567*
tectum *571*
thalamus *571*
white matter *569*

Readings

Barlow, R., Jr. April 1990. "What the Brain Tells the Eye." *Scientific American* 262(4):90–95.

Bloom, F., and A. Lazerson. 1988. *Brain, Mind, and Behavior.* Second edition. New York: Freeman.

Churchland, P., and P. Churchland. January 1990. "Could a Machine Think?" *Scientific American* 262(1):32–37.

Julien, R. 1985. *A Primer of Drug Action.* Fourth edition. New York: Freeman. Effectively fills the gap between popularized (and often superficial or misleading) accounts of drug action and the upper-division books in pharmacology. Paperback.

Kalil, R. December 1989. "Synapse Formation in the Developing Brain." *Scientific American* 261(6):76–85.

Romer, A., and T. Parsons. 1986. *The Vertebrate Body.* Sixth edition. Philadelphia: Saunders. Insights into the evolution of vertebrate nervous systems.

Shepherd, G. 1988. *Neurobiology.* Second edition. New York: Oxford. Paperback.

Springer, S., and Deutsch, G. 1985. *Left Brain, Right Brain.* Revised edition. New York: Freeman.

Hormone Jamboree

In the early 1960s, at her camp in the forests along the shores of Lake Tanganyika in Africa, the primatologist Jane Goodall let it be known that bananas were available. Among the first chimpanzees attracted to the delicious new food was a female—Flo, as she came to be called. Flo brought along her offspring, an infant female and a juvenile male, and tended carefully to them. Three years later, Flo's preoccupation with being a mother gave way to a preoccupation with sex. Male chimpanzees followed Flo to the camp, and stayed for more than the bananas.

Sex, Goodall discovered, is the premier force in the social life of chimpanzees. These primates do not mate for life, as eagles do, or wolves. Before the rainy season begins, the mature females that are undergoing a fertile cycle become sexually active. In response to changing blood concentrations of sex hormones, their external genitalia become enormously swollen and pink, a visual signal that is magnetic to males. Their swellings are the flags of sexual jamborees, of great gatherings of highly

stimulated chimps in which any males present may copulate in sequence with the same female.

Although a swollen bottom makes it rather difficult to sit and is vulnerable to being torn, it has its advantages. The gathering of many flag-waving females in the same place draws together individuals that otherwise would spend most of the year foraging alone or in small family groups. Now they spend time together, reestablishing bonds that hold them together in their rather fluid community. Infants and juveniles interact with one another and with the adults. Future dominance hierarchies have their foundations in the playful and aggressive jostlings. Consider Flo, a high-ranking member of the social hierarchy. Through her sexual attractiveness and direct solicitations, she built alliances with many male chimps. Through her high

Figure 34.1 (**a**) Primatologist Jane Goodall in Gombe National Park, near the shores of Lake Tanganyika, scouting for chimpanzees. (**b**) The socially dominant Flo, center, and three of her offspring.

a

b

status and aggressive behavior, she helped her male offspring win confrontations with other young male chimps.

Intriguingly, the sexual swelling lasts somewhere between ten and sixteen days—yet the female actually is fertile for only one to five days. Furthermore, swellings may occur during nonfertile periods and at irregular times even after a female has become pregnant. It is not difficult to imagine why prolonged swelling has been favored during the course of chimpanzee evolution. The males groom a sexually attractive female more often and give her a larger share of their food. She is allowed to travel with males to new, peripheral food sources and is protected by them. The greater her acceptance by males, the higher up she goes in the social hierarchy—and the more her offspring benefit.

Through their effects, hormones help orchestrate the reproductive cycle of chimpanzees. They do the same for nearly all animals, from invertebrate worms to humans. Besides this, hormones help orchestrate growth and development. They help control minute-to-minute and day-to-day metabolic functions. Through their interplays with one another and with the nervous system, hormones have major influence over the physical appearance of individuals, their well being, and how they will behave. How individuals behave has major influence on whether they survive, either on their own or as members of social groups.

This chapter is about hormones and other signaling molecules—their sources, targets, and interactions. It is about their effects and the molecular mechanisms governing their secretion and action. If the details start to seem remote, remember that this is the stuff of life. Hormones underwrote Flo's appearance, behavior, and rise through the chimpanzee's social hierarchy—and just imagine what they have been doing for you.

KEY CONCEPTS

1. Hormones and other signaling molecules help integrate cell activities in ways that benefit the whole body. Some hormones help the body adjust to short-term changes in diet and levels of activity. Other hormones have roles in long-term adjustments underlying growth, development, and reproduction.

2. The hypothalamus and pituitary gland interact in ways that coordinate the stimulation and inhibition of many endocrine glands. Together they control many of the body's functions.

3. Neural signals, hormonal signals, chemical changes in the blood, and environmental cues trigger hormone secretions. Only cells with receptors for a specific hormone are its targets. Steroid hormones trigger the activation of genes and protein synthesis in target cells. Protein hormones alter the activity of existing enzymes in target cells.

"THE ENDOCRINE SYSTEM"

The word "hormone" dates back to the early 1900s, when W. Bayliss and E. Starling were trying to figure out what triggers the secretion of pancreatic juices that act on food traveling through the canine gut. At the time, it was known that acids mix with food in the stomach, and that the pancreas secretes an alkaline solution when the acidic mixture has passed into the small intestine. Was the nervous system or something else stimulating the pancreatic response?

To find the answer, Bayliss and Starling cut off the nerve supply to the upper small intestine but left its blood vessels intact. When an acid was introduced into the intestinal region, the pancreas still was able to make the secretory response. More telling, extracts of cells taken from the epithelial lining of the small intestine also induced the response. Glandular cells in the lining had to be the source of a substance that stimulated the pancreas into action.

The substance itself came to be called secretin. Demonstration of its existence and its mode of action was the first confirmation of an idea that had been around for centuries: *Internal secretions released into the bloodstream influence the activities of tissues and organs.* Starling coined the word hormone for such internal glandular secretions (the Greek *hormon* means to set in motion).

Later work led to the discovery of many hormones and their sources (Figure 34.2). The sources include but are not limited to the following array, which is characteristic of most vertebrates:

Pituitary gland
Pineal gland
Thyroid gland
Parathyroid glands (number varies; four in humans)
Adrenal glands (two)
Gonads (two)
Thymus gland
Pancreatic islets (multiple)
Endocrine cells of gut, liver, kidneys, placenta

These scattered sources of hormones came to be viewed as "the endocrine system." The phrase implied that they formed a separate means of control within the body, apart from the nervous system. (The Greek *endon* means within; *krinein* means to separate.)

In time, however, refined biochemical studies and improvements in electron microscopy revealed that the boundaries between the nervous and endocrine systems are not sharply defined. Glands thought to be free of neural control turned out to be well supplied with nerves. Some neurons actually secrete "hormones." The hypothalamus, a brain region, exerts major control over the pituitary—the so-called master gland of the endocrine system.

We no longer can say that a natural division exists between the nervous and endocrine systems, given their intricate overlapping. For that reason, we will begin with the secretions that serve as agents of integration, regardless of their neural or endocrine origin.

HORMONES AND OTHER SIGNALING MOLECULES

Cells have built-in means of responding to chemical changes in their immediate surroundings, including mechanisms for the uptake and release of substances. In complex animals, the individual responses of thousands, millions, even many billions of cells must be integrated in ways that benefit the whole body.

Integration depends on *signaling molecules*. These are hormones and other chemical secretions that alter the behavior of target cells. Any cell is a "target" if it has receptors to which specific signaling molecules can bind and elicit a cellular response. A target may or may not be adjacent to the secreting cell.

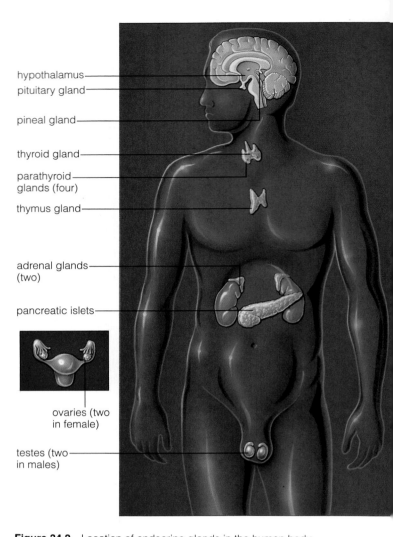

Figure 34.2 Location of endocrine glands in the human body. The liver and kidneys do other things besides secreting hormones, but these organs do have component endocrine cells. Intestinal epithelium also has endocrine cells; so does the heart.

Different cells and glands produce at least four types of signaling molecules, which may be defined as follows:

1. Hormones are secreted from endocrine glands, endocrine cells, and some neurons, then are transported by the bloodstream to nonadjacent targets.

2. Transmitter substances are released from neurons, act on immediately adjacent target cells, then are rapidly degraded or recycled (page 555).

3. Local signaling molecules are secreted from cells in many different tissues. They alter chemical conditions in the immediate vicinity, then are degraded swiftly.

4. Pheromones, secreted by some exocrine glands, have targets outside the body. They diffuse through water or air and act on cells of other animals of the same species. They integrate social activities between animals.

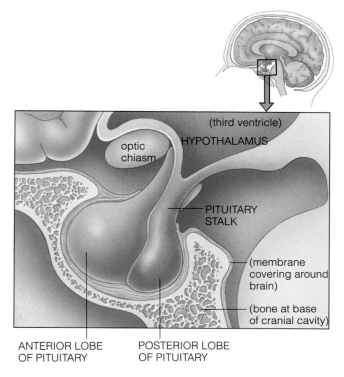

Figure 34.3 Location of the pituitary gland in an adult human.

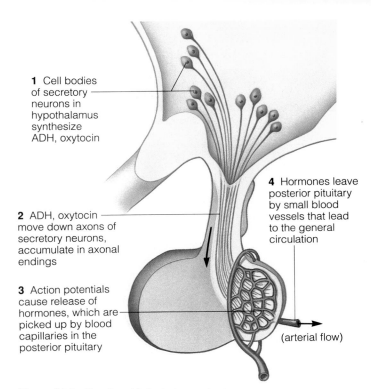

1 Cell bodies of secretory neurons in hypothalamus synthesize ADH, oxytocin

2 ADH, oxytocin move down axons of secretory neurons, accumulate in axonal endings

3 Action potentials cause release of hormones, which are picked up by blood capillaries in the posterior pituitary

4 Hormones leave posterior pituitary by small blood vessels that lead to the general circulation

(arterial flow)

Figure 34.4 Functional links between the hypothalamus and the posterior lobe of the pituitary in humans.

THE HYPOTHALAMUS-PITUITARY CONNECTION

The hypothalamus and pituitary work together to integrate many of the body's activities, and so are called the **neuroendocrine control center**. The hypothalamus, recall, monitors internal organs and emotional states. The human pituitary weighs 0.5 gram and is about the size of a garden pea. It is suspended from a slender stalk that extends downward from the base of the hypothalamus (Figure 34.3).

The pituitary gland is lobed. Its *posterior lobe* stores and secretes two hormones, which are actually produced in the hypothalamus. Its *anterior lobe* produces and secretes six hormones, some of which control the release of several more hormones from other endocrine glands. The pituitary of many vertebrates (not humans) also has an *intermediate lobe*. Often the secretions from this third lobe induce changes in skin color.

Posterior Lobe Secretions

Figure 34.4 shows the links between the hypothalamus and the posterior lobe of the pituitary. Notice the cell bodies of neurons in the hypothalamus and their axons, which extend down the pituitary stalk and into the posterior lobe. Antidiuretic hormone (ADH) and oxytocin are produced in the neuron cell bodies and stored in the axon endings, next to a capillary bed. In such "beds," thin-walled blood vessels (capillaries) thread profusely through the surrounding interstitial fluid. ADH or oxy-

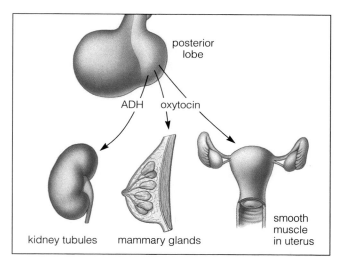

Figure 34.5 Secretions of the posterior lobe of the pituitary and some of their targets.

tocin released from the axon endings diffuse into the capillaries. They are transported throughout the body by the bloodstream, and they act on specific target cells (Figure 34.5).

Cells in the walls of kidney tubules are targets for ADH. This hormone helps control the volume of extracellular fluid by preventing excess water loss in the urine. At high concentration, ADH also has a role in adjustments over the volume of blood flowing through

Table 34.1	Effect of Releasing and Inhibiting Hormones on Anterior Pituitary	
Hormone	Influences Secretion of:	Effect*
Corticotropin-releasing hormone (CRH)	Corticotropin (ACTH)	+
Thyrotropin-releasing hormone (TRH)	Thyrotropin (TSH)	+
Gonadotropin-releasing hormone (GnRH)	Follicle-stimulating hormone (FSH)	+
	Luteinizing hormone (LH)	+
STH-releasing hormone (STHRH)	Somatotropin (STH); also called growth hormone (GH)	+
Somatostatin	Somatotropin, TSH	−
Dopamine	Prolactin (PRL), LH, FSH	−

*Stimulatory (+) or inhibitory (−).

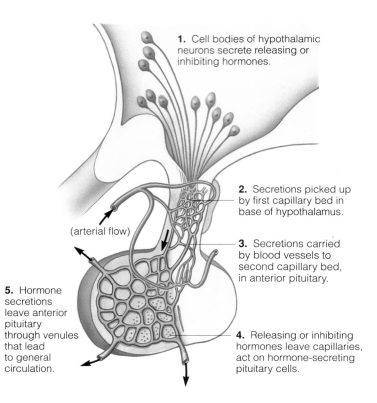

1. Cell bodies of hypothalamic neurons secrete releasing or inhibiting hormones.

2. Secretions picked up by first capillary bed in base of hypothalamus.

(arterial flow)

3. Secretions carried by blood vessels to second capillary bed, in anterior pituitary.

5. Hormone secretions leave anterior pituitary through venules that lead to general circulation.

4. Releasing or inhibiting hormones leave capillaries, act on hormone-secreting pituitary cells.

Figure 34.6 Functional links between the hypothalamus and the anterior lobe of the pituitary.

different regions of the body. It causes arterioles in some tissues to constrict and so helps divert blood flow to other tissues, where metabolic demands are greater. Oxytocin has roles in reproduction. For example, it triggers muscle contractions in the uterus during labor and causes milk to be released when the young are being nursed.

Anterior Lobe Secretions

Role of the Hypothalamus. The hypothalamus produces and secretes **releasing hormones** and **inhibiting hormones**. Both are signaling molecules that act on specific cells of the anterior lobe of the pituitary. Most stimulate target cells to secrete other hormones; others slow down the secretions (Table 34.1). As Figure 34.6 shows, the releasing or inhibiting hormones travel through two capillary beds before leaving the blood and binding to receptors on target cells in the anterior lobe.

Releasing hormones were identified by monumental research efforts that began in 1955, most notably by Roger Guillemin and Andrew Schally. Over one four-year period, Guillemin's team purchased 500 tons of sheep brains from meat processing plants and extracted 7 tons of hypothalamic tissue from them. They eventually ended up with a single milligram of a substance that stimulates the release of "TSH," a hormone that controls the functions of the thyroid gland.

Anterior Pituitary Hormones. In response to commands from the hypothalamus, different cells of the anterior pituitary secrete the following hormones of their own:

Corticotropin	ACTH
Thyrotropin	TSH
Follicle-stimulating hormone	FSH
Luteinizing hormone	LH
Prolactin	PRL
Somatotropin (or growth hormone)	STH (or GH)

The first four hormones listed act on endocrine glands, which in turn produce other hormones. The effect of ACTH on adrenal glands and TSH on the thyroid will be described shortly. FSH and LH play elegant roles in reproductive function, a topic that will occupy our attention in Chapter 43.

The last two hormones listed, prolactin and somatotropin, have effects on body tissues in general (Figure 34.7 and Table 34.2). Prolactin influences a variety of activities among vertebrate species ranging from primitive fishes to humans. One of its functions is to stimulate and sustain milk production in mammary glands during lactation (page 775). Prolactin does this only when the tissues in those glands have been primed by other hormones. Prolactin also affects the production of hormones by ovaries.

Table 34.2 Hormones Released from the Mammalian Pituitary Gland

Pituitary Lobe	Secretions	Abbreviation	Main Targets	Primary Actions
Posterior				
Nervous tissue (extension of hypothalamus)	Antidiuretic hormone	ADH	Kidneys	Induces water conservation required in control of extracellular fluid volume (and, indirectly, solute concentrations)
	Oxytocin		Mammary glands Uterus	Induces milk movement into secretory ducts Induces uterine contractions
Anterior				
Mostly glandular tissue	Corticotropin	ACTH	Adrenal cortex	Stimulates release of adrenal steroid hormones
	Thyrotropin	TSH	Thyroid gland	Stimulates release of thyroid hormones
	Gonadotropins: Follicle-stimulating hormone	FSH	Ovaries, testes	In females, stimulates follicle growth, helps stimulate estrogen secretion, ovulation; in males, promotes spermatogenesis
	Luteinizing hormone	LH	Ovaries, testes	In females, stimulates ovulation, corpus luteum formation; in males, promotes testosterone secretion, sperm release
	Prolactin	PRL	Mammary glands	Stimulates and sustains milk production
	Somatotropin (also called growth hormone)	STH (GH)	Most cells	Has growth-promoting effects in young; induces protein synthesis, cell division; has role in glucose, protein metabolism in adults
Intermediate *				
Mostly glandular tissue	Melanocyte-stimulating hormone	MSH	Pigmented cells in skin, other surface coverings	Induces color changes in response to external stimuli; affects behavior

*Present in most vertebrates (not adult humans). MSH is associated with the anterior lobe in humans.

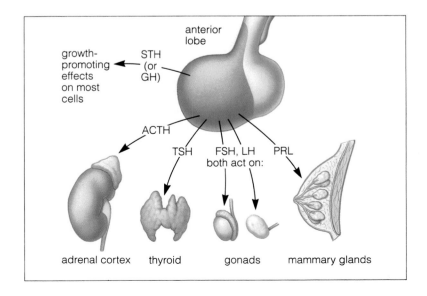

Figure 34.7 Secretions of the anterior lobe of the pituitary and some of their targets.

Figure 34.8 (a) Manute Bol, an NBA center, is 7 feet 6-3/4 inches tall owing to excessive STH production during childhood.

(b) Effect of somatotropin (STH) on overall body growth. The person at the center is affected by gigantism, which resulted from excessive STH production during childhood. The person at right displays pituitary dwarfism, which resulted from underproduction of STH during childhood. The person at the left is average in size.

a

b

Figure 34.9 Acromegaly, which resulted from excessive production of STH during adulthood. Before this female reached maturity, she was symptom-free.

age nine   sixteen

thirty-three   fifty-two

Somatotropin, or growth hormone, stimulates protein synthesis and cell division in target cells. It profoundly influences overall growth, especially of cartilage and bone. Figure 34.8 shows what can happen with too little or too much somatotropin. *Pituitary dwarfism* results when not enough somatotropin was produced during childhood. The adult is similar in proportion to a normal person but much smaller. *Gigantism* results when excessive amounts of somatotropin were produced during childhood. The adult is similar in proportion to a normal person but is much larger.

Acromegaly results from excessive secretion of somatotropin during adulthood, when long bones no longer can lengthen. Cartilage, bone, and other connective tissues of the hands, feet, and jaws thicken, as do epithelial tissues of the skin, nose, eyelids, lips, and tongue (Figure 34.9). Skin thickening is pronounced on the forehead and soles of the feet.

SELECTED EXAMPLES OF HORMONAL ACTION

Table 34.3 lists hormones from endocrine glands other than the pituitary. Here we will focus on a few examples of endocrine activity to show how hormonal controls work. These examples will lay the groundwork for understanding how hormones affect digestion, circulation, reproduction, and other activities described in chapters to come.

Table 34.3 Hormone Sources Other Than the Mammalian Hypothalamus and Pituitary

Source	Its Secretion(s)	Main Targets	Primary Actions
Adrenal cortex	Glucocorticoids (including cortisol)	Most cells	Promote protein breakdown and conversion to glucose
	Mineralocorticoids (including aldosterone)	Kidney	Promote sodium reabsorption; control salt, water balance
Adrenal medulla	Epinephrine (adrenalin)	Liver, muscle, adipose tissue	Raises blood level of sugar, fatty acids; increases heart rate, force of contraction
	Norepinephrine	Smooth muscle of blood vessels	Promotes constriction or dilation of blood vessel diameter
Thyroid	Triiodothyronine, thyroxine	Most cells	Regulates metabolism; has roles in growth, development
	Calcitonin	Bone	Lowers calcium levels in blood
Parathyroids	Parathyroid hormone	Bone, kidney	Elevates calcium levels in blood
Gonads:			
Testis (in males)	Androgens (including testosterone)	General	Required in sperm formation, development of genitals, maintenance of sexual traits; influences growth, development
Ovary (in females)	Estrogens	General	Required in egg maturation and release; prepares uterine lining for pregnancy; other actions same as above
	Progesterone	Uterus, breast	Prepares, maintains uterine lining for pregnancy; stimulates breast development
Pancreatic islets	Insulin	Muscle, adipose tissue	Lowers blood sugar level
	Glucagon	Liver	Raises blood sugar level
	Somatostatin	Insulin-secreting cells of pancreas	Influences carbohydrate metabolism
Endocrine cells of stomach, gut	Gastrin, secretin, etc.	Stomach, pancreas, gallbladder	Stimulates activity of stomach, pancreas, liver, gallbladder
Liver	Somatomedins	Most cells	Stimulates overall growth, development
Kidney	Erythropoietin*	Bone marrow	Stimulates red blood cell production
	Angiotensin*	Adrenal cortex, arterioles	Helps control blood pressure, aldosterone secretion
	Vitamin D_3*	Bone, gut	Enhances calcium resorption and uptake
Heart	Atrial natriuretic hormone	Kidney, blood vessels	Increases sodium excretion; lowers blood pressure
Thymus	Thymosin, etc.	Lymphocytes	Has roles in immune responses
Pineal	Melatonin	Gonads (indirectly)	Influences daily biorhythms, seasonal sexual activity

*These hormones are not produced in the kidneys but are formed when *enzymes* produced in kidneys activate specific substances in the blood.

There are three points to keep in mind about hormonal action. *First*, the response of target cells depends on the number of receptors they have and on the concentration of hormone molecules at a given time. *Second*, many hormones are linked to the neuroendocrine control center by **homeostatic feedback loops**. In such loops, the hypothalamus, pituitary, or both detect a change in the concentration of a hormone in some body region, then respond by inhibiting or stimulating the gland that secretes the hormone. *Third*, hormones interact to produce some effect on body functions.

As you will see, three kinds of hormonal interactions are common:

1. *Antagonistic interaction*. The effect of one hormone may oppose the effect of another. Insulin, for example, promotes a decrease in the glucose level in the blood and glucagon promotes an increase.

2. *Synergistic interaction*. The sum total of the action of two or more hormones is necessary to produce the required effect on target cells. Mammals, for example, cannot produce and secrete milk without the synergistic interaction of the hormones prolactin, oxytocin, estrogen, and progesterone.

3. *Permissive interaction*. One hormone exerts its effect only when a target cell has become "primed" to respond in an enhanced way to that hormone. The priming is accomplished by previous exposure to another hormone. Getting pregnant, for example, depends on the lining of the uterus being exposed first to estrogens, then to progesterone.

Adrenal Glands

Adrenal Cortex. Humans have a pair of adrenal glands, one above each kidney (Figure 34.10). We call the outer portion of each gland the **adrenal cortex**. Glucocorticoids are among the hormones secreted by the adrenal cortex. And homeostatic feedback loops to the neuroendocrine control center govern those secretions.

Glucocorticoids influence metabolic reactions that help maintain the glucose level in the blood. They also suppress inflammatory responses to tissue injury or infection. Cortisol is an example. Cortisol blocks the uptake and use of glucose by muscle cells, stimulates liver cells to store glucose (as glycogen), and makes more glucose in blood available to the brain. It also stimulates liver cells to form glucose from amino acids when the blood glucose level falls.

When the blood level of glucose falls below a set point, a condition called *hypoglycemia*, the hypothala-

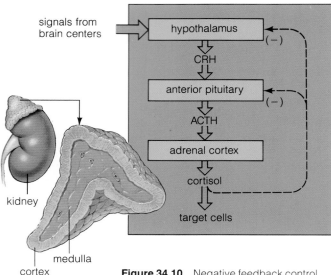

Figure 34.10 Negative feedback control of cortisol secretion.

mus is called into action. The hypothalamus initiates a *stress response* by secreting a releasing hormone (CRH). As Figure 34.10 suggests, the CRH prods the anterior pituitary into secreting corticotropin (ACTH). The ACTH stimulates the adrenal cortex into secreting cortisol—which works to prevent muscle cells from withdrawing glucose from the blood. As you can see, control of cortisol secretion is based on *negative* feedback mechanisms.

The feedback control of cortisol secretion is overridden by the nervous system when the body is abnormally stressed. A painful injury or severe illness may trigger shock, tissue inflammation, or both. Then, increased secretion of cortisol and other signaling molecules is essential to recovery. That is why cortisol-like drugs, such as cortisone, are administered to persons suffering from asthma and serious inflammatory disorders. Massive doses of such drugs also block certain immune reactions and so help prevent the body from rejecting surgically transplanted organs.

Adrenal Medulla. Hormone-secreting neurons are located in the **adrenal medulla**, the inner region of the adrenal gland (Figure 34.10). Their secretions, epinephrine and norepinephrine, help regulate blood circulation and carbohydrate metabolism. The hypothalamus and other brain centers govern their secretion by issuing commands along sympathetic nerves that service the adrenal glands.

The adrenal medulla helps mobilize the body's defenses during times of excitement or stress. For example, in response to epinephrine and norepinephrine, the heart beats faster and harder, blood flow is diverted to

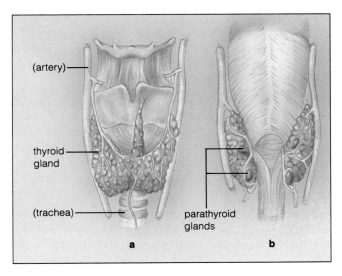

Figure 34.11 (**a**) Anterior view of the human thyroid gland. (**b**) Posterior view, showing the location of the four parathyroid glands adjacent to it.

Figure 34.12 A mild case of goiter, as displayed by Maria de Medici in 1625. During the late Renaissance, a rounded neck was considered a sign of beauty; it occurred regularly in parts of the world where iodine supplies were insufficient for normal thyroid function.

heart and muscle cells from other body regions, airways in the lungs dilate, and more oxygen is delivered to cells throughout the body. These are features of the "fight-flight" response described on page 568.

Thyroid Gland

Earlier we saw how excessive or insufficient output from the pituitary gland affects body functioning. Abnormal secretions from other endocrine glands also can have profound effects on the body. Consider the human **thyroid gland**. Figure 34.11 shows its location at the base of the neck, in front of the trachea (windpipe). The main secretions of this gland, thyroxine and triiodothyronine, influence overall metabolic rates, growth, and development.

Thyroid hormones contain iodine and cannot be produced without it. In the absence of iodine, thyroid hormone levels in the blood decrease. The anterior pituitary responds by secreting thyroid-stimulating hormone (TSH). Excess TSH overstimulates the thyroid gland and causes it to enlarge. The resulting tissue enlargement is a form of *goiter* (Figure 34.12). Goiter caused by iodine deficiency is no longer common in countries where iodized table salt is used. Elsewhere, hundreds of thousands of people still suffer from the disorder, which is easily preventable.

Insufficient thyroid output is called *hypothyroidism*. Hypothyroid adults are sluggish, dry-skinned, and intolerant of cold. If the disorder is present at birth and is not detected early in infants, *cretinism* may result. Cretinism may arise from a genetic disorder that affects the thyroid

gland in the fetus. Affected children show mental retardation, and if they do not receive treatment, their growth will be stunted. When cretinism is not identified in time, the mental retardation cannot be reversed.

Excessive thyroid output can lead to *hyperthyroidism*. The most common hyperthyroid disorder is called *Graves' disease*. Affected people show increases in metabolic rates, heart rate, and blood flow, and they lose weight even when they take in normal or increased amounts of food. They are excessively nervous, agitated, and typically have trouble sleeping. They are intolerant of heat and sweat profusely. Apparently, Graves' disease is an *autoimmune disorder* (in which the immune system mounts an attack against the body's own cells). In this case, cells of the immune system attack thyroid cells in much the same way that TSH stimulates them. Often individuals have a genetic predisposition to the disorder, which may be triggered by some environmental event. Treatment involves the surgical removal of some (or all) of the thyroid gland or treatment with drugs that suppress the synthesis of thyroid hormones.

Parathyroid Glands

Some glands are not stimulated directly by hormones or nerves. Rather, they respond homeostatically to a chemical change in their immediate surroundings. The **parathyroid glands** are like this. (Figure 34.11 shows how four of these glands are adjacent to the back of the human thyroid.) In response to a drop in extracellular levels of calcium ions, the glands secrete parathyroid hormone (PTH), which helps restore blood calcium levels. By its

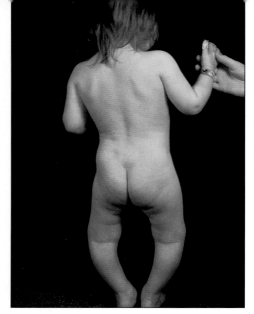

Figure 34.13 A child with rickets. Notice the bowed legs characteristic of the disorder.

action, then, the parathyroid glands influence the availability of calcium ions for enzyme activation, muscle contraction, blood clotting, and many other tasks.

PTH stimulates calcium and phosphate removal from bone and its movement into extracellular fluid. It stimulates the kidneys to conserve calcium. It also helps activate vitamin D. The activated form, which is a type of hormone, enhances calcium absorption from food moving through the gut. Vitamin D deficiency leads to *rickets*, a disorder arising from insufficient calcium and phosphorus for proper bone development (Figure 34.13).

Gonads

Homeostatic feedback loops also govern the function of gonads—the primary reproductive organs. Male gonads are called testes (singular, testis) and female gonads, ovaries. Gonads produce gametes. They also secrete estrogens, progesterone, and androgens (including testosterone). These are sex hormones that control reproductive function and the development of secondary sexual traits, as they did for Flo and others in the chimpanzee community described at the start of this chapter. How they do this is an elegant but intricate story, and for that reason we postpone our discussion of them until Chapter 43.

Pancreatic Islets

In one respect, the pancreas is an exocrine gland associated with the digestive system; it secretes digestive enzymes. (Exocrine gland products, recall, are not picked up by the bloodstream; they are secreted onto a free epithelial surface.) But the pancreas also has small clusters of endocrine cells scattered through it. There are about

2 million of these clusters, which are called the **pancreatic islets**. The following are three types of hormone-secreting cells in the islets:

1. *Alpha cells* secrete the hormone **glucagon**. Between meals, cells use the glucose delivered to them by the bloodstream. The blood glucose level decreases, at which time glucagon secretions cause glycogen (a storage polysaccharide) and amino acids to be converted to glucose in the liver. In such ways, *glucagon raises the glucose level in the blood*.

2. *Beta cells* secrete the hormone **insulin**. After meals, when the blood glucose level is high, insulin stimulates uptake of glucose by liver, muscle, and adipose cells especially. It also promotes synthesis of proteins and fats, and inhibits protein conversion to glucose. Thus *insulin lowers the glucose level in the blood*.

3. *Delta cells* secrete **somatostatin**, a hormone with regulatory functions in the digestive system. Somatostatin also can block the secretion of insulin and glucagon.

Figure 34.14 shows how interplays among the pancreatic hormones help keep blood glucose levels fairly constant despite great variation in when—and how much—we eat. The importance of this function is clear if we consider what happens when the body cannot produce enough insulin or when insulin's target cells cannot respond to it. Disorders in carbohydrate, protein, and fat metabolism occur.

Insulin deficiency can lead to *diabetes mellitus*. Blood glucose levels rise, and glucose accumulates in the urine. This promotes urination to the extent that the body's water-solute balance is disrupted. Affected persons become dehydrated and excessively thirsty. Their insulin-deprived (glucose-starved) cells start degrading proteins and fats for energy, and this leads to weight loss. Ketones, which are normal acidic products of fat breakdown, accumulate in the blood and urine, and they promote excess water loss. The imbalance disrupts brain function and, in extreme cases, death may follow.

In "type 1 diabetes," the body mounts an autoimmune response against its own insulin-secreting beta cells and destroys them. Genetic susceptibility and environmental triggers combine to produce the disorder, which is the less common but more immediately dangerous of the two types of diabetes. Symptoms usually appear during childhood and adolescence, hence the disorder also is called "juvenile-onset diabetes." Type 1 diabetic patients survive with insulin injections.

In "type 2 diabetes," insulin levels are close to or above normal—but target cells cannot respond to the hormone. As affected persons grow older, their beta

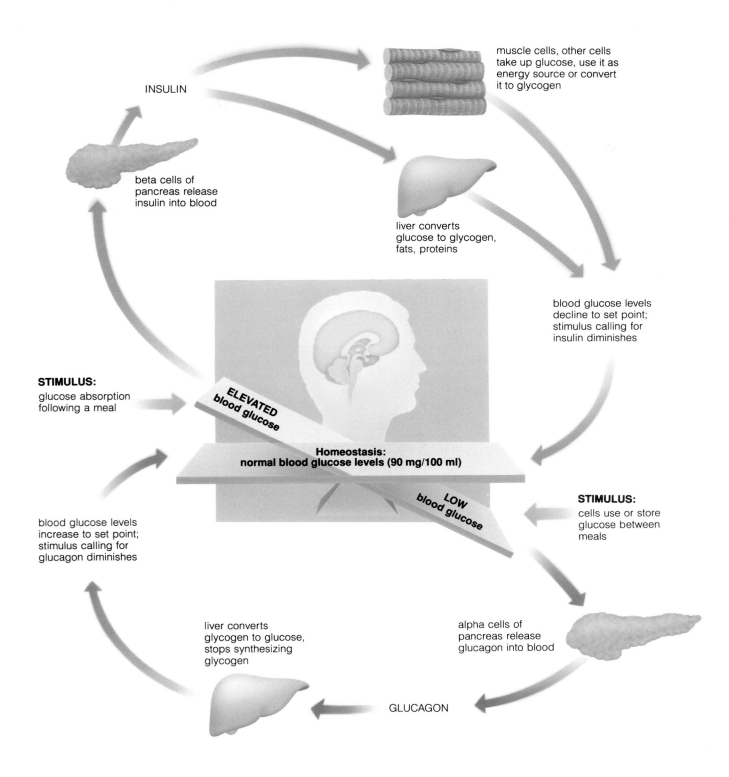

INSULIN

muscle cells, other cells take up glucose, use it as energy source or convert it to glycogen

beta cells of pancreas release insulin into blood

liver converts glucose to glycogen, fats, proteins

blood glucose levels decline to set point; stimulus calling for insulin diminishes

STIMULUS:
glucose absorption following a meal

ELEVATED blood glucose

Homeostasis:
normal blood glucose levels (90 mg/100 ml)

LOW blood glucose

STIMULUS:
cells use or store glucose between meals

blood glucose levels increase to set point; stimulus calling for glucagon diminishes

liver converts glycogen to glucose, stops synthesizing glycogen

alpha cells of pancreas release glucagon into blood

GLUCAGON

Figure 34.14 Some of the homeostatic controls over glucose metabolism. *Following* a meal, glucose enters the bloodstream faster than cells can use it. Blood glucose levels rise. Pancreatic beta cells are stimulated to secrete insulin. The hormonal targets (mainly liver, fat, and muscle cells) use glucose or store it as glycogen.

Between meals, blood glucose levels drop. Pancreatic alpha cells are stimulated to secrete glucagon. The hormonal targets convert glycogen back to glucose, which enters the blood. Also, the hypothalamus prods the adrenal medulla into secreting other hormones that slow down the conversion of glucose to glycogen in liver, fat, and muscle cells.

Rhythms and Blues

The pineal gland, a bump of tissue in your brain, secretes the hormone melatonin. When the brain receives sensory signals from your eyes about the waning light at sunset, the pineal gland steps up its melatonin secretion, which is picked up by the bloodstream, which transports it to target cells—in this case, certain brain neurons. Those neurons are involved in sleep behavior, a lowering of body temperature, and possibly other physiological events. At sunrise, when the eye detects the light of a new day, melatonin production slows down. Your body temperature increases, and you wake up and become active.

The cycle of sleep and arousal is evidence of an internal biological clock that seems to tick in synchrony with daylength. Think about what happens when circumstances disturb the clock. There is the night worker who tries to sleep in the morning but ends up staring groggily at sunbeams on the ceiling. There is the traveler from the United States to Paris who starts off a vacation with four days of disoriented "jet lag." Two hours past midnight he is sitting upright in bed, wondering where the coffee and croissants are. Two hours past noon he is ready for bed. His body will gradually shift to a new routine when melatonin's signals begin arriving at their target neurons on Paris time.

Maybe you have heard of people affected by severe *winter blues*—depression, carbohydrate binges, and an overwhelming desire to sleep. Their discomfort may result from a biological clock that is out of synchrony with the changes in daylength in winter (days are shorter and nights longer). Their symptoms worsen when they are given doses of melatonin. And they improve dramatically when they are exposed to intense light—which shuts down pineal activity.

Melatonin is a long-time player in the course of vertebrate evolution. It is known to affect physiological rhythms in trout, alligators, sea turtles, sparrows, armadillos, hamsters, and rats. Researchers have identified its action in sexual behavior and other aspects of the reproductive cycle. They have identified its role in the seasonal deposition of body fat. Melatonin's roles may be varied, but they are part and parcel of life's tempos.

cells deteriorate and they produce less and less insulin. Type 2 diabetes usually occurs in middle age and is less dramatically dangerous than the other type. Affected persons can lead a normal life by controlling their diet, controlling their weight, and sometimes taking drugs that enhance insulin action or secretion.

Thymus Gland

The lobed **thymus gland** is located behind the breastbone and between the lungs. Certain lymphocytes (white blood cells) multiply, differentiate, and mature in this gland, which secretes a group of hormones collectively called thymosins. These hormones affect the functioning of lymphocytes that defend the body against disease (Chapter 39).

Pineal Gland

So far, we have seen how endocrine glands and endocrine cells respond to other hormones, to signals from the nervous system, and to chemical changes in their surroundings. Now we can start thinking about a larger picture, in which reproduction and development of the body are controlled by hormonal responses to environmental cues.

Until about 240 million years ago, it seems, vertebrates commonly had a third eye, on top of the head. Lampreys still have one, beneath the skin. A modified form of this photosensitive organ persists in nearly all vertebrates. We call it the **pineal gland** (Figure 34.2). The pineal gland secretes melatonin, a hormone that functions in the development of gonads and in reproductive cycles.

As described in the *Commentary*, melatonin is secreted in the absence of light. This means melatonin levels vary from day to night. The levels also change with the seasons, as when winter days are shorter than summer days. The hormonal effects can be observed during hamster reproductive cycles. High melatonin levels in winter suppress sexual activity, and in summer, when melatonin levels are low, sexual activity peaks. In humans, decreased melatonin secretion might help trigger the onset of *puberty*, the age at which reproductive organs and structures start to mature. If disease destroys the pineal gland, puberty may begin prematurely.

LOCAL SIGNALING MOLECULES

In mammals, cells with mediating functions detect changes in the surrounding chemical environment and alter their activity, often in ways that either counteract or amplify the change. The cells secrete local signaling molecules, the action of which is confined to the immediate vicinity of change. Most of the signaling molecules are taken up so rapidly that not many are left to enter the general circulation. Prostaglandins and growth factors are examples of such secretions.

Prostaglandins

More than sixteen different prostaglandins have been identified in tissues throughout the body. They are released continually, but the rate of synthesis often increases in response to local chemical changes. The stepped-up secretion can influence neighboring cells as well as the prostaglandin-releasing cells themselves.

At least two prostaglandins help adjust blood flow through local tissues. When their secretion is stimulated by epinephrine and norepinephrine, they cause smooth muscle in the walls of blood vessels to constrict or dilate. Prostaglandins have similar effects on smooth muscle of airways in the lungs. Allergic responses to airborne dust and pollen may be aggravated by prostaglandins (page 693).

Prostaglandins have major effects on some mammalian reproductive events. When women menstruate, many of them experience painful cramping and excessive bleeding—both of which have been traced to prostaglandin action. (Aspirin and other anti-prostaglandin drugs block synthesis of this local signaling molecule and alleviate the discomfort.) Prostaglandins also influence the corpus luteum, a glandular structure that develops from cells that earlier surrounded a developing ovum in the ovary. When pregnancy does not follow ovulation (that is, the release of an ovum from the ovary), a corpus luteum self-destructs. It produces copious amounts of prostaglandins that interfere with its own function. Prostaglandins also have roles in stimulating uterine contractions during labor.

Growth Factors

Signaling molecules called **growth factors** influence growth by regulating the rate at which certain cells divide. Epidermal growth factor (EGF), discovered by Stanley Cohen, influences the growth of many cell types. Nerve growth factor (NGF) is another example. NGF, discovered by Rita Levi-Montalcini, promotes survival and growth of neurons in the developing embryo. One experiment demonstrated that certain immature neurons survive indefinitely in tissue culture when NGF is present but die within a few days if it is not. NGF also may define the direction of growth for these embryonic neurons, laying down a chemical path that leads the elongating processes to target cells.

SIGNALING MECHANISMS

Hormones and other signaling molecules induce diverse responses in target cells. They can trigger the entry of substances into cells. They can alter the rate of protein synthesis, cause modification in proteins already present in the cell, and induce changes in the cell's shape and internal structure. What dictates the nature of the target cell's response?

That depends largely on two things. First, different signals activate different cellular mechanisms. Second, not all cells *can* respond to all types of signals. Many types of cells have receptors for cortisol and some other hormones; that is why those hormones have such widespread effects. But only a few cell types have receptors for more specific hormones, which have highly directed effects.

Let's think about some responses to just two of the main categories of hormones, as listed below and in Table 34.4:

1. *Steroid hormones*, which are synthesized from cholesterol. Steroid hormones are lipid-soluble and readily cross plasma membranes.

2. *Nonsteroid hormones*, which are synthesized from amino acids. They include protein and peptide hormones, and catecholamines. All are water-soluble and cannot cross the lipid bilayer of the plasma membrane.

Table 34.4 Two Main Categories of Hormones	
Type of Hormone	Examples
Steroid	Estrogens, testosterone, aldosterone, cortisol
Nonsteroid:	
Amines	Norepinephrine, epinephrine
Peptides	ADH, oxytocin, TRH
Proteins	Insulin, somatotropin, prolactin
Glycoproteins	FSH, LH, TSH

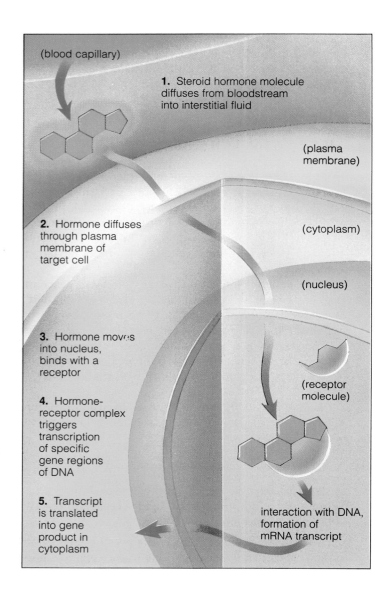

(blood capillary)

1. Steroid hormone molecule diffuses from bloodstream into interstitial fluid

(plasma membrane)

2. Hormone diffuses through plasma membrane of target cell

(cytoplasm)

(nucleus)

3. Hormone moves into nucleus, binds with a receptor

4. Hormone-receptor complex triggers transcription of specific gene regions of DNA

(receptor molecule)

5. Transcript is translated into gene product in cytoplasm

interaction with DNA, formation of mRNA transcript

(blood capillary)

1. Protein hormone molecule diffuses from bloodstream into interstitial fluid

(unoccupied receptor site for hormone on plasma membrane of target cell)

cyclic AMP + P_i

ATP

2. Binding of the hormone to receptor activates adenyl cyclase (a membrane-bound enzyme system that catalyzes cyclic AMP formation)

3. Cyclic AMP (a second messenger) activates many enzymes in cytoplasm

4. Enzymes cause alteration in some cell activity (such as a change in active transport of a specific substance across a membrane)

Figure 34.15 Proposed mechanism of steroid hormone action on a target cell. This same type of mechanism is also thought to occur for thyroid hormones, with one qualification. Recent studies suggest that membrane proteins facilitate the movement of thyroid hormones across the plasma membrane.

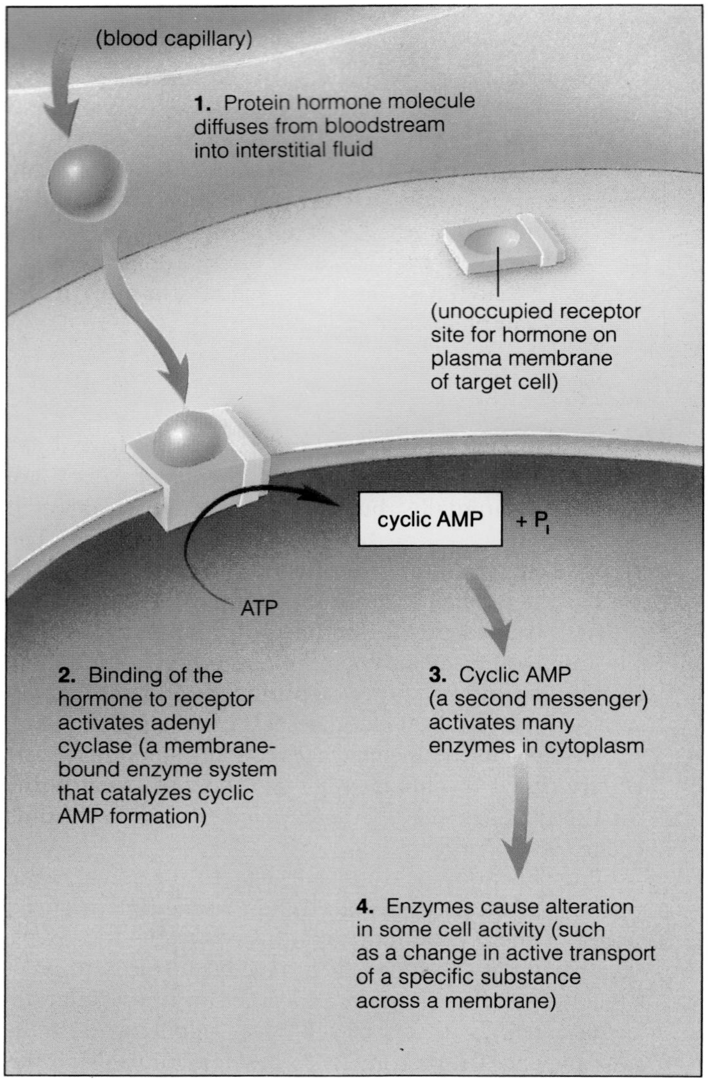

Figure 34.16 Proposed mechanism of protein hormone action on a target cell. The response is mediated by a second messenger inside the cell—in this case, cyclic AMP. Other chemical messengers may be involved, depending on the particular hormone and its particular target cell.

Steroid Hormone Action

Steroid hormones stimulate or inhibit protein synthesis by switching certain genes on or off. They do not alter the activity of already existing proteins. Being lipid-soluble, steroid hormones diffuse easily across the plasma membrane of target cells (Figure 34.15). Once inside, they move into the nucleus, where they bind with receptors for them.

The three-dimensional shape of the hormone-receptor complex allows it to interact with the DNA. The complex triggers the transcription of certain gene regions. Translation of the transcripts into specific proteins follows.

Testosterone is an example of a steroid hormone. It influences the development of male sexual traits. In *testicular feminization syndrome*, the receptor to which testosterone binds is defective. The affected individuals are males (XY), and they have functional testes that secrete testosterone. But none of the target cells can respond to the hormone, so the secondary sexual traits that do develop are like those of females.

Nonsteroid Hormone Action

Protein hormones and other water-soluble signaling molecules cannot cross the plasma membrane of target cells without assistance. First they bind to receptors at the plasma membrane. In some cases, the hormone-receptor complex moves into the cytoplasm by way of endocytosis, then further action occurs inside the cell. Figure 5.13 shows photomicrographs of this type of endocytosis. Some hormones bind to receptors that recruit transport proteins into action or trigger the opening of channel proteins across the membrane. Certain ions or other substances move inward, and their cytoplasmic concentration changes in ways that affect cell activities.

Most peptide and protein hormones, including glucagon, activate **second messengers**. These are molecules inside the cell that mediate the response to a hormone. An example is *cyclic AMP*. (The full name is cyclic adenosine monophosphate.) First a hormone binds to a membrane receptor on a target cell. Binding alters the activity of a membrane-bound enzyme system (Figure 34.16). An enzyme, adenylate cyclase, is prodded into action. This enzyme speeds the conversion of ATP to cyclic AMP.

The hormone-receptor complex activates many molecules of the enzyme, not just one. Each enzyme molecule increases the rate at which many ATP molecules are converted to cyclic AMP. Each cyclic AMP molecule so formed then activates many enzyme molecules. Each of the enzyme molecules so activated can convert a very large number of substrate molecules into activated enzymes, and so on. Soon the number of molecules representing the final cellular response to the initial signal is enormous. Thus, second messengers *amplify* the response to a signaling molecule.

SUMMARY

This chapter concludes our survey of controls over the integration of body activities in multicelled animals. Throughout the remainder of this unit, we will be looking at specific examples of these controls, so keep the following key concepts in mind:

1. For metabolic activity to proceed smoothly, the chemical environment of a cell must be maintained within fairly narrow limits.

2. In complex animals, thousands to billions of cells continually take up some substances from the extracellular fluid and secrete other substances into it. The nature and amount of the substances can change with the diet or level of activity; they inevitably change during the course of development.

3. It follows that the myriad withdrawals and secretions must be integrated in ways that ensure cell survival through the whole body.

4. Integration is accomplished by signaling molecules: chemical secretions by one cell that adjust the behavior of other, target cells. A target is one having receptors to which specific signaling molecules can bind and elicit a cellular response. It may or may not be adjacent to the signaling cell.

5. Signaling molecules include hormones, transmitter substances, local signaling molecules, and pheromones.

6. A neuroendocrine control center integrates many activities for the vertebrate body. This center consists of the hypothalamus and pituitary gland.

7. Two hypothalamic hormones—ADH and oxytocin—are stored in and released from the posterior lobe of the pituitary. ADH influences extracellular fluid volume. Oxytocin influences contraction of the uterus and milk release from mammary glands.

8. Six other hypothalamic hormones, called releasing or inhibiting hormones, control the secretions by cells of the anterior lobe of the pituitary.

9. The anterior lobe of the pituitary produces and secretes six hormones. Two (prolactin and somatotropin, or growth hormone) have general effects on body tissues. The remainder (ACTH, TSH, FSH, and LH) act on specific endocrine glands.

10. Hormone secretion is influenced by neural signals, hormonal interactions, local chemical changes in the surrounding tissues, and changes in the external environment.

11. Antagonistic, synergistic, and permissive interactions occur among hormones. The secretion of many hormones is controlled by homeostatic feedback of the neuroendocrine control center.

12. Fast-acting hormones such as parathyroid hormone (PTH) or insulin generally come into play when the extracellular concentration of a substance must be homeostatically controlled. Slow-acting hormones such as somatotropin have more prolonged, gradual, and often irreversible effects, such as those on development.

13. Cells respond to specific hormones or other signaling molecules only if they have receptors for them. Steroid hormones have receptors in the nucleus of target cells. Nonsteroid hormones (the amines, peptides, proteins, and glycoproteins) have receptors on the plasma membrane of target cells; responses to them are often mediated by a second messenger (such as cyclic AMP) inside the cell.

14. Steroid hormones trigger gene activation and protein synthesis. Most nonsteroid hormones alter the activity of proteins already present in target cells. These cellular responses contribute in some way to maintaining the internal environment or to the developmental or reproductive program.

Review Questions

1. Which secretions of the posterior and anterior lobes of the pituitary glands have the targets indicated? (Fill in the blanks; *see pages 583 and 585.*)

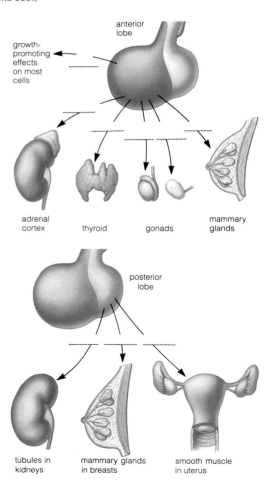

anterior lobe

growth-promoting effects on most cells

adrenal cortex thyroid gonads mammary glands

posterior lobe

tubules in kidneys mammary glands in breasts smooth muscle in uterus

2. Name the main endocrine glands and state where each is located in the human body. *582*

3. Define hormone. What functions do hormones serve? How do these functions differ from those of transmitter substances? 582

4. How do steroid and polypeptide hormones act on a target cell? *593–595*

5. The hypothalamus and pituitary are considered to be a neuroendocrine control center. Can you describe some of the functional links between these two organs? *583*

6. How does the hypothalamus control secretions of the posterior lobe of the pituitary? The anterior lobe? *583, 584*

7. Name three endocrine glands and a substance that each one secretes. What are the main consequences of their secretion? *585, 587*

8. Which hormone secreted by the anterior pituitary has an effect on most body cells rather than on a specific cell type? What are the clinical consequences of too little or too much secretion of this hormone? *584, 586*

Self-Quiz *(Answers in Appendix IV)*

1. _____, _____, _____, and _____ are types of signaling molecules that help integrate cell activities in ways that benefit the whole body.

2. The _____ and _____ gland interact as a neuroendocrine control center to integrate endocrine gland secretions.

3. Stimulation or inhibition of hormone secretions often involves _____ loops between the neuroendocrine control center and glands.

4. A target cell for a specific hormone or some other signaling molecule has _____ for that molecule.

5. The hypothalamus produces two hormones that are released from the posterior lobe of the pituitary gland. One hormone,

_____, affects kidney function; the other, _____, affects some reproductive events.
   a. ADH; oxytocin
   b. prolactin; ADH
   c. oxytocin; ADH
   d. ADH; prolactin

6. The anterior lobe of the pituitary gland produces two hormones, _____ and _____, that have general effects on body tissues in general.
   a. ACTH; somatotropin
   b. prolactin; FSH
   c. ACTH; FSH
   d. prolactin; somatotropin

7. Which of the following does *not* stimulate hormone secretion?
   a. neural signals
   b. local chemical changes
   c. hormonal signals
   d. environmental cues
   e. all of the above can stimulate hormone secretion

8. Insulin is an example of a _____ hormone which must work to accomplish homeostatic control of an extracellular substance; somatotropin is an example of a _____ hormone which operates during body development.
   a. growth; metabolic
   b. fast-acting; slow-acting
   c. metabolic; growth
   d. slow-acting; fast-acting

9. Which of the following statements is true?
   a. Steroid hormones have receptors on the plasma membrane of target cells.
   b. Protein hormones have receptors in target cell nuclei.
   c. Most protein hormones alter the activity of genes.
   d. Steroid hormones activate genes and protein synthesis.

10. Match the endocrine control concepts
   _____ oxytocin
   _____ ADH
   _____ steroid hormone
   _____ somatotropin (GH)
   _____ hypothalamus/ pituitary
   a. neuroendocrine control center
   b. affects kidney function
   c. has general effects on growth
   d. affects reproductive events
   e. triggers protein synthesis

Readings

Cantin, M., and J. Genest. February 1986. "The Heart As an Endocrine Gland." *Scientific American* 254(2):76–81.

Fellman, B. May 1985. "A Clockwork Gland." *Science 85* 6(4):76–81. Describes some of the known functions of the pineal gland.

Goodall, J. 1986. *The Chimpanzees of Gombe.* Cambridge, Massachusetts: Belknap Press of Harvard University Press.

Hadley, M. 1988. *Endocrinology.* Second edition. Englewood Cliffs, New Jersey: Prentice-Hall.

Sapolsky, R. January 1990. "Stress in the Wild." *Scientific American* 262(1):116–123. Study of hormonal effects on stress responses in baboons.

Snyder, S. October 1985. "The Molecular Basis of Communication Between Cells." *Scientific American* 253(4):132–141.

35 SENSORY RECEPTION

Nobody Calls the Bat Man's Best Friend

How many of your friends have pet bats? None, most likely. Something about bats makes them unloved, the models for gargoyles and other imagined monsters. Nearly all bats sleep by day and spread their webbed wings at dusk, when different species take to the air in search of nectar, fruit, frogs, or insects. Probably it is the few blood-sucking species among them that have given the entire order of bats a bad name. Yet bats are rather close to us on the family tree for animals. They are every bit as mammalian as a dog or cat, and even though you might be reluctant to scratch one behind the ears, you have to give them credit for being relatives with some distinguishing sensory traits.

Consider that many of the sensory receptors in the eyes, nose, ears, mouth, and skin of bats are not that different from your own. When male frogs attempt to attract females by croaking as they float about in a pond, sensory receptors in your ears allow you to hear the relatively low-pitched sounds and use them to locate the frog's general whereabouts. The frog-eating bat shown in Figure 35.1 can do the same thing—with greater accuracy. Such bats routinely zero in on vocal but unlucky male frogs.

Other kinds of bats hunt in different ways, with a sense of hearing that we cannot begin to match. Think about the species of bats with tiny eyes, nearly blind, yet able to navigate with ease through forests and capture flying insects with great precision, even in the dark! Such bats are masters of *echolocation*. They emit calls, and when the sound waves of those calls bounce off objects in the environment—insects, tree branches—the echoes are assessed by the bat brain and used to judge the precise location and shape of those objects.

As an echolocating bat flies, it emits a steady stream of about ten clicking sounds every second—sounds *you* cannot hear at all. The clicks are intense. But they are "ultrasounds," meaning they are above the range of sound waves that sensory receptors in human ears can detect. Bats can hear even the extremely faint echoes of ultrasounds that are returning from distant objects.

Figure 35.1 (**a**) Having used its well-developed sense of hearing to track down a singing tropical frog, this frog-eating bat is about to scoop dinner from the water. (**b**) Other bats, including this one, detect prey by listening to echoes of their own high-frequency cries. The echoes bounce back from objects in the environment, and the bat brain deciphers them in ways that provide the bat with a "sound map." With this map, the bat easily captures mosquitoes and moths in midair, without even seeing them.

When an echolocating bat hears a pattern of distant echoes from, say, an airborne mosquito or moth, it increases the rate of ultrasonic clicks to as many as 200 per second—which is faster than a machine gun fires bullets. There are only a few milliseconds of silence between clicks—but in that blip of silence the bat's receptors detect the echoes. Sensory nerves carry signals to the bat's brain, where the signals are decoded and processed. The brain somehow constructs a "map" of sounds that the bat uses in its maneuvers through the night world. With this map, the bat can swoop in on its insect snack without ever having caught a glimpse of it.

With this chapter, we turn to the means by which animals receive signals from the external and internal environments—and then decode those signals in ways that give rise to awareness of sounds, sights, odors, pain, and other sensations. Sensory neurons, nerve pathways, and brain regions are required for these tasks. Together, they represent the portions of the nervous system that are called *sensory systems*.

KEY CONCEPTS

1. Sensory systems are part of the nervous system. Each consists of specific types of sensory receptors, nerve pathways from those receptors to the brain, and brain regions that deal with sensory information.

2. Sensory receptors at or near the body surface detect specific stimuli that can give rise to sensations of hearing, taste, smell, touch, externally applied pressure, temperature, pain, and sight.

3. Sensory receptors associated with skeletal muscles provide information about body position and movement. Those associated with internal organs other than skeletal muscles have roles in sensations of fatigue, hunger, thirst, nausea, and internal pressure and pain.

4. A sensation is a conscious awareness of change in external or internal conditions. It takes four steps to reach such awareness. First, sensory receptors detect a specific stimulus. Second, the stimulus energy is converted to local signals that can give rise to action potentials. Third, information about the stimulus becomes encoded in the number and frequency of action potentials sent to the brain along particular nerve pathways. Fourth, specific brain regions translate the information into a sensation.

SENSORY SYSTEMS: AN OVERVIEW

Sensory systems are the front doors of the nervous system, the parts that let in information about changing conditions and allow the brain to become aware of pertinent changes that are going on outside and inside the body. Sponges are among the very few animals that get along without sensory systems—but then, not much changes from one day to the next in the life of a sponge. You as well as most other animals would never even survive without elaborate sensory systems, given the sheer volume of information that you deal with every day. (Think about what it takes simply to cross a busy street during rush hour.)

Each sensory system has three component parts. These are (1) sensory receptors, (2) nerve pathways

599

Table 35.1   Receptors Associated with the Major Senses

Category of Receptor	Examples	Stimulus
Chemoreceptors:		
Internal chemical senses	Carotid bodies in blood vessel wall	Substances (CO_2, etc.) dissolved in extracellular fluid
Taste	Taste receptors of tongue	Substances dissolved in saliva, etc.
Smell	Olfactory receptors of nose	Odors in air, water
Mechanoreceptors:		
Touch, pressure	Pacinian corpuscles in skin	Mechanical pressure against body surface
Stretch	Muscle spindle in skeletal muscle	Stretching
Auditory	Hair cells within ear	Vibrations (sound or ultrasound waves)
Balance	Hair cells within ear	Fluid movement
Photoreceptors:		
Visual	Rods, cones of eye	Wavelengths of light
Thermoreceptors	Cold or warm receptors in skin; central thermoreceptors in hypothalamus, etc.	Presence of or change in radiant energy (heat)
Nociceptors	Free nerve endings in skin	Any stimulus that causes tissue damage and leads to sensation of pain

leading from the receptors to the brain, and (3) brain regions where sensory information is processed.

Sensory receptors are finely branched endings of sensory neurons or specialized cells adjacent to them, and they detect specific stimuli. Think of a *stimulus* as light, heat, mechanical pressure, or some other form of energy that is capable of eliciting a response from a sensory receptor. Once detected, the stimulus energy is converted to action potentials—the form in which information travels along nerve pathways to the brain. Finally, specific brain regions translate information about the stimulus into a sensation.

What we call a **sensation** is a conscious awareness of a stimulus. It is not the same thing as **perception**, or understanding what the sensation means. Plunge your finger accidentally into boiling water and you are conscious of the stimulus—but beyond this, you understand acutely that the intense heat is hurting you. Plunge a live lobster into a cookpot and its reactions tell us that the stimulus (boiling water) has indeed registered. But lobsters and other invertebrates probably do not have the neural means to understand the meaning of their predicament.

A sensory system consists of sensory receptors for specific stimuli, nerve pathways that conduct information from those receptors to the brain, and brain regions where the information is processed.

Information flowing through a sensory system may or may not lead to conscious awareness of the stimulus.

Types of Sensory Receptors

Different sensory receptors are specialized to detect different kinds of stimuli. Many sensory receptors are positioned individually, like sentinels, in the skin and other body tissues. Others are part of sensory organs, such as eyes, that amplify or focus the stimulus energy.

By using the different types of stimulus energy as a guide, we can define five major types of sensory receptors, as listed below and in Table 35.1:

1. **Chemoreceptors** detect the chemical energy of specific substances dissolved in the fluid surrounding them.

2. **Mechanoreceptors** detect mechanical energy associated with changes in pressure, changes in position, or acceleration.

3. **Photoreceptors** detect visible and ultraviolet light.

4. **Thermoreceptors** detect infrared energy (heat).

5. **Nociceptors** (pain receptors) detect tissue damage.

Keep in mind that different animals do not have the same kinds or numbers of sensory receptors, so they sample the environment in different ways and differ in their awareness of it. You don't have photoreceptors for ultraviolet light, as bees do—so you do not "see" many flowers the way they do (page 506). Unlike the bat in Figure 35.1b, you do not have mechanoreceptors for ultrasound. Unlike the python in Figure 35.2, you do not have thermoreceptors that detect warm-blooded prey in the dark.

Figure 35.2 A python of southern Asia, equipped with thermoreceptors inside the pits, shown here, above and below its mouth. The python eats small, night-foraging mammals. Its thermoreceptors are sensitive to body heat (infrared energy) of its prey, which are much warmer than the night air. They notify the snake brain, which has a program for assessing signals about the location of objects. The program works very well—the snake's strike may be only a few degrees off-center. Yet the same snake might slither past a motionless, edible frog. Frog skin is cool and blends with background colors. The snake does not have receptors for detecting it or a neural program for responding to it.

Sensory Pathways

Each sensory pathway starts at receptors of sensory neurons that are sensitive to the same type of stimulus, such as light, cold, or pressure. As Figure 35.3 shows, the axons of sensory neurons lead into the central nervous system. There they might converge on a single interneuron or branch out and connect with several to many interneurons.

Sensory nerve pathways from different receptors lead to different parts of the cerebral cortex, the outermost layer of the brain. For example, signals from receptors in the skin and joints travel to the *somatic sensory cortex*. Cells of this cortical region are laid out like a map corresponding to the body surface. Some map regions are larger than others; the body regions they represent are functionally more important and have more

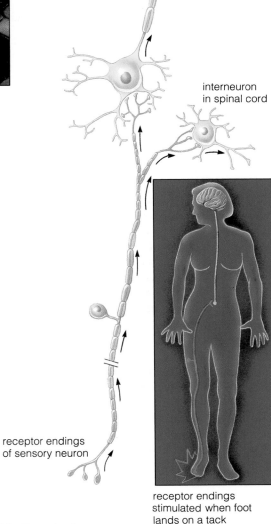

message sent on to interneurons in brain

interneuron in spinal cord

receptor endings of sensory neuron

receptor endings stimulated when foot lands on a tack

Figure 35.3 Example of a sensory nerve pathway leading from a sensory receptor to the brain. The sensory neuron is coded red; interneurons are coded yellow.

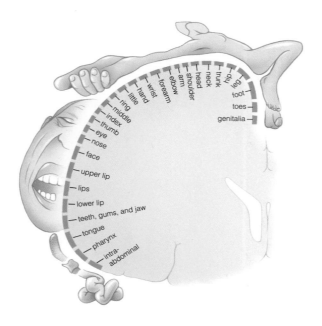

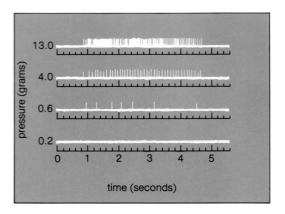

Figure 35.5 Action potentials recorded from a single pressure receptor of the human hand. The recordings correspond to variations in stimulus strength. A thin rod was pressed against the skin with the pressure indicated on the vertical axis of this diagram. Vertical bars above each thick horizontal line represent individual action potentials. Notice the increases in frequency, which correspond to increases in stimulus strength.

Figure 35.4 Body regions represented in the primary somatic sensory cortex. This region is a strip a little more than an inch (2.5 centimeters) wide, running from the top of the head to just above the ear on the surface of *each* cerebral hemisphere. The diagram is a cross-section through the right hemisphere of someone facing you. Compare this with Figure 33.14, which shows the motor cortex.

receptors. In humans, a large part of the primary somatic sensory cortex responds to receptors located in the fingers, thumbs, and lips, as shown by the diagram in Figure 35.4.

Similarly, the visual world is mapped onto the *visual cortex* (Figures 33.13 and 35.19). The map region corresponding to the center of the visual field is larger than other regions.

Information Flow Along Sensory Pathways

All sensory receptors convert stimulus energy to a form that can travel along communication lines to the brain. When the plasma membrane of a receptor ending is stimulated, the disturbance allows ions to move through channels across a local patch of the membrane—that is, a local, graded potential. Recall that ion distributions contribute to a difference in charge across the unstimulated plasma membrane of a neuron (page 550). Local signals do not spread far from the point of stimulation, and they vary in magnitude, depending on the strength of the stimulus. If the stimulus is intense or repeated

fast enough for the local signals to be added together, action potentials may be triggered.

Action potentials are propagated rapidly from the receptors to the axon endings of the sensory neuron. There, the sensory neuron releases a transmitter substance that affects the activity of interneurons (or motor neurons) adjacent to it. The arrival of transmitter substances may trigger action potentials in the interneurons, which may be part of nerve tracts leading to the brain.

Action potentials being propagated along sensory neurons are not like a wailing ambulance siren; they don't vary in amplitude. What exactly do they "tell" the brain about the stimulus?

First, the genetically determined network of neurons in the brain of each animal can interpret incoming action potentials only in certain ways. That is why you "see stars" when you accidentally poke your eye in the dark. Your brain always interprets action potentials arriving from your eyes, by way of a particular sensory pathway, as "light."

Second, strong stimulation of a receptor causes action potentials to fire more frequently. Thus, even though the same receptor will detect the sound of a throaty whisper as well as that of a wild screech, the brain senses the difference through frequency variations in the signals that the receptor sends to it. Figure 35.5 shows frequency variations corresponding to differences in sustained pressure on human skin.

Third, strong stimulation "recruits" more receptors in a larger area. You activate some receptors when you lightly touch skin on your arm, but you activate many more when you press hard on the same area. The increased disturbance sets off action potentials in many sensory axons at the same time. And the brain interprets the combined activity as an increase in stimulus intensity.

Each type of sensation is triggered when action potentials arriving from *particular nerve pathways* activate specific neurons in the brain.

Besides sensing a stimulus, the brain also interprets variations in stimulus intensity. Interpretation is based on the *frequency* of action potentials propagated along single axons and the *number* of axons carrying action potentials from a given tissue.

Sometimes the frequency of action potentials decreases or stops even when a stimulus is being maintained at a constant strength. This occurrence is called **adaptation**. After you have put on clothing, for example, you no longer are aware of its pressure against your skin. Mechanoreceptors (Pacinian corpuscles) in your skin have fired only at the onset of the stimulus. These receptors are a rapidly adapting type. By contrast, when you pick up a warm muffin, thermoreceptors in your skin rapidly fire off action potentials. Then the frequency drops to a lower level that is maintained for as long as you hold the muffin. These receptors are a slowly adapting type.

Let's now focus on a few kinds of sensory receptors to illustrate how changing conditions in the internal and external environments can be detected in the first place. Some kinds occur in more than one location in the body. They contribute to *somatic sensations*. Other kinds are restricted to special locations, such as the eyes or ears. They contribute to the *special senses*.

SOMATIC SENSATIONS

The **somatic sensations** include awareness of touch, pressure, heat, cold, and pain. They also include awareness of limb movements and the body's positions in space. Such sensations start with receptor endings that are embedded in skin and other tissues at the body's surfaces, in skeletal muscles, and in the walls of internal organs.

Touch and Pressure

Sensations at the body surface depend largely on mechanoreceptors. These receptors are widespread, and they are more abundant in some regions than others. Fingertips and the tip of the tongue have the most; these regions show the greatest sensitivity to stimulation. The back of the hand and neck do not have nearly as many, and these regions show far less sensitivity to stimulation.

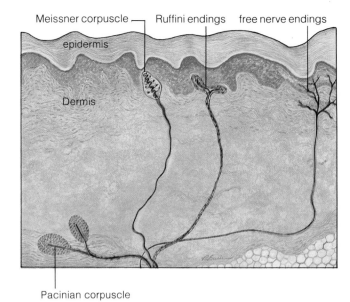

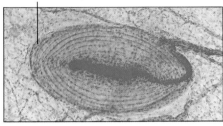

Figure 35.6 Tactile receptors in human skin. Free nerve endings contribute to sensations of temperature, light pressure, and pain. Pacinian corpuscles contribute to sensations of vibration. Meissner corpuscles are stimulated at the onset and end of sustained pressure; the Ruffini endings react continually to ongoing stimuli.

Mechanoreceptors in the skin are the receiving end of sensory neurons, with or without a surrounding capsule of epithelial or connective tissue. The free nerve endings and Pacinian corpuscles shown in Figure 35.6 are examples. All skin mechanoreceptors respond easily to deformation of the skin resulting from pressure. But about half of them fire off action potentials only when the stimulus is first encountered and when it ends. They make you aware of touch, even light tickles and vibrations. The other half fire off action potentials as long as the stimulus is present. These are the mechanoreceptors that make you aware of pressure.

Temperature Changes

When the temperature near the body surface remains much the same, thermoreceptors in skin fire off a steady barrage of action potentials to the brain. With increases in temperature, the firing frequency increases. We know that free nerve endings serve as the "heat" receptors. Researchers still have not figured out what the "cold" receptors are.

Pain

Pain is the *perception* of injury to some body region. That perception begins with signals from nociceptors, which occur in nearly all tissues of the body. Nociceptors include free nerve endings that detect any stimulus causing tissue damage. Signals from these receptors travel

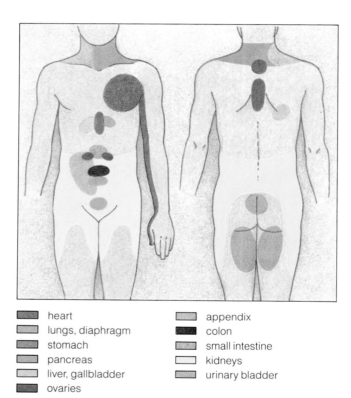

▮ heart	▯ appendix
▮ lungs, diaphragm	▮ colon
▮ stomach	▮ small intestine
▮ pancreas	▯ kidneys
▯ liver, gallbladder	▮ urinary bladder
▮ ovaries	

Figure 35.7 Regions of referred pain. Sensations of certain internal disorders are localized to the skin areas indicated.

along spinal or cranial nerves to the thalamus. They are relayed to the part of the cerebrum called the parietal lobe, where they are processed (Figure 33.12).

Free nerve endings apparently detect any stimulus that is intense enough to damage the surrounding tissue. They respond to strong mechanical stimulation, intense heat or cold, and chemical irritation, as when prostaglandins flood a local tissue region (page 593). When the stimulus intensity passes a certain threshold, the signals generated become translated into sensations of pain.

Responses to pain often depend on the brain's ability to associate the pain with the affected tissue. Get hit in the face with a snowball and you "feel" the contact on facial skin. You "feel" pain in your chest when the double-layered membrane surrounding your lung cavity becomes inflamed and the layers rub against each other. (This is a symptom of the early stages of a respiratory disorder called *pleurisy*.)

For reasons that are not fully understood, the brain does not always associate internal injury with the tissue that is actually injured. Instead, it may associate the pain with a tissue that is located some distance away from the real stimulation point. This phenomenon is called "referred pain."

A heart attack, for example, might be felt as pain in the skin above the heart and along the left shoulder and arm (Figure 35.7). Generally speaking, nerve pathways to both the injured tissue and the tissue where pain is referred pass through the same segment of the spinal cord.

Muscle Sense

Mechanoreceptors in skeletal muscle, joints, tendons, ligaments, and skin are responsible for awareness of the body's position in space and of limb movements. Stretch receptors are foremost among them. As described on page 558, stretch receptors are part of sensory organs called muscle spindles. Such organs are embedded in skeletal muscle tissue and run parallel with the muscle itself. The response of stretch receptors to stimulation depends on how much and how fast the muscle is stretched.

THE SPECIAL SENSES

Taste and Smell

Different animals "taste" substances with their mouth, antennae, legs, tentacles, or fins. It all depends on where the appropriate chemoreceptors are distributed. Whatever their location, *taste receptors* detect differences among substances, including nutritious and noxious ones, that become dissolved in fluid next to some body

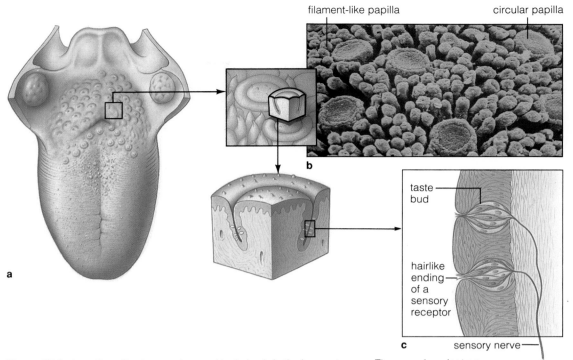

filament-like papilla   circular papilla

b

taste bud

hairlike ending of a sensory receptor

c   sensory nerve

a

Figure 35.8 Location of taste receptors and taste buds in the human tongue. The scanning electron micrograph shows filamentlike papillae that help move food; these do not contribute to the sense of taste. The taste receptors are organized inside taste buds, which are ringed by circular papillae.

surface. Those on animal tongues often are part of sensory organs, the taste buds (Figure 35.8).

Similarly, animals "smell" substances with *olfactory receptors*, such as the ones in your nose (Figure 35.9). Olfaction, the sense of smell, is one of the most ancient senses—and for good reason. Food and predators both give off chemical substances that can diffuse through water or air, and so give clues or advance warning of their whereabouts. Even humans, with their relatively insensitive sense of smell, have about 10 million olfactory receptors in the nose. The nose of a bloodhound has more than 200 million. Sharks have an olfactory sense that strikes terror in the heart of human divers who have accidentally cut or scraped themselves in the water.

Potential mates and rivals also give off odors. These include *pheromones*, the signaling molecules that are used as communication signals among members of the same species. Among other things, pheromones can raise an alarm, mark territorial boundaries, and attract a mate. Olfactory receptors on a male silk moth can detect one molecule of bombykol, a sex-attracting pheromone, in 10^{15} molecules of air! Each receptor, making contact with one molecule per second, can trigger an action potential. The signals reaching the moth brain are enough to help a male locate a female, even in the dark, and even more than a kilometer upwind from him.

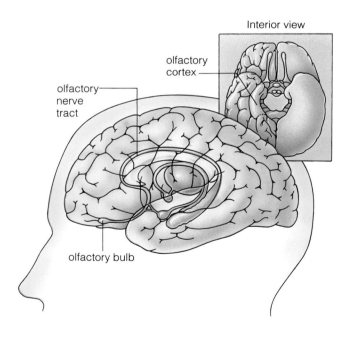

Interior view

olfactory cortex

olfactory nerve tract

olfactory bulb

Figure 35.9 Sensory nerve pathway leading from olfactory receptors in the nasal cavity to primary receiving centers in the brain.

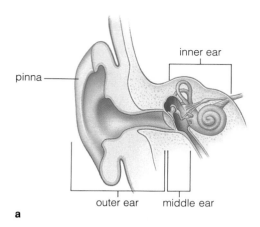

a

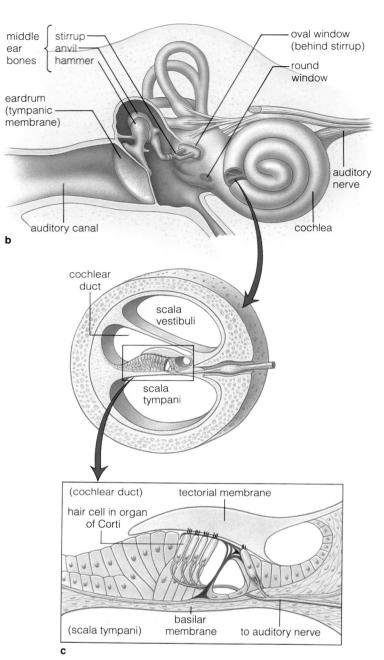

b

c

Figure 35.10 Sensory receptors in the human ear.

(**a, b**) Pressure waves funneled through the ear canal strike the eardrum (tympanic membrane), which bows in and out at the same frequency as the waves. This activates the middle earbones, a lever system that amplifies the stimulus by transmitting the force of pressure waves to the smaller surface of the oval window. The oval window is an elastic membrane over the entrance to the coiled inner ear.

The oval window bows in and out, producing fluid pressure waves in two ducts in the inner ear (scala vestibuli and scala tympani). The waves reach a membrane (round window) that bulges under pressure. Without this bulging, fluid would not be able to move back and forth in the inner ear.

Pressure waves are sorted out at the third duct (cochlear duct) of the coiled inner ear. The duct's basement membrane (basilar membrane) starts out narrow and stiff, but it becomes broader and flexible deep in the coil. High-frequency waves set up membrane vibrations in the stiff region. So do low-frequency waves, but the vibrations are lower in amplitude and continue on, into the more elastic regions.

(**c**) Perched on the basilar membrane is the organ of Corti, which contains hair cells. Vibrations of different regions of the membrane push different patches of hair cells against an overhanging flap (the tectorial membrane). Signals from disturbed hair cells initiate action potentials that are carried to the brain by the auditory nerve.

Hearing

Many arthropods and nearly all vertebrates have an acoustical sense, meaning that they can hear sounds. Crickets and birds sense one another singing; frogs sense one another croaking. Not all animals have ears like yours, of course, and their "ears" are not always located on the head. In insects and amphibians, acoustical receptors are located on thin membranes that are flush with the body surface. They are located on the front legs of crickets and near the hind legs of grasshoppers.

In all cases, however, the sense of hearing starts with mechanoreceptors that can detect vibrations. A vibration is a wavelike form of mechanical energy. For example, clapping produces waves of compressed air. Each time your hands clap together, molecules are forced outward and a low-pressure state is created in the region they vacated. The pressure variations can be depicted as a wave form, and the amplitude of its peaks corresponds to loudness:

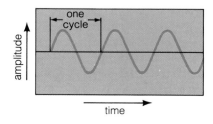

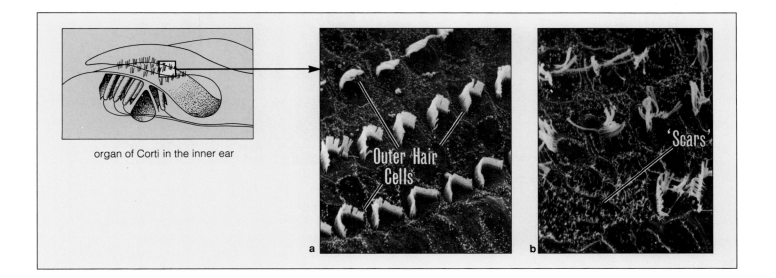

organ of Corti in the inner ear

'Scars'

Outer Hair Cells

a

b

The frequency of a sound is the number of wave cycles per second. Each "cycle" extends from the start of one wave peak to the start of the next. The more cycles per second, the higher the frequency—and the higher the perceived pitch of the sound.

When sound waves travel into an ear, they soon reach a membrane that is stretched across their path and they make it vibrate. In invertebrates, the vibrations directly stimulate mechanoreceptors attached to the membrane. In vertebrates, the membrane vibrations cause a fluid inside the ear to be displaced. The fluid movement causes mechanoreceptors to bend. With enough deformation, action potentials are produced. They travel along an auditory nerve leading from the receptors to the brain.

The Mammalian Ear. Land-dwelling mammals have a pair of ears, each with three regions that receive, amplify, and sort out signals (Figure 35.10). The *outer ear* collects sound waves with its external flaps, then channels them inward through a canal to an eardrum. The waves cause the eardrum to vibrate, and the vibrations are picked up and transferred inward by small bones of the *middle ear*. The commotion causes pressure waves in the coiled tube of the *inner ear*. There, different parts of a membrane vibrate most strongly to sounds of different frequencies. Stimulation of mechanoreceptors called *hair cells* on that membrane gives rise to action potentials that travel along an auditory nerve to the brain.

Some sounds are so intense that prolonged exposure to them can cause permanent damage to the inner ear, as Figure 35.11 suggests. Amplified music and the thundering of jet planes taking off can do this. Such recent developments exceed the functional range of the evolutionarily ancient hair cells in the ear.

Figure 35.11 Effect of intense sound on the inner ear. (**a**) Normal organ of Corti from a guinea pig, showing two rows of outer hair cells. (**b**) Organ of Corti after twenty-four-hour exposure to noise levels approached by loud rock music (2,000 cycles per second at 120 decibels).

Echolocation. The example at the start of this chapter introduced us to a sense of hearing that is associated with echolocation. Bats are not the only echolocating animals. Dolphins and whales also emit high-frequency sound waves (ultrasounds), and echoes from the waves bounce back to them from objects in their surroundings. By perceiving frequency variations in the echoes, all of these mammals can pinpoint the distance and direction of movement of predators, prey, and the like.

Although *we* cannot hear bat cries, the sound waves being produced are not weak. The cries have been measured at 100 decibels—about the same intensity as thunder booming overhead or a freight train rumbling past.

Balance

Almost all animals must have a sense of the "natural" position for their body, given the way they return to it after being tilted or turned upside-down. The baseline against which animals assess displacement from their natural position is called the "equilibrium position." Then, the body is balanced in relation to gravity, velocity, acceleration, and other forces that influence position and movement. The same balancing act occurs during the movement of wings and other body parts. The sense of balance depends on organs of equilibrium, which depend on the bending of hair cells in response to gravity and other forces.

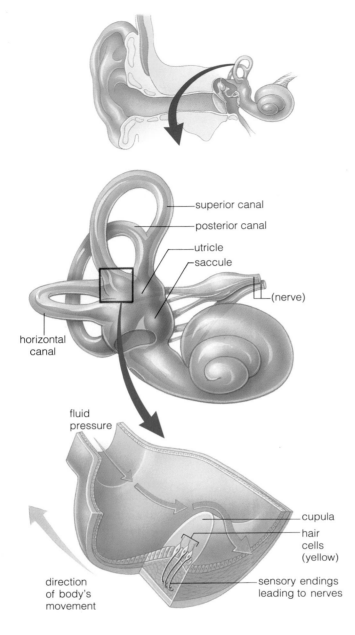

Figure 35.12 Vestibular apparatus, an organ of equilibrium that is given a workout when you ride a rollercoaster. Its three semicircular canals, positioned at angles corresponding to three planes of space, detect changes in movements. Its two sacs (utricle and saccule) each have an otolith organ that detects tilts in the head's position.

Body movement in a given direction displaces fluid in the canal corresponding to that direction. The fluid pressure bends hair cells inside the canal, and these produce signals that travel to the brain. Tilting the head affects the otolith organs (not visible here). Inside each organ, a membrane slides in response to gravity, like a baked cookie on a greased cookie sheet. It slides right over hair cells and so activates them.

Vertebrates rely partly on input from receptors in the eyes, skin, and joints for their sense of balance. They also rely on the **vestibular apparatus**, a closed system of fluid-filled sacs and canals inside the ear. The *otolith organs* of this system detect changes in the head's orien-

tation relative to gravity. The *semicircular canals* of this system detect changing movements. Amphibians, birds, and mammals have three semicircular canals positioned at angles corresponding to three planes of space. Figure 35.12 shows how fluid pressure inside the canals deforms hair cells and so produces action potentials.

Motion sickness may occur when monotonous linear, angular, or vertical motion overstimulates hair cells in the vestibular apparatus. Visual sensations often contribute to the sickness; fear and anxiety also play roles. Action potentials triggered by the sensory input reach a brain center that governs the vomiting reflex. Nausea and vomiting are the chief symptoms in individuals who get carsick, airsick, or seasick.

Vision

All organisms, whether they see or not, are sensitive to light. Even a single-celled amoeba will stop abruptly when you shine a light on it. What we call **vision**, however, requires more than a complex system of photoreceptors. It also requires a neural program in the brain that can interpret the patterns of action potentials arriving from different parts of the photoreceptor system. Those incoming signals encode information about the position, shape, brightness, distance, and movement of a visual stimulus.

Eyes are photoreceptor organs that contribute to image formation. Some types of eyes are better than others at doing this. For example, most types incorporate an adjustable *lens*, a transparent cone or sphere that focuses incoming light onto a dense layer of photoreceptor cells behind it. Yet even with lens-equipped eyes, some invertebrates still cannot see as we do. Their lenses channel light either in front of or behind their photoreceptors, the result being a very diffuse kind of stimulation. These invertebrates detect a general change in light intensity, as when another animal passes overhead in the water. But in many cases they cannot discern the size or shape of objects.

Many invertebrates have **eyespots**, not eyes. These are nothing more than clusters of photosensitive cells arranged in a cuplike depression in the epidermis. Mol-

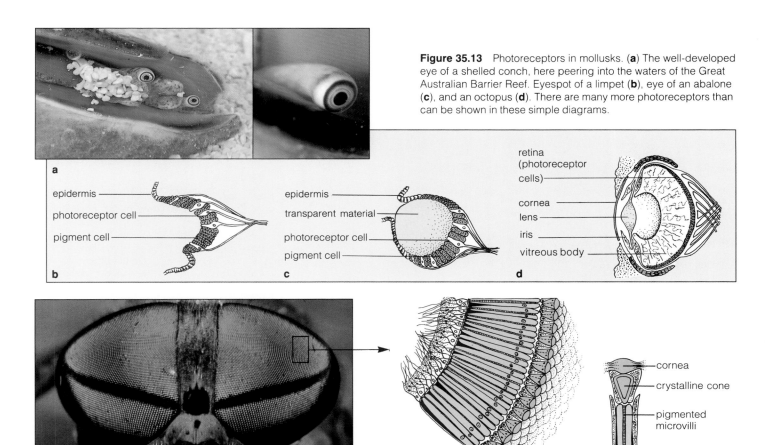

Figure 35.13 Photoreceptors in mollusks. (**a**) The well-developed eye of a shelled conch, here peering into the waters of the Great Australian Barrier Reef. Eyespot of a limpet (**b**), eye of an abalone (**c**), and an octopus (**d**). There are many more photoreceptors than can be shown in these simple diagrams.

Figure 35.14 (**a**) Compound eyes of a deerfly. The crystal-like cones of its photoreceptor units (ommatidia) act like a lens that focuses light on pigmented photoreceptor cells below (**b**).

lusks are the animals with the simplest eyes (Figure 35.13). Some molluscan eyes have a transparent lens with a transparent cover, a *cornea*, over it. They also have a *retina*, a light-sensitive tissue with densely packed photoreceptors. Squids, octopuses, and other splendidly complex mollusks have a pair of eyes that contribute to the formation of clear visual images. Their eyes contain an *iris*, a ring of contractile tissue that can be adjusted to admit more or less light. The *pupil* is the opening at the center of this ring.

Insects and crustaceans such as crabs have **compound eyes** that contain closely packed photosensitive units, of the sort shown in Figure 35.14. Some compound eyes have many thousands of these units, which are called *ommatidia* (singular, *ommatidium*). According to the **mosaic theory** of image formation, each ommatidium samples only a small part of the overall visual field. An image is built up according to signals about differences in light intensities across the field, with each unit contributing a separate bit to a visual mosaic (Figure 35.15). The units must be good at detecting movement; have you ever tried to sneak up on a fly?

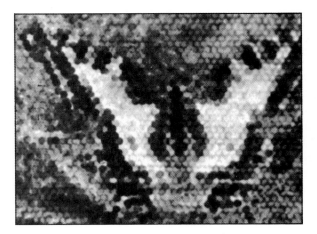

Figure 35.15 An approximation of light reception in the insect eye. This image of a butterfly was actually formed when a photograph was taken through the outer surface of a compound eye that had been detached from an insect. However, it may not be what the insect actually "sees." Integration of signals sent to the brain from photoreceptors in the eye may produce a more crisply defined image. The representation shown here is useful insofar as it suggests how the overall visual field may be *sampled* by separate ommatidia.

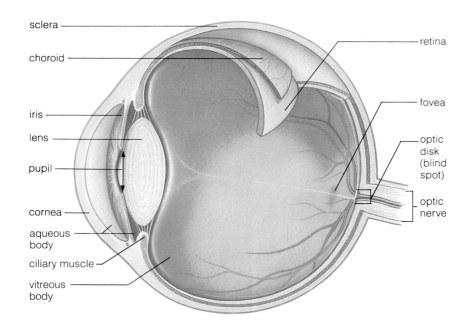

Figure 35.16 Structure of the human eye.

Table 35.2 Vertebrate Eye Components	
Eye Region	Functions
Outer Layer:	
Sclera	Protect eyeball
Cornea	Focus light
Middle Layer:	
Choroid	
Pigmented tissue	Prevent light scattering
Iris	Control amount of light
Pupil	Entrance for light
Lens	Finely focus light on photoreceptors
Aqueous body	Transmit light, maintain pressure
Vitreous body	Transmit light, support lens, eye
Inner Layer:	
Retina	Absorb, convert light
Fovea	Increase visual acuity
Start of optic nerve	Transmit signals to brain

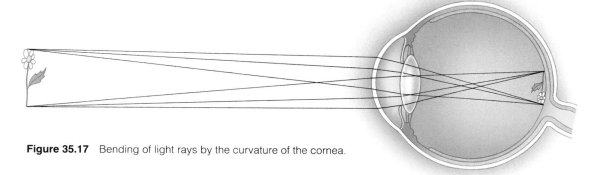

Figure 35.17 Bending of light rays by the curvature of the cornea.

The Vertebrate Eye

Eye Structure. Nearly all vertebrates have eyes capable of forming clear images. Think of the vertebrate eye as having three layers. The outer layer consists of a sclera and transparent cornea. The middle layer consists of a choroid and lens, as well as semifluid and jellylike substances. The key feature of the inner layer is the retina (Figure 35.16 and Table 35.2).

The *sclera*—the dense, fibrous "white" of the eye, protects most of the eyeball, and the cornea covers the rest. In the eye's middle layer, a clear fluid (aqueous humor) bathes both sides of the lens and a jellylike substance (vitreous body) fills the chamber behind the lens. The *choroid*, a dark-pigmented tissue, prevents light scattering inside the eyeball. Suspended beneath the transparent circle of the cornea is a doughnut-shaped, pigmented *iris* (after *irid*, meaning colored circle). The dark "hole" in the center of the doughnut is the *pupil*,

the entrance for light. When bright light hits the eye, circular muscles in the iris contract and so shrink the size of the pupil. In dim light, radial muscles contract and so enlarge the pupil. Behind the iris is the *lens*, with its onionlike layers of transparent proteins.

The surface of the cornea is curved, so light rays coming from the same source hit it at different angles and their trajectories change. (Light rays bend at boundaries between two different substances, and this sends them in new directions; refer to Figure 4.4.) The rays converge at the back of the eyeball. There, because of the way the rays were bent at the curved cornea, they form a pattern of stimulation that is upside-down and reversed left to right, relative to the original light source (Figure 35.17).

Light rays from sources at different distances from the eye strike the cornea at different angles and will be focused at different distances behind it. Therefore, adjustments must be made so that the light will be

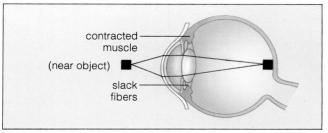

a Accommodation for near objects (lens bulges)

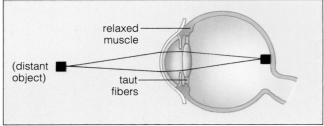

b Accommodation for distant objects (lens flattens)

Figure 35.18 Visual accommodation in the human eye. (**a**) Close objects are brought into focus when eye muscles contract enough to slacken certain fibers interposed between them and the lens, and this causes the lens to thicken at its equator. (**b**) Distant objects are brought into focus when eye muscles relax, thereby putting tension on the fibers and stretching the lens into a flatter shape.

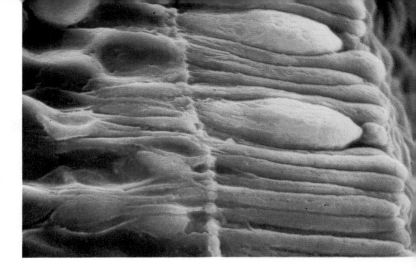

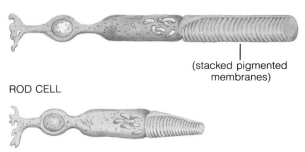

Figure 35.19 Structure of rods and cones, the photoreceptors of the vertebrate eye.

focused precisely onto the retina. Such adjustments are called *accommodation*. If adjustments are not made, rays from very distant objects will be focused in front of the retina. And rays from very close objects will be focused in back of the retina.

Normally, the lens can be adjusted so that the focal point coincides exactly with the retina. In fish and reptiles, eye muscles move the entire lens forward or back, like the focusing apparatus inside a camera. Increasing the distance moves the focal point forward; decreasing the distance moves it back.

By contrast, in birds and mammals, a "ciliary muscle" adjusts the *shape* of the lens. The ciliary muscle rings the lens and attaches to it by fiberlike ligaments (Figures 35.16 and 35.18). When the muscle contracts, the lens flattens, so the focal point moves farther back. When the muscle relaxes, the lens bulges, so the focal point moves forward.

Sometimes the lens cannot be adjusted enough to match the focal point with the retina. Sometimes also, the eyeball is not shaped quite right. The lens is too close or too far away from the retina, so accommodation alone cannot bring about a precise match. As indicated in the *Commentary* on the page that follows, eyeglasses can correct both problems, which are called nearsightedness and farsightedness.

Photoreception. Most birds and mammals have a keen sense of vision and a well-developed retina. In these animals, the retina is a thin, complex layer of neural tissue at the back of the eyeball. It has a basement layer—a pigmented epithelium that covers the choroid. Resting on the basement layer are densely packed photoreceptors that are functionally linked with a variety of neurons. Axons from these neurons converge to form the optic nerve at the back of the eyeball. The optic nerve is the trunk line to the thalamus, which sends information on to the visual cortex.

The photoreceptors of the retina are called **rod cells** and **cone cells** because of their shapes (Figure 35.19). Rods are sensitive to very dim light. They contribute to coarse perception of movements by detecting changes in light intensity across the field of vision. Typically they are abundant in the periphery of the retina. Cones respond to bright light. They contribute to sharp daytime vision and color perception. Pigments in different cone cells are selectively sensitive to wavelengths corresponding to red, green, or blue colors. Cones of the human eye are densely packed in the *fovea*, a funnel-shaped depression near the center of the retina, where the overlying nerve tissue is thinner. Cones at the fovea contribute most to visual acuity—that is, to precise discrimination between adjacent points in space.

Disorders of the Eye

Two-thirds of all the sensory receptors your body requires are located in your eyes. Those photoreceptors do more than detect light. They also allow you to see the world in a rainbow of colors. Your eyes are the single most important source of information about the outside world.

Injuries, disease, inherited abnormalities, and advancing age can disrupt functions of the eyes. The consequences range from relatively harmless conditions, such as near-sightedness, to total blindness. Each year, many millions of people must deal with such consequences.

Color Blindness

Consider a common heritable abnormality, *red-green color blindness*. It is an X-linked, recessive trait that shows up most often in males. The retina lacks some or all of the cone cells with pigments that normally respond to light of red or green wavelengths. Most of the time, color-blind persons merely have trouble distinguishing red from green in dim light. However, some cannot distinguish between the two even in bright light. The rare few who are totally color blind have only one of three kinds of pigments that selectively respond to red, green, or blue wavelengths. They see the world only in shades of gray.

Focusing Problems

Other heritable abnormalities arise from misshapen features of the eye that affect the focusing of light. *Astigmatism*, for example, results from corneas with an uneven curvature; they cannot bend incoming light rays to the same focal point.

Nearsightedness (myopia) commonly occurs when the vertical axis of the eyeball is longer than the horizontal axis. It also occurs when the ciliary muscle responsible for adjustments in the lens contracts too strongly. The outcome is that images of distant objects are focused in front of the retina instead of on it (Figure *a*). *Farsightedness* (hyperopia) is the opposite problem. The horizontal axis of the eyeball is longer than the vertical axis (or the lens is "lazy"), so close images are focused behind the retina (Figure *b*).

Even a normal lens loses some of its natural flexibility as a person grows older. That is why people over forty years old often start wearing eyeglasses.

Eye Diseases

The structure of the eye and its functions are vulnerable to infection and disease. Especially in the southeastern United States, for example, a fungal infection of the lungs (histo-

plasmosis) can lead to retinal damage. This complication can cause partial or total loss of vision. As another example, *Herpes simplex*, a virus that causes skin sores, also can infect the cornea and cause it to ulcerate.

Trachoma is a highly contagious disease that has blinded millions, mostly in North Africa and the Middle East. The culprit is a bacterium that also is responsible for the sexually transmitted disease chlamydia (page 782). The eyeball and the lining of the eyelids (conjunctiva) become damaged. The damaged tissues are entry points for bacteria that can cause secondary infections. In time the cornea can become so scarred that blindness follows.

Age-Related Problems

Cataracts, a gradual clouding of the lens, is a problem associated with aging, although it also may arise through injury or diabetes. Possibly the condition arises when the transparent proteins making up the lens undergo structural changes. The clouding may skew the trajectory of incoming light rays. If the lens becomes totally opaque, light cannot enter the eye at all.

Glaucoma results when excess aqueous humor accumulates inside the eyeball. Blood vessels that service the retina collapse under the increased fluid pressure. Vision deteriorates as neurons of the retina and optic nerve die off. Although chronic glaucoma often is associated with advanced age, the problem actually starts in middle age. If detected early, the fluid pressure can be relieved by drugs or surgery before the damage becomes severe.

Eye Injuries

Retinal detachment is the eye injury read about most often. It may follow a physical blow to the head or an illness that tears the retina. As the semifluid vitreous body oozes through the torn region, the retina becomes lifted from the underlying choroid. In time it may peel away entirely, leaving its blood supply behind. Early symptoms of injury include blurred vision, flashes of light that occur in the absence of outside stimulation, and loss of peripheral vision. Without medical intervention, the injured person may become totally blind in the injured eye.

New Technologies

Today a variety of tools are used to correct some eye disorders. In *corneal transplant surgery*, the defective cornea is removed, then an artificial cornea (made of clear plastic) or

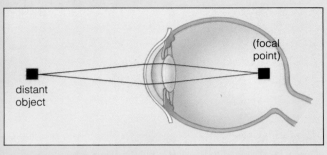

a Focal point in nearsighted vision. The example shows flamingos in Tanzania, East Africa.

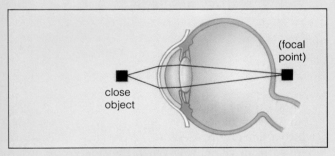

b Focal point in farsighted vision.

a natural cornea from a donor is stitched in place. Within a year, the patient is fitted with eyeglasses or contact lenses. Similarly, cataracts can be surgically corrected by removing the lens and replacing it with an artificial one, although the operation is not always successful.

Severely nearsighted people may opt for *radial keratotomy*, a still-controversial surgical procedure in which tiny, spokelike incisions are made around the edge of the cornea

to flatten it more. When all goes well, the adjustment eliminates the need for corrective lenses. Sometimes, however, the result is overcorrected or undercorrected vision.

Retinal detachment may be treatable with *laser coagulation*, a painless technique in which a laser beam seals off leaky blood vessels and "spot welds" the retina to the underlying choroid.

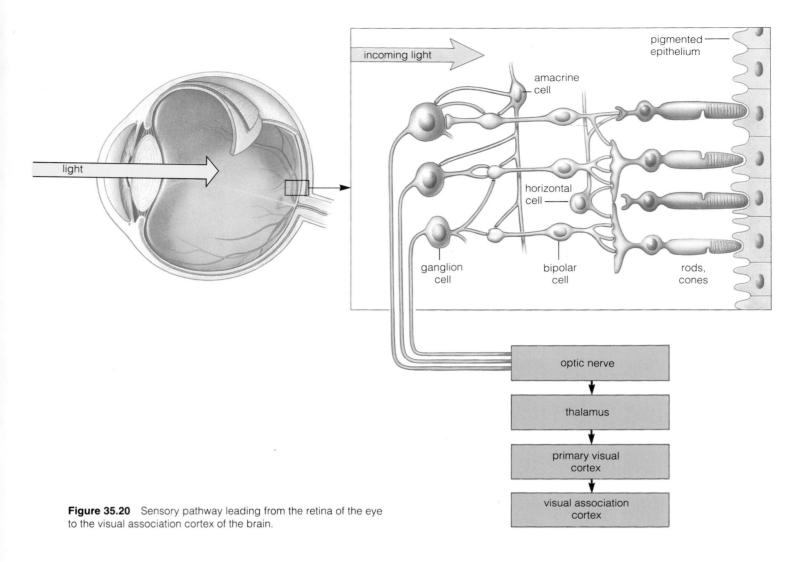

Figure 35.20 Sensory pathway leading from the retina of the eye to the visual association cortex of the brain.

Rods contain molecules of rhodopsin in their membranes. Each molecule consists of a protein (opsin) to which a side group (*cis*-retinal) is attached. The retinal is derived from vitamin A. When the side group absorbs light energy, it is temporarily converted to a slightly different form (*trans*-retinal):

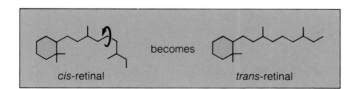

In this altered form, the side group initiates a series of chemical reactions within the photoreceptor. The reactions lead to a voltage change across the photoreceptor's plasma membrane. This local, graded potential affects the release of a transmitter substance from the photoreceptor. That substance acts on neighboring neurons, in ways that will now be described.

Processing Visual Information. How does the information from photoreceptors become translated into the sense of vision? Part of the answer is that the information moves in increasingly organized ways through *levels* of synapsing neurons—first in the retina, then in different parts of the brain. That information includes signals about form, movement, depth, color, and texture. Each type of signal seems to be processed along a separate communication channel. And many different channels run in parallel to the brain.

Only the rods and cones respond directly to light. When stimulated, they pass graded signals horizontally

STIMULI:

RESPONSES:

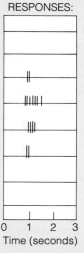

Time (seconds)

Orientation of stimulus (bar of light) in visual field analyzed by one kind of neuron

Recordings of action potentials in response to changing stimulus orientation

Figure 35.21 From signaling to visual perception. Neurons in the visual cortex are stacked in columns at right angles to the brain's surface. Connections run between neurons in each column and between different columns. Each column apparently deals with only one kind of stimulus, received from only one location. Visual perception seems to be based on the organization and synaptic connections between neurons in these columns. The neurons fall into a few categories. In each category, they seem to be tripped into action the same way. For instance, excitatory signals traveling up through the cortex activate certain neurons, which then send out inhibitory signals to other neurons. The excitatory and inhibitory signals between neurons form narrow bands of electrical activity.

Experiments show that the pattern of excitation through specific columns of neurons is highly focused. For example, David

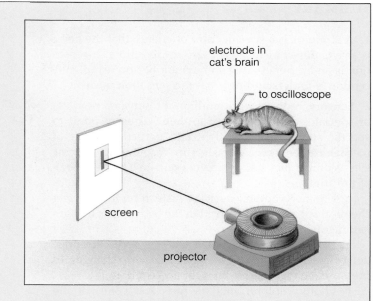

Hubel and Torsten Wiesel implanted electrodes in individual neurons in the brain of an anesthetized cat. Then they positioned the cat in front of a small screen. They projected images of different shapes (including a bar) onto the screen. When the bar was tilted at different angles, changes in electrical activity that corresponded to the different angles were recorded.

The strongest activity was recorded for one type of neuron when the bar image was vertical (numbered 5 in the sketch). When the bar image was tilted slightly, the signals were less frequent. When the image was tilted past a certain angle, the signals stopped. In other experiments, a certain neuron fired only when an image of a block was moved from left to right across the screen; another fired when the image was moved from right to left.

to one another as well as to adjacent neurons, including the bipolar cells shown in Figure 35.20. Action potentials start with functional groups of bipolar cells, which differ from one another in their sensitivity to contrast and color. Those neurons pass signals to ganglion cells, the axons of which converge to form the optic nerve leading to the brain. The optic nerves from both eyes converge at the base of the brain. A portion of the axons of each nerve cross over here before continuing onward. (This partial crossover, the optic chiasm shown on page 573, ensures that information from both eyes will reach both cerebral hemispheres.) Most axons of the optic nerves lead into the thalamus, which passes on information to the visual cortex.

The visual cortex has several subdivisions, but each has the whole visual field mapped onto it. The *visual field* is the portion of the outside world that is being detected by photoreceptors at any given time. A particular portion of each subdivision thus receives input from a particular portion of the visual field. Even within such portions, some neurons are sensitive to stimuli in the visual field that are oriented in one direction only—a line in the outside world that is horizontal, vertical, tilted left, or tilted right. (Figure 35.21 shows an example of this.) And nearby neurons may respond to stimuli oriented in a different direction.

Now different bits of information that have reached the visual cortex are sent to different parts of the cerebral cortex. Some parts analyze what the stimulus might be, another part analyzes where it is located in the visual field, and so on. All the information is processed rapidly, at the same time, in different cortical regions. Finally, signals are integrated to produce the organized electrical activity that gives rise to the sensation of sight.

SUMMARY

1. A stimulus is a specific form of energy that the body is able to detect by means of sensory receptors. A sensation is an awareness that stimulation has occurred. Perception is understanding what the sensation means.

2. Sensory receptors are endings of sensory neurons or specialized cells adjacent to them. They respond to specific stimuli, such as light and particular forms of mechanical energy. Animals can respond to specific events only if they have receptors sensitive to the energy of the stimulus.

    a. Chemoreceptors, such as taste receptors, detect chemical substances dissolved in the body fluids that are bathing them.

    b. Mechanoreceptors, such as free nerve endings, detect mechanical energy associated with changes in pressure, changes in position, or acceleration.

    c. Photoreceptors, such as rods and cones of the retina, detect light.

    d. Thermoreceptors detect the presence of or changes in radiant energy from heat sources.

3. At receptor endings, the stimulus triggers local, graded signals. When the stimulus is strong enough, summation of graded signals may produce action potentials. The action potentials travel on particular nerve pathways from the receptors to parts of the cerebral cortex.

4. Variations in stimulus intensity are encoded in (1) the frequency of action potentials propagated along an information-carrying neuron and (2) the number of action potentials generated in a given tissue.

5. Somatic sensations include touch, pressure, temperature, pain, and muscle sense. The receptors associated with these sensations are not localized in a single organ or tissue. Stretch receptors, for example, occur in skeletal muscles throughout the body.

6. The special senses include taste, smell, hearing, balance, and vision. The receptors associated with these senses typically reside in sensory organs or some other particular region.

Review Questions

1. Label the component parts of the human eye: *610*

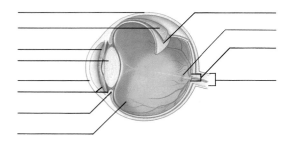

2. What is a stimulus? Receptor cells detect specific kinds of stimuli. When they do, what happens to the stimulus energy? *600*

3. Give some examples of chemoreceptors and mechanoreceptors. *600, 603–607*

4. What is sound? How are amplitude and frequency related to sound? Give some examples of animals that apparently perceive sounds. *606–607*

5. What is pain? Can you name one of the receptors associated with pain? *604*

6. How does vision differ from photoreception? What sensory apparatus does vision require? *608*

7. How does the vertebrate eye focus the light rays of an image? What is meant by *nearsighted* and *farsighted*? *610–611, 612*

Self-Quiz *(Answers in Appendix IV)*

1. A _____ is a specific form of energy that is capable of eliciting a response from a sensory receptor.

2. Conscious awareness of a stimulus is called a _____ .

3. _____ is understanding what particular sensations mean.

4. Each sensory system is composed of _____ .
    a. nerve pathways from specific receptors to the brain
    b. sensory receptors
    c. brain regions that deal with sensory information
    d. all of the above are components of sensory systems

5. _____ detect mechanical energy associated with changes in pressure, in position, or acceleration.
    a. Chemoreceptors   c. Photoreceptors
    b. Mechanoreceptors   d. Thermoreceptors

6. Detecting chemical substances present in the body fluids that bathe them is the function of _____ .
    a. thermoreceptors   c. mechanoreceptors
    b. photoreceptors   d. chemoreceptors

7. Which of the special senses is based on the following events: Membrane vibrations cause fluid movements, which bend mechanoreceptors and so trigger action potentials.
    a. taste
    b. smell
    c. hearing
    d. vision

8. The outer layer of the human eye includes the _____.
    a. lens and choroid
    b. sclera and cornea
    c. retina
    d. both a and c are correct

9. The middle layer of the human eye includes the _____.
    a. lens and choroid
    b. sclera and cornea
    c. retina
    d. start of optic nerve

10. Match each term with the appropriate description.
    _____ somatic senses
    _____ stimulus
    _____ special senses
    _____ variations in stimulus intensity
    _____ action potential
    _____ sensory receptor

    a. produced by strong stimulation and summation of graded signals
    b. endings of sensory neurons or specialized cells next to them
    c. taste, smell, hearing, balance, and vision
    d. a specific form of energy that can elicit a response from a sensory receptor
    e. frequency and number of action potentials
    f. touch, pressure, temperature, pain, and muscle sense

Selected Key Terms

adaptation *603*
chemoreceptors *600*
compound eye *609*
cone cell *611*
ear *607*
echolocation *599*
eye *608*
eyespot *608*
hair cell *607*
iris *609*
lens *608*
mechanoreceptor *600*
mosaic theory *609*
nociceptor *604*
olfactory receptor *605*

pain *604*
perception *600*
pheromone *605*
photoreceptor *600*
rod cell *611*
sensation *600*
sensory receptor *600*
sensory system *599*
somatic sensation *603*
somatic sensory cortex *601*
taste receptor *604*
thermoreceptor *600*
vision *608*
visual cortex *602*
visual field *615*

Readings

Eckert, R., D. Randall, and G. Augustine. 1988. *Animal Physiology: Mechanisms and Adaptations.* Third edition. New York: Freeman.

Hubel, D. H., and T. N. Wiesel. September 1979. "Brain Mechanisms of Vision." *Scientific American* 241(3):150–162. Describes studies on information processing in the primary visual cortex.

Hudspeth, A. January 1983. "The Hair Cells of the Inner Ear." *Scientific American* 248(1):54–66.

Jacobs, G. 1983. "Colour Vision in Animals." *Endeavour* 7(3): 137–140.

Kandel, E., and J. Schwartz. 1985. *Principles of Neural Science.* Second edition. New York: Elsevier. Advanced reading, but good coverage of sensory perception.

Newman, E. A., and P. H. Hartline. March 1982. "The Infrared 'Vision' of Snakes." *Scientific American* 246(3):116–127.

Parker, D. November 1980. "The Vestibular Apparatus." *Scientific American* 243(5):118–130.

Stryer, L. July 1987. "The Molecules of Visual Excitation." *Scientific American* 257(1)42–50. Well-written description of the cascade reactions that give rise to nerve signals in the retina.

Vander, A., J. Sherman, and D. Luciano. 1990. *Human Physiology.* Fifth edition. New York: McGraw-Hill. Chapter 9 is a clear introduction to sensory systems.

Wu, C. H. November–December 1984. "Electric Fish and the Discovery of Animal Electricity." *American Scientist* 72(6):598–607.

Young, J. 1978. *Programs of the Brain.* New York: Oxford University Press. An extraordinary book, beautifully written.

36 PROTECTION, SUPPORT, AND MOVEMENT

The Challenge of the Iditarod

"All right—GO!"

Once again Susan Butcher is mushing out. On command, her trained Alaskan huskies leap forward, gathering momentum for the long haul across 1,157 miles of the Iditarod Trail. They will be towing a 200-pound sled in a race that will take them on a frozen, isolated route between Anchorage and Nome, Alaska. At the minimum, they will face eleven days and nights of snow, ice, and treacherous river crossings. Butcher is confident; many consider her to be the finest long-distance sled-dog racer of all time.

When it comes to speed, stamina, and built-in protection against the elements, we humans are not the superstars of the animal kingdom. Long before the Iditarod race began, Butcher began following a marathoner's regimen of diet and exercise to put her arm and leg muscles in peak condition. Through physical workouts, she increased the capacity of her skeletal-muscular system to support and help move her body. Lacking the fur coat of mammals that are native to the Far North, Butcher selected clothing that would insulate and protect her while still allowing freedom for strenuous movements.

Protection, support, movement—these aspects of animal anatomy and physiology are the topics of this chapter. Some basic rules apply for all of the examples we will be using. For example, the body must have a system of structural support—some type of skeleton. The skeleton must have fairly rigid parts that muscles can work against and so transmute force into body movement. Those skeletal parts must be lightweight as well as strong, thus minimizing the amount of energy required for movement. The muscle cells and tissues must be organized to work with one another as well as with the skeleton. Only then will the animal be able to execute the movements and positional changes required for a particular life-style in a particular environment.

Consider Granite, the lead dog of Butcher's team when she won the Iditarod in 1986, then in 1987, and again in 1988. Through artificial selection practices, huskies have been bred for remarkable strength and endurance. They can haul a light load at moderate speed over great distances, and most of the time in bitter cold. The long bones of Granite's legs are large in diameter and quite sturdy, yet much of the structural material inside them is lightweight. Granite's chest is

deep but not too broad. Breeders of huskies favor dogs with a rib cage that is a bit flattened on the sides, so the forelegs will have full freedom of movement. Although lean overall, Granite has well-developed muscles that ripple across the upper bones of his hind legs. These are not the muscles of a greyhound, cheetah, or some other sprinter. They are the muscles of a load-pulling, long-distance runner. The pads on Granite's feet are notably thick and tough. Packed with the protein collagen, they are built-in cushions against sharp ice and frozen rock. Granite also has a superior fur coat. Dense, fine hair serves as an insulative layer next to the skin. Above it is a slightly oily layer of tougher, longer hairs that take the first brunt of biting winds and near-freezing moisture.

Huskies, humans, and other animals show considerable variation in their systems of support, movement, and protection. When thinking about those variations, keep in mind that they are evolutionary responses to life in a particular environment. From this standpoint, the Iditarod is a testing ground—not only for variations that enhance speed and endurance, but for the functioning of all the interacting systems.

KEY CONCEPTS

1. The body of nearly all animals has an outer covering (integument), muscle cells, and some form of skeleton. The contractile force of the muscle cells acts against a fluid medium, rigid structures (such as bones), or both.

2. The vertebrate integumentary system (skin and its component structures) protects the body from dehydration, ultraviolet radiation, abrasion, bacterial attack, and other environmental insults. It contributes to overall body functioning, as when it helps control moisture loss and when some of its cells synthesize the vitamin D required for calcium metabolism.

3. Smooth, cardiac, and skeletal muscle tissues occur in vertebrates. When adequately stimulated, the cells of all three types of muscle tissue can contract (shorten), then return to the resting position.

4. In vertebrates, the force of skeletal muscle contraction acts against an internal skeleton of bone and cartilage. Together, the skeletal and muscular systems change the positions of body parts and move the body through the environment. The bones function not only in movement but also in protection and support for soft organs, in mineral storage, and in blood cell formation.

Figure 36.1 To the left, Susan Butcher and her Alaskan huskies, superbly illustrating their systems of support, movement, and protection along the Iditarod Trail.

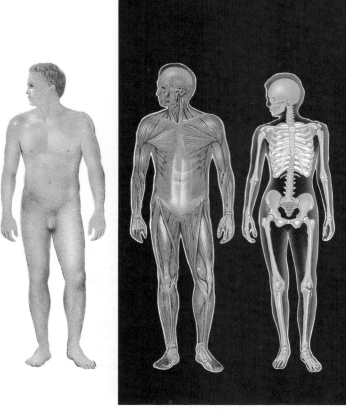

Figure 36.2 Starting point for a tour of three types of organ systems, using the human body as an example. Traveling from the outside in, these diagrams show the integumentary system (skin and its derivatives), muscle system, and skeletal system.

In the image you hold of yourself, you are tall or short, pale or dark, sparsely or profusely haired, taut-skinned or flabby, slow-moving or always on the move (or somewhere in between). Like most other animals, you have three organ systems to thank for your body's shape, superficial features, and capacity for movement. These are the integumentary, skeletal, and muscular systems. Figure 36.2 gives a general picture of what the systems look like for humans.

INTEGUMENTARY SYSTEM

Animals ranging from worms to humans have an outer cover for the body. It is called an **integument** (after the Latin *integere*, meaning "to cover"). In most cases, the integument is tough yet pliable, a barrier against a great variety of environmental insults. For insects, crabs, and other arthropods, the integument is a hardened covering called a *cuticle*. Figure 36.3 shows examples. Arthropod cuticles consist of chitin, protein, and sometimes lipid secretions. For vertebrates, the integument is skin and the structures derived from epidermal cells of its outer layers of tissue.

Integuments vary among vertebrates. Depending on the vertebrate group, the skin may be decked out with scales, feathers, hair, beaks, hooves, horns, claws, nails, quills, and other structures (Figure 36.4). Variation exists within groups as well. Thus, as you read in Chapter 26,

Figure 36.3 Off with the old, on with the new. (**a**) A green cicada (*Tibicen superbus*) and (**b**) a centipede (*Lithobius*), each shedding its outgrown cuticle during a molting cycle. The new cuticle is pale and soft, but soon will harden and darken.

a

b

some fishes have hard scales, others have bare skin coated with slimy mucus. Here our focus will be on the properties of human skin.

Functions of Skin

No garment ever made approaches the qualities of the one covering your body—your skin. What besides skin maintains its shape in spite of repeated stretchings and washings, kills many bacteria on contact, screens out harmful rays from the sun, is waterproof, repairs small cuts and burns on its own, and with a little care, will last as long as you do?

Skin does more than protect the rest of the body from dehydration, abrasion, and bacterial attack. It helps control the body's internal temperature. It has so many small blood vessels that it serves as one of the reservoirs for blood. The reservoirs can be tapped and shunted to metabolically active regions, such as leg muscles during strenuous pushes along the Iditarod Trail. Skin produces vitamin D, which is required for calcium metabolism. And signals from sensory receptors in skin help the brain assess what is happening in the outside world.

Structure of Skin

Assuming you are an average-sized adult, your skin weighs about 9 pounds. Stretched out, it would have a surface area of 15 to 20 square feet. For the most part, your skin is as thin as a paper towel. It thickens only on

the soles of your feet and in other regions subjected to pounding or abrasion.

As is the case for other vertebrates, skin has two distinct regions. As Figure 36.5 shows, the outermost region is the **epidermis** and the underlying region, the **dermis**. Beneath this is the *hypodermis*, a tissue that anchors the skin and yet allows it some freedom of movement. Fat stored in the hypodermis insulates the body against cold. Strictly speaking, the hypodermis is not part of skin.

Figure 36.4 (*Above*) Feathers—one of the diverse kinds of structures arising from cell differentiations in the epidermis.

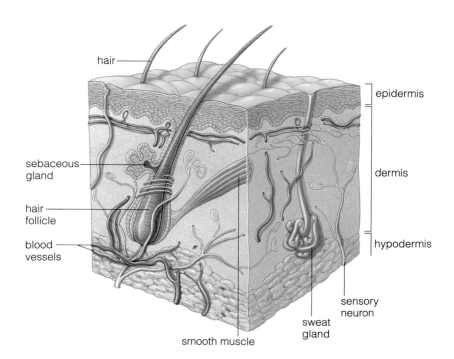

Figure 36.5 The two-layered structure of human skin. The hypodermis is a subcutaneous layer, not part of skin.

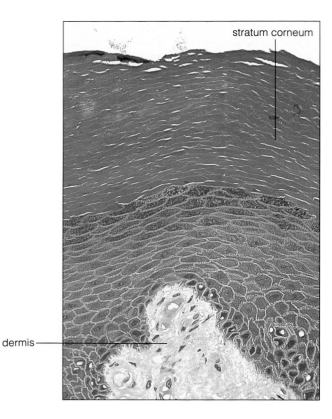

stratum corneum

dermis

Figure 36.6 Section through human skin, showing the uppermost layer of epidermis (stratum corneum), the deeper epidermal layers, and the underlying dermis.

Epidermis. The epidermis consists mostly of stratified epithelium, a tissue described earlier on page 535. An abundance of cell junctions of the sort shown in Figure 31.6 knits the epithelial cells together. The cells themselves arise within the epidermis but are pushed toward its free surface, as rapid and ongoing mitotic divisions produce new cells beneath them.

Most cells of the epidermis are keratinocytes. Each is a tiny factory for manufacturing *keratin*, a tough, water-insoluble protein. Those cells start producing keratin when they are in mid-epidermal regions. By the time they reach the skin's free surface, they are dead and flat-tened. All that remain are fibers of keratin, packed inside plasma membranes. This is the composition of the outermost layer of skin—the tough, waterproof "stra-tum corneum" (Figure 36.6). Millions of the flattened keratin packages are worn off daily, but cell divisions con-tinually push up replacements. The rapid divisions also contribute to skin's capacity to mend itself quickly after cuts or burns.

In the deepest epidermal layer, cells called melano-cytes produce *melanin*, a brownish-black pigment. This pigment is transferred to keratin-producing cells and accumulates inside them, forming a shield against ultra-violet radiation. Melanin also contributes to skin color.

Humans generally have the same number of melano-cytes. Variations in skin color arise through differences in melanocyte distribution and activity. Albinos, for example, lack melanin; their melanocytes cannot produce all of the enzymes required for its production (page 174). Melanocyte activity is increased in suntanned skin (Figure 36.7).

Skin color also is influenced by *hemoglobin* (the oxygen-carrying pigment of red blood cells) and *carotene* (a yellow-orange pigment). Pale skin, for example, has a pinkish cast. It does not have much melanin, so the presence of hemoglobin is not masked. Hemoglobin's red color shows through thin-walled blood vessels and the epidermis, both of which are transparent.

With its multiple layers of keratinized, melanin-shielded epidermal cells, skin helps the body conserve water, avoid damage from ultraviolet radiation, and resist mechanical stress.

Rapid, continuous cell divisions in deep epidermal layers underlie skin's capacity to heal itself after being abraded, burned, or cut.

Dermis. Dense connective tissue makes up most of the dermis, and it fends off damage from everyday stretch-ing and other mechanical insults. There are limits to this protection. For example, the dermis tears when skin over the abdomen is stretched too much during preg-nancy, leaving white scars ("stretch marks"). With per-sistent abrasion, the epidermis separates from the dermis and you get a "blister."

Blood vessels, lymph vessels, and the receptor end-ings of sensory nerves thread through the dermis. Nutri-ents from the bloodstream reach epidermal cells by diffusing through the dermal tissue. Sweat glands, oil glands, and the husklike cavities called hair follicles reside mostly in the dermis, even though they are derived from epidermal tissue.

The fluid secreted from *sweat glands* is 99 percent water, along with dissolved salts, traces of ammonia and other metabolic wastes, vitamin C, and other substances. You have about 2.5 million sweat glands, which are con-trolled by sympathetic nerves. One type abounds in the palms of the hands, soles of the feet, forehead, and armpits. They function mainly in temperature regulation (page 733). They also function in "cold sweats," one of the responses you make when you are frightened, ner-vous, or merely embarrassed. Another type of sweat gland prevails in skin around the sex organs. Their secre-tion steps up during stress, pain, sexual foreplay, and estrus. Do they have functions similar to those of scent glands in other animals? No one knows.

Figure 36.7 Sunlight and the skin. Melanin-producing cells of the epidermis are stimulated by exposure to ultraviolet radiation. With prolonged sun exposure, melanin levels increase and light-skinned people become tanned (visibly darkened). Tanning provides some protection against ultraviolet radiation, but prolonged exposure can damage the skin. Over the years, tanning causes elastin fibers of the dermis to clump together. The skin loses its resiliency and begins to look like old leather.

Prolonged exposure to ultraviolet radiation also suppresses the immune system. Certain phagocytes and other specialized cells in the epidermis defend the body against specific viruses and bacteria. Sunburns interfere with the functioning of these cells. This may be why sunburns can trigger the small, painful blisters called "cold sores." The blisters are a symptom of a viral infection. Nearly everyone harbors this virus (*Herpes simplex*); usually it becomes localized in a nerve ending near the skin surface, where it remains dormant. Stress factors—including sunburn—can activate the virus and trigger the skin eruptions.

Ultraviolet radiation from sunlight or from the lamps of tanning salons also can activate proto-oncogenes in skin cells (page 241). Epidermal skin cancers start out as scaly, reddened bumps. They grow rapidly and can spread to

adjacent lymph nodes unless they are surgically removed. Basal cell carcinomas start out as small, shiny bumps and slowly grow into ulcers with beaded margins. Their threat to the individual ceases when they are surgically removed, provided they are removed in time.

Oil glands (also called sebaceous glands) are everywhere except on the palms and soles. They function to soften and lubricate both the hair and the skin—and to kill surface bacteria. *Acne* is a skin condition in which the ducts of oil glands have become infected by bacteria, followed by inflammation of the glands.

Hairs are flexible structures, composed mostly of keratinized cells. Each has a root embedded in skin and a shaft that projects above the skin's surface. As living cells divide near the base of the root, older cells are pushed upward, then flatten and die. The outermost layer of the shaft consists of flattened cells that overlap one another like roof shingles (Figure 36.8). The most abused of these cells tend to frizz out near the end of the hair shaft; we call these "split ends."

The average scalp has about 100,000 hairs, but a person's genes, nutrition, and hormones influence hair growth and density. Protein deficiency causes hair to thin, for hair cannot grow without the amino acids required for keratin synthesis. Severe fever, emotional stress, and excessive vitamin A intake also cause hair thinning. Excessive hairiness (hirsutism) may result when the body produces abnormal amounts of testosterone. This hormone influences patterns of hair growth and other secondary sexual traits.

As we age, epidermal cells divide less often, and our skin becomes thinner and more susceptible to injury. Glandular secretions that kept the skin soft and moistened start dwindling. Collagen and elastin fibers in the

Figure 36.8 Close look at a hair. This scanning electron micrograph shows overlapping cells of the outer layers of a hair shaft, here emerging from the epidermal surface of skin. Compare Figure 3.20, which shows a hair's molecular structure.

dermis break down and become sparser, so the skin loses its elasticity and wrinkles deepen. Excessive tanning, prolonged exposure to drying winds, and tobacco smoke accelerate the skin aging processes.

Hairs, oil glands, sweat glands, and other structures associated with skin are derived from epidermal cells, but they are largely embedded in skin's underlying region, the dermis.

Blood and lymph vessels as well as receptor endings of sensory neurons also reside in the dermis.

Figure 36.9 Effects of muscle contractions on the hydrostatic skeleton of sea anemones. Compare this photograph with Figure 33.2. In (**a**), radial muscles that ring the gut cavity are relaxed, and longitudinal muscles running parallel with the body axis are contracted. Anemones typically look like this at low tide, when currents are not bringing in morsels of food. In (**b**), the radial muscle cells are contracted and the longitudinal muscles are stretched, so that the body is stretched into an upright position. This is the way sea anemones look when they attempt to gather food.

a Relaxed position   **b** Feeding position

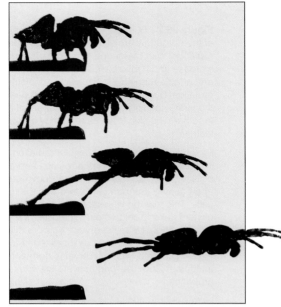

Figure 36.11 Leap of the jumping spider *Sitticus pubescens*, based on the hydraulic extension of the hind legs when blood surges into them under high pressure.

Figure 36.10 Example of a hinged region of an insect exoskeleton. This hinge is composed of layers of chitin and a highly elastic protein that withstands rapid, sustained movement characteristic of insect flight.

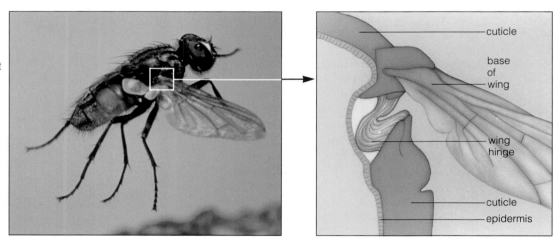

cuticle

base of wing

wing hinge

cuticle

epidermis

SKELETAL SYSTEMS

So far in this unit, we have considered how the nervous system samples the external and internal environments with great precision and keeps informed of change. Many responses to those changes require movements—either of the whole body or some parts of it. Those movements occur by the activation, contraction, and relaxation of muscle cells. But muscle cells alone cannot produce movement. They require the presence of some medium or structural element against which the force

of contraction can be applied. An internal or external skeleton fulfills this requirement. Three kinds of skeletal systems predominate in the animal world:

hydrostatic skeleton	*Force of contraction applied against internal body fluids, which transmit the force*
exoskeleton	*Force of contraction applied against rigid external body parts, such as shells or armor plates*
endoskeleton	*Force of contraction applied against rigid internal body parts (cartilage and bone)*

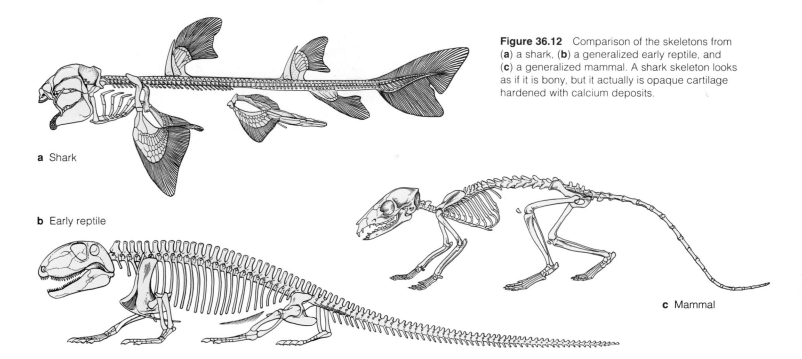

Figure 36.12 Comparison of the skeletons from (**a**) a shark, (**b**) a generalized early reptile, and (**c**) a generalized mammal. A shark skeleton looks as if it is bony, but it actually is opaque cartilage hardened with calcium deposits.

a Shark

b Early reptile

c Mammal

Invertebrate Skeletons

As we saw in Chapter 25, many soft-bodied invertebrates have hydrostatic skeletons. In all hydrostatic skeletons, some type of fluid is confined to a limited space. Like a fully filled waterbed, the hydrostatic skeleton resists compression, thereby serving as the medium against which muscles can work.

Sea anemones, for example, apply the force of contraction against their fluid-filled gut cavity. Between meals, longitudinal muscles are contracted, rings of muscles in the body wall are relaxed, and the animal looks rather like a flattened blob (Figure 36.9). When the radial muscles contract, fluid in the gut cavity is forced out, and longitudinal muscles in the body wall are stretched. Thus the body lengthens into its upright feeding position.

Animals with hydrostatic skeletons do not move with Olympian precision, although invertebrates with segmented bodies show more complex body movements than sea anemones do. Recall that earthworms have a series of coelomic compartments, each with its own muscles and nerves (page 426). Each segment has its own nerve supply, as well as longitudinal and radial muscles in the body wall. This means that the body can lengthen and shorten a few segments at a time in controlled ways. By coordinating muscle contractions on one side or the other of different segments, the earthworm can thrash from side to side as well as move forward and back.

Invertebrates with rigid exoskeletons lack the flexibility of soft bodies, but they benefit in other ways. Hard external parts work like armor against predators. They also provide support for increased body size, especially on land, where animals are deprived of water's buoyancy. Because hard parts can be moved like levers by the sets of muscles attached to them, they afford more precise and often more rapid movements. With an exoskeleton, large movements of body parts (such as wings) can result from small muscle contractions. Think about the cuticle of flying insects. It extends over all body segments *and* over the gaps between segments (Figure 36.10). The cuticle remains pliable at these gaps and acts like a hinge when muscles raise and lower either the wing or the body parts to which wings are attached.

Jumping spiders have hinged exoskeletons, but they also use body fluids to transmit force when they leap at prey. Muscle contractions cause blood inside the spider's body tissues to surge rapidly into the hind leg spines. It's something like giving a water-filled rubber glove a quick, hard squeeze, so that the glove's skinny fingers become rigidly erect. Figure 36.11 shows the outcome of this hydraulic pressure ("hydraulic" meaning fluid pressure inside tubes).

Vertebrate Skeletons

Humans and other vertebrates have an endoskeleton of bone and cartilage (or cartilage alone). Some fishes have a flexible skeleton of an elastic, translucent form of cartilage that almost looks like glass. Sharks have a skeleton of an opaque form of cartilage, hardened with calcium deposits (Figure 36.12). However, most vertebrate skeletons are constructed primarily of bone. Let's turn now to the functions and characteristics of bones, using the human skeletal system as our example.

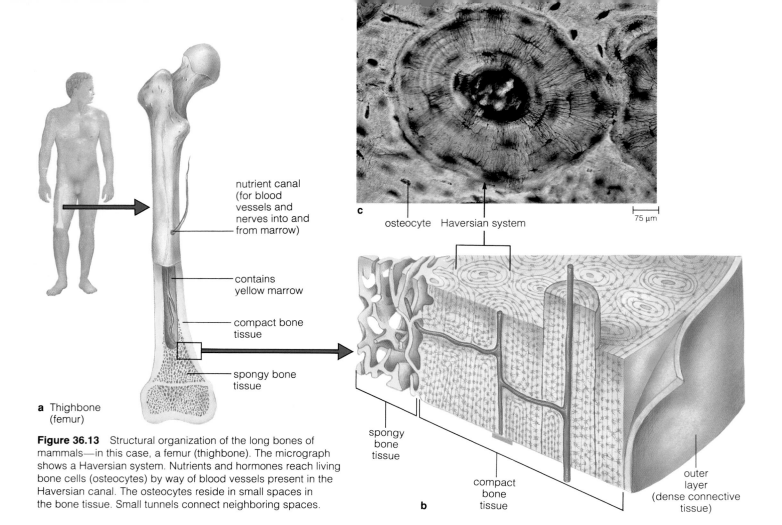

Figure 36.13 Structural organization of the long bones of mammals—in this case, a femur (thighbone). The micrograph shows a Haversian system. Nutrients and hormones reach living bone cells (osteocytes) by way of blood vessels present in the Haversian canal. The osteocytes reside in small spaces in the bone tissue. Small tunnels connect neighboring spaces.

Labels in figure:
- nutrient canal (for blood vessels and nerves into and from marrow)
- contains yellow marrow
- compact bone tissue
- spongy bone tissue
- **a** Thighbone (femur)
- osteocyte Haversian system
- 75 µm
- **c**
- spongy bone tissue
- compact bone tissue
- outer layer (dense connective tissue)
- **b**

Functions of Bone

Just as skin is more than a baglike covering, so is the skeletal system more than a frame to hang muscles on. Its major component parts, called **bones**, are complex organs composed of a number of tissues. Those organs function in movement, protection, support, mineral storage, and blood cell formation:

1. *Movement:* Through interactions with skeletal muscle, bones maintain or change the position of body parts.

2. *Protection:* Bones are hard compartments that enclose and protect the brain, lungs, and other vital organs.

3. *Support:* Bones of the skeletal system support and anchor muscles and soft organs.

4. *Mineral Storage:* Bone tissue serves as a "bank" for calcium, phosphorus, and other mineral ions. The body makes deposits and withdrawals of these reserves, depending on metabolic needs.

5. *Blood Cell Formation:* Parts of some mature bones (such as the breastbone) are sites of blood cell production.

Characteristics of Bone

In size and shape, human bones range from tiny ear-bones to pea-size wrist bones to strong, clublike thighbones. Bones are classified as long, short (or cubelike), flat, and irregular. Here we will focus mainly on long bones that occur in the body's limbs.

Bone Structure. Like other organs, bones are made of tissues, including epithelium and various connective tissues, but they alone incorporate *bone tissue*. Bone tissue consists of living cells and collagen fibers distributed through a ground substance. Both the fibers and the ground substance are hardened by deposits of calcium salts.

Take a look at Figure 36.13, which shows the internal organization of a thighbone. The tissue that forms the bone's shaft and outer portion of its two ends is dense, or *compact bone tissue*. Such tissue forms the shaft of all long bones and allows them to withstand mechanical shocks. Notice, in Figure 36.13b, how the tissue is organized as thin, concentric layers around small canals. These "Haversian canals" are interconnected channels for blood vessels and nerves that service the living bone cells that reside in compact bone tissue.

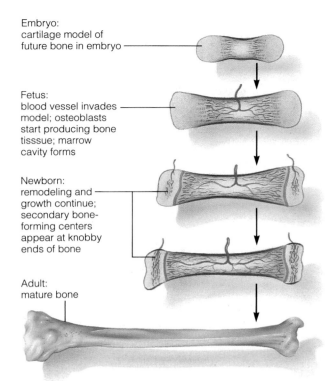

Embryo:
cartilage model of
future bone in embryo

Fetus:
blood vessel invades
model; osteoblasts
start producing bone
tisssue; marrow
cavity forms

Newborn:
remodeling and
growth continue;
secondary bone-
forming centers
appear at knobby
ends of bone

Adult:
mature bone

Figure 36.14 Long bone formation, starting with osteoblast activity in a cartilage model (here, already formed in the animal embryo). Bone-forming cells are active in the shaft region first. Their activities are repeated in the knobby bone ends until only cartilage is left in the joints at both ends of the shaft.

The bone tissue *inside* the shaft and the ends is less packed; it has a spongelike appearance. Tiny, flattened parts make up this *spongy bone tissue*, which actually is quite firm and strong. In many bones, **red marrow** fills the spaces in the spongy tissue, which serves as a major site of blood cell formation. Most mature bones have **yellow marrow** in interior cavities. Yellow marrow is mostly fat. It converts to red marrow and produces red blood cells if blood loss from the body is severe.

Bones are complex organs composed of living cells and various tissues, the most notable of which is bone tissue.

A distinguishing feature of bone tissue is its extracellular matrix of collagen fibers and ground substance, both of which are mineralized.

How Bones Develop. Long bones form in cartilage models that develop in the embryo. Bone-forming cells (osteoblasts) secrete material inside the shaft and on the surface of the cartilage model. Gradually, the cartilage breaks down in the shaft region and the marrow cavity forms (Figure 36.14). The bone-forming cells continue to secrete bone tissue and eventually become trapped by

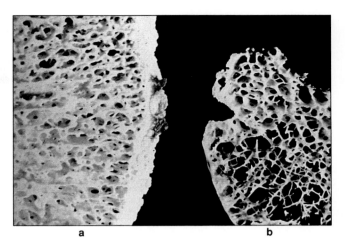

a   b

Figure 36.15 Effect of osteoporosis on bone tissue. In normal tissue (**a**), mineral deposits continually replenish mineral withdrawals, so the tissue is maintained. (**b**) After the onset of osteoporosis, mineral replacements cannot keep pace with the withdrawals, and the tissue gradually erodes. Bones become progressively hollow and brittle.

their own secretions. Then, they are called **osteocytes** (living bone cells). Figure 36.13 shows some of these cells, which are responsible for maintaining mature bones.

Bone Tissue Turnover. Bone tissue is like a bank from which minerals are constantly deposited and withdrawn. The turnover occurs when adult bone is subjected to exercise, which generally increases bone density. It occurs also after stress or injury. In bone remodeling programs that occur when young individuals are growing, this turnover is especially important. For example, the diameter of the thighbone increases as certain bone cells deposit minerals at the surface of the shaft. At the same time, other bone cells destroy a small amount of bone tissue inside the shaft. Thus the thighbone becomes thicker and stronger—but not too heavy.

Bone turnover also helps maintain calcium levels for the body as a whole. Consider how the body resorbs calcium. First, bone cells secrete enzymes that break down bone tissue. As the component minerals dissolve, the released calcium enters interstitial fluid; from there, it is taken up by the blood. This resorption activity is central to the hormonal control of calcium balance, as described on page 589.

Bone turnover can deteriorate with increasing age, especially among older women. The bone mass decreases in the backbone, hips, and elsewhere (Figure 36.15). The backbone can collapse and curve so much that the ribcage position is lowered, leading to complications in internal organs. The syndrome is called *osteoporosis*. Decreasing osteoblast activity, calcium and sex hormone deficiencies, excessive protein intake, and decreased physical activity are suspected of contributing to osteoporosis.

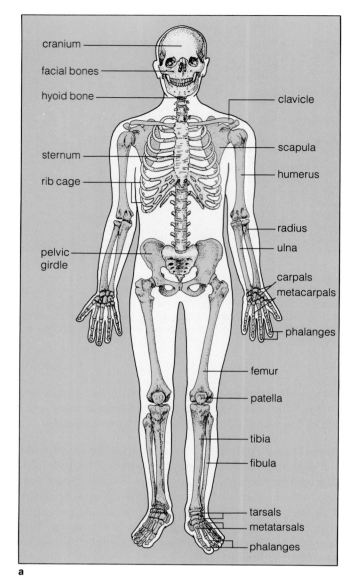

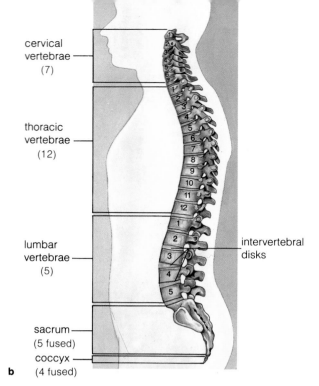

Figure 36.16 (**a**) Human skeleton, with the axial portion color-coded in yellow and the appendicular portion, in tan. Can you identify similar structures in the endoskeleton of the mammal diagrammed in Figure 36.12? (**b**) Side view of the vertebral column, or backbone. The cranium balances on the column's uppermost vertebra. Compare Figure 33.8.

HUMAN SKELETAL SYSTEM

Skeletal Structure

Humans started walking on their hind legs about 3 million years ago and haven't stopped since. As Figure 21.3 suggests, the upright posture puts the backbone into an S-shaped curve, which is not an ideal arrangement. (The older we get, the longer we have resisted the pull of gravity in an imperfect way, and the more lower back pain we have.) But evolution occurs through modifications of preexisting structures, and the skeleton of ancestral four-legged vertebrates was one of them.

Your body has 206 bones, in the two skeletal regions shown in Figure 36.15. The **axial skeleton** includes the skull, vertebral column (backbone), ribs, and sternum (the breastbone). The **appendicular skeleton** includes bones of the pectoral girdles (at the shoulders), arms, hands, pelvic girdle (at the hips), legs, and feet. Straps of dense, regular connective tissue called **ligaments** connect the bones at joints. Cords or straps of dense, regular connective tissue called **tendons** attach muscles to the bones (or to other muscles).

The flexible, curved backbone extends from the base of the skull to the pelvic girdle, where it transmits the weight of your torso to the lower limbs. The delicate spinal cord threads through a cavity formed by bony parts of the vertebrae, which are arranged one above the other (Figure 33.8).

Intervertebral disks, which contain cartilage, occur between the vertebrae (Figure 36.16b). They serve as shock absorbers and flex points. However, severe or rapid shocks may cause a disk to herniate. A *herniated disk* has slipped out of place and possibly may rupture. The protruding disk may press against neighboring nerves or the spinal cord and cause excruciating pain.

Each pectoral girdle has a large, flat shoulder blade and a long, slender collarbone that connects to the breastbone. It is not a sturdy arrangement. Fall on an outstretched arm and you might end up with a fractured clavicle or dislocated shoulder. Of all bones, the collarbone is the one most frequently broken.

Joints

"Joints" are areas of contact or near-contact between bones. The most familiar type, the **synovial joint**, is

On Runner's Knee

When you run, one foot and then the other is pounding hard against the ground. Each time a foot hits the ground, the knee joint above it must absorb the full force of your body weight. The knee joint allows us to do many things. It allows the leg bones beneath it to swing and, to some degree, to bend and twist. And the joint can absorb a force nearly seven times the body's weight—but there is no guarantee that it can do so repeatedly. Nearly 5 million of the 15 million joggers and runners in the United States alone suffer from "runner's knee," which refers generally to various disruptions of the bone, cartilage, muscle, tendons, and ligaments at the knee joint.

Like most joints, the knee joint permits considerable movement. The two long bones joined here (the femur and tibia) are actually separated by a cavity. They are held together by ligaments, tendons, and a few fibers that form a capsule around the joint. A membrane that lines the capsule produces a fluid that lubricates the joint, and where the bone ends meet, they are capped with a cushioning layer of cartilage.

Between the femur and tibia are wedges of cartilage that add stability and act like shock absorbers for the weight placed on the joint. Here also are thirteen fluid-filled sacs (bursae) that help cut down friction.

When the knee joint is hit hard or twisted too much, its cartilage can be torn. Once cartilage is torn, the body often cannot repair the damage. Orthopedic surgeons usually recommend removing most or all of the torn tissue; otherwise it can cause arthritis. Each year, more than 50,000 pieces of torn cartilage are surgically removed from the knees of football players alone. Football players,

tennis players, basketball players, weekend joggers—all are helping to support the burgeoning field of "sports medicine."

The seven ligaments that strap the femur and tibia together are also vulnerable to injury. A ligament is not meant to be stretched too far, and blows to the knee during collision sports (such as football) can tear it apart. A ligament is composed of many connective tissue fibers. If only some of the fibers are torn, it may heal itself. If the ligament is severed, however, it must be surgically repaired. (Edward Percy likens the surgery to sewing two hairbrushes together.) Severed ligaments must be repaired within ten days. The fluid that lubricates the knee joint happens to contain phagocytic cells that remove the debris resulting from day-to-day wear and tear in the joint. The cells will also go to work indiscriminately on torn ligaments and turn the tissue to mush.

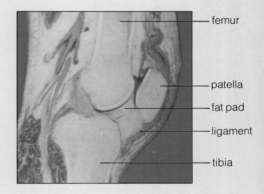

femur
patella
fat pad
ligament
tibia

Longitudinal section through the knee joint.

freely movable. Such joints are stabilized in part by straplike ligaments that are capable of stretching. A flexible capsule of dense connective tissue surrounds the bones of a synovial joint. Cells of a membrane that lines the interior of the capsule secrete a lubricating fluid into the joint.

Unfortunately, freely movable joints sometimes move too freely and their structural organization is disrupted (see *Commentary*).

As a person ages, the cartilage covering the bone ends of freely movable joints may simply wear away, a condition called *osteoarthritis*. In contrast, *rheumatoid arthritis* is a degenerative disorder with a genetic basis. The synovial membrane becomes inflamed and thick-

ened, cartilage degenerates, and bone becomes deposited in the joint.

At **cartilaginous joints**, cartilage fills the space between bones and permits only slight movement. Such joints occur between vertebrae and between the breastbone and ribs. At **fibrous joints**, fibrous tissue unites the bones and no cavity is present. Fibrous joints loosely connect the flat skull bones of a fetus. During childbirth, the loose connections allow the bones to slide over each other and so prevent skull fractures. The skull of a newborn still has fibrous joints and membranous areas that are known as "soft spots" (fontanels). But the fibrous tissue hardens completely during childhood, so the skull bones become fused into a single unit.

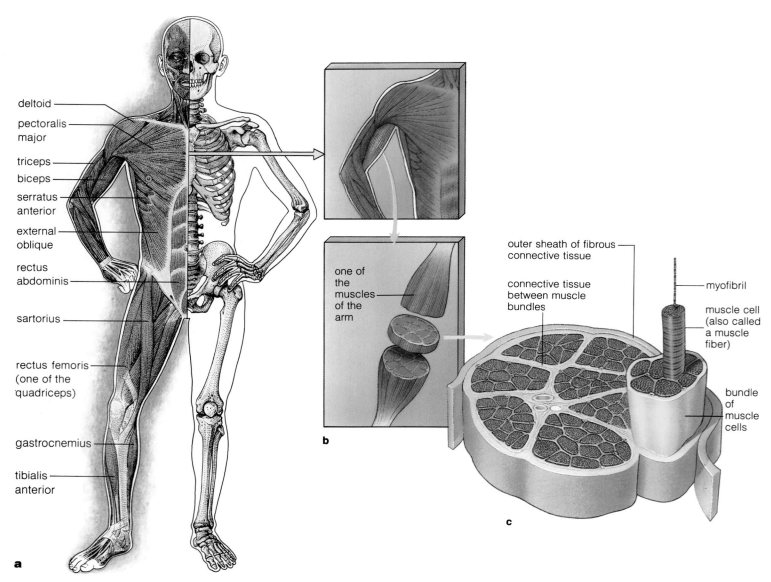

deltoid

pectoralis
major

triceps

biceps

serratus
anterior

external
oblique

rectus
abdominis

sartorius

rectus femoris
(one of the
quadriceps)

gastrocnemius

tibialis
anterior

a

one of
the
muscles
of the
arm

b

outer sheath of fibrous
connective tissue

connective tissue
between muscle
bundles

myofibril

muscle cell
(also called
a muscle
fiber)

bundle
of
muscle
cells

c

Figure 36.17 (**a**) Some of the major skeletal muscles of the human skeletal-muscular system. (**b**) Closer view of the fine structure of an individual skeletal muscle. (**c**) Location of myofibrils, the threadlike structures inside each muscle cell. As shown in Figure 36.18, each myofibril has units of contraction (sarcomeres) arranged one after another along its length.

MUSCULAR SYSTEM

Comparison of Muscle Tissues

We turn now to the skeletal muscle—the functional partner of bone. Recall that there are three types of muscle tissues: skeletal, cardiac, and smooth (page 538). Although skeletal and cardiac muscle are quite different in appearance from smooth muscle, they are all alike in three respects.

First, muscle cells show *excitability*. All cells have a voltage difference across the plasma membrane (the outside is more positively charged than the inside). In excit-

able cells, the voltage reverses suddenly and briefly in response to adequate stimulation. This sudden reversal in charge, called an action potential, was described in earlier chapters. Second, muscle cells can *contract* (shorten) in response to action potentials. Third, muscle cells are *elastic*; after contracting, they return to their original, relaxed position.

Skeletal muscle is the only type of muscle tissue that interacts with the skeleton to bring positional changes of body parts and locomotion. In vertebrates, smooth muscle occurs mostly in the wall of internal organs. (For example, smooth muscle in the stomach and intestinal walls helps

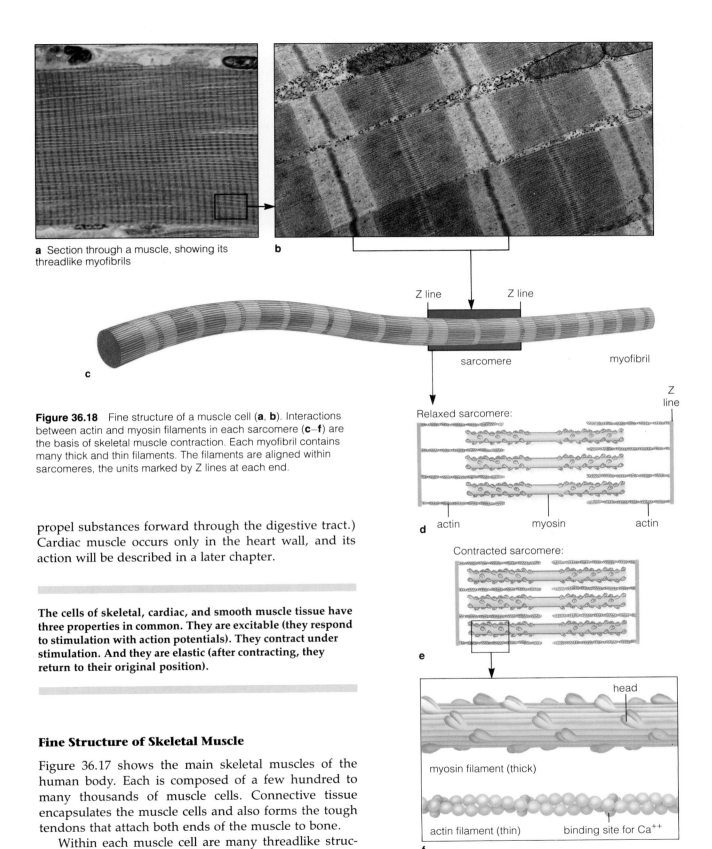

a Section through a muscle, showing its threadlike myofibrils

b

Z line   Z line

sarcomere   myofibril

c

Figure 36.18 Fine structure of a muscle cell (**a**, **b**). Interactions between actin and myosin filaments in each sarcomere (**c–f**) are the basis of skeletal muscle contraction. Each myofibril contains many thick and thin filaments. The filaments are aligned within sarcomeres, the units marked by Z lines at each end.

Z line

Relaxed sarcomere:

d   actin   myosin   actin

Contracted sarcomere:

e

head

myosin filament (thick)

actin filament (thin)   binding site for Ca^{++}

f

propel substances forward through the digestive tract.) Cardiac muscle occurs only in the heart wall, and its action will be described in a later chapter.

The cells of skeletal, cardiac, and smooth muscle tissue have three properties in common. They are excitable (they respond to stimulation with action potentials). They contract under stimulation. And they are elastic (after contracting, they return to their original position).

Fine Structure of Skeletal Muscle

Figure 36.17 shows the main skeletal muscles of the human body. Each is composed of a few hundred to many thousands of muscle cells. Connective tissue encapsulates the muscle cells and also forms the tough tendons that attach both ends of the muscle to bone.

Within each muscle cell are many threadlike structures called **myofibrils**. You can see these in Figures 36.17 and 36.18. In turn, within each myofibril are many thin and thick filaments, side by side in parallel array. Close

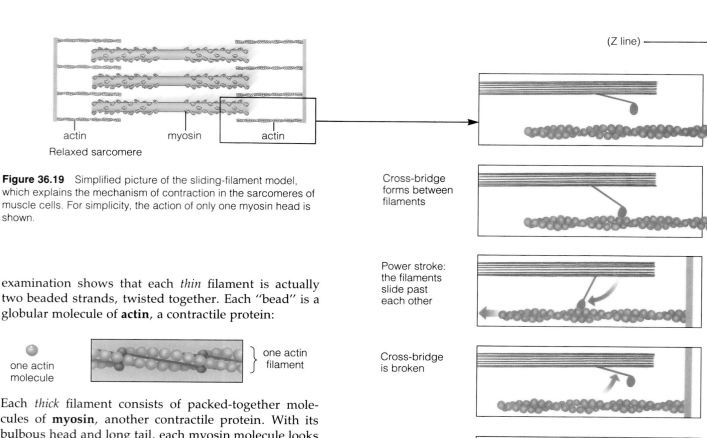

Figure 36.19 Simplified picture of the sliding-filament model, which explains the mechanism of contraction in the sarcomeres of muscle cells. For simplicity, the action of only one myosin head is shown.

Labels in figure (top to bottom): (Z line); actin; myosin; actin; Relaxed sarcomere; Cross-bridge forms between filaments; Power stroke: the filaments slide past each other; Cross-bridge is broken; Another cross-bridge forms; Another power stroke (toward center of sarcomere)

examination shows that each *thin* filament is actually two beaded strands, twisted together. Each "bead" is a globular molecule of **actin**, a contractile protein:

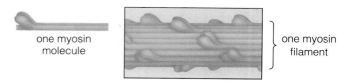

one actin molecule

one actin filament

Each *thick* filament consists of packed-together molecules of **myosin**, another contractile protein. With its bulbous head and long tail, each myosin molecule looks rather like a double-headed golf club. In thick filaments, the myosin tails are packed together in parallel, and the heads stick out to the sides:

one myosin molecule

one myosin filament

The actin and myosin filaments are components of **sarcomeres**, the basic units of muscle contraction. The organization of actin and myosin filaments in sarcomeres is so highly ordered, it gives skeletal and cardiac muscles a striped appearance (Figure 36.18c).

Mechanism of Skeletal Muscle Contraction

The only way that skeletal muscles can move the body parts to which they are attached is to shorten. When a skeletal muscle shortens, its cells are shortening. And when a muscle cell shortens, its component sarcomeres are shortening. *The combined decreases in length of the individual sarcomeres account for contraction of the whole muscle.*

How does a sarcomere contract? According to the **sliding-filament model**, myosin filaments physically slide along and pull the actin filaments toward the center of a sarcomere during contraction.

The sliding movement depends on the formation of cross-bridges between adjacent actin and myosin filaments. A cross-bridge forms when the "head" of a myosin molecule attaches to binding sites on actin (Figure 36.19). An ATP molecule is associated with each myosin head. When some of its energy is released, the myosin head tilts in a short power stroke, toward the center of the sarcomere. As actin filaments become attached to the myosin heads, they also move toward the center. Now another energy input (from ATP) causes each myosin head to detach, reattach at the next actin binding site in line, and move the actin filaments a bit more. A single contraction takes a whole series of power strokes by myosin heads in each sarcomere.

In the absence of ATP, the cross-bridges never do detach. Following death, for instance, ATP production stops along with other metabolic activities. Cross-

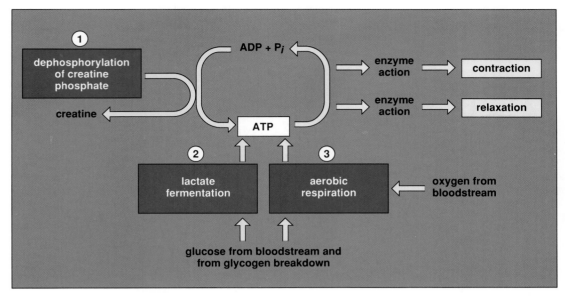

Figure 36.20 Three possible metabolic pathways for producing ATP in muscles.

bridges remain locked in place and all skeletal muscles in the body become rigid. This condition, *rigor mortis*, lasts up to sixty hours after death.

In skeletal muscle, contraction occurs in sarcomeres, which are *contractile units* organized one after another in the myofibrils of muscle cells. Each sarcomere contains parallel arrays of actin filaments and myosin filaments.

Each sarcomere *shortens* when its actin and myosin filaments slide past each other, propelled by cross-bridge formation.

The combined decrease in length of the individual sarcomeres accounts for contraction of the muscle.

Energy Metabolism in Muscles

How are muscle cells assured of getting enough ATP? As Figure 36.20 and the following list indicate, three metabolic supply routes are available to them:

1. Creatine phosphate metabolism

2. Lactate fermentation

3. Aerobic respiration

Which metabolic pathway dominates at a given time depends on the demands being placed on the muscle. During the sudden onset of contraction, many ATP molecules are stripped of a phosphate group (during the formation and detachment of cross-bridges). Creatine phosphate instantaneously restores the ATP by donat-

ing a phosphate group to ADP. Supplies of creatine phosphate are so limited, however, that they would dwindle within a few seconds unless synthesis reactions met the demands of the contracting cells.

When the demand for muscle action is *intense but brief* (say, a 100-meter race), muscles use an anaerobic pathway. As described in Chapter 8, lactate fermentation is an anaerobic route by which glucose is broken down to lactate, with a net yield of two ATP molecules. (Lactate is the ionized form of lactic acid.) Muscle cells get the glucose from blood and by tapping into their "storage form" of glucose—glycogen molecules. This pathway also produces energy very quickly, but lactate builds up in the muscles.

If the demand for muscle action is *moderate*, it also can be *prolonged*, as during the Iditarod sled-dog race. At such times, most of the required ATP forms by electron transport phosphorylation, the final stage of the aerobic pathway. As described on page 126, this event proceeds within mitochondria. And it has a net energy yield of 36 ATP per glucose molecule.

In the muscles of humans and other complex animals, the rate of ATP production by the aerobic pathway is linked to the rate at which the circulatory system is delivering oxygen to mitochondria. It is linked also to the number of mitochondria in muscle cells. When world-class marathoners engage in rigorous training, one of the physiological results is an increase in the number of mitochondria in their muscle cells. (Remember that mitochondria can divide independently of the cell in which they are located.) This is one outcome of the training that Butcher and her king-of-the-sled, Granite, undergo before a race.

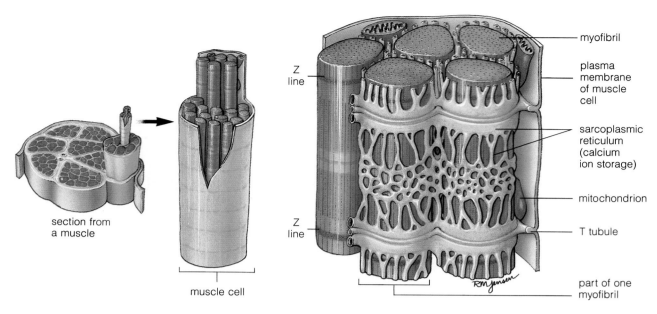

Figure 36.21 Membrane systems of a muscle cell. The plasma membrane surrounds the myofibrils. The plasma membrane is continuous with membranous tubes (T tubules) that thread inward. The tubes are located very close to a calcium-storing system (sarcoplasmic reticulum). Signals travel along the plasma membrane and the T tubules, and then trigger calcium release from the sarcoplasmic reticulum. Without calcium ions, actin and myosin filaments in the myofibrils cannot interact to bring about contraction.

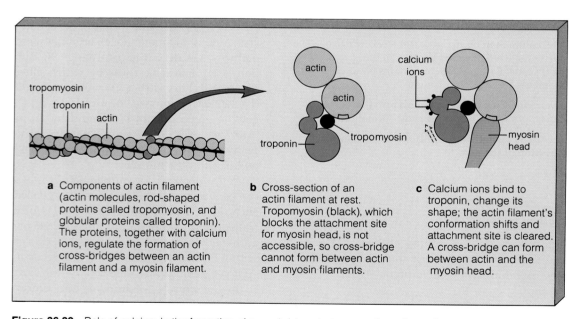

a Components of actin filament (actin molecules, rod-shaped proteins called tropomyosin, and globular proteins called troponin). The proteins, together with calcium ions, regulate the formation of cross-bridges between an actin filament and a myosin filament.

b Cross-section of an actin filament at rest. Tropomyosin (black), which blocks the attachment site for myosin head, is not accessible, so cross-bridge cannot form between actin and myosin filaments.

c Calcium ions bind to troponin, change its shape; the actin filament's conformation shifts and attachment site is cleared. A cross-bridge can form between actin and the myosin head.

Figure 36.22 Role of calcium in the formation of cross-bridges between actin and myosin.

Control of Contraction

Skeletal muscle contracts under commands from motor neurons. Appropriate stimulation from those neurons can trigger action potentials that travel along the plasma membrane of a muscle cell. They continue along infoldings of the plasma membrane that form many small tubes (the transverse tubule system). Those tubes reach the **sarcoplasmic reticulum**, a continuous system of membranous chambers around the myofibrils (Figure 36.21). The chambers actually are a compartment inside the cell that serves as a storehouse for calcium ions. The arrival of action potentials causes the compartment's membrane to become more permeable to calcium ions. After they

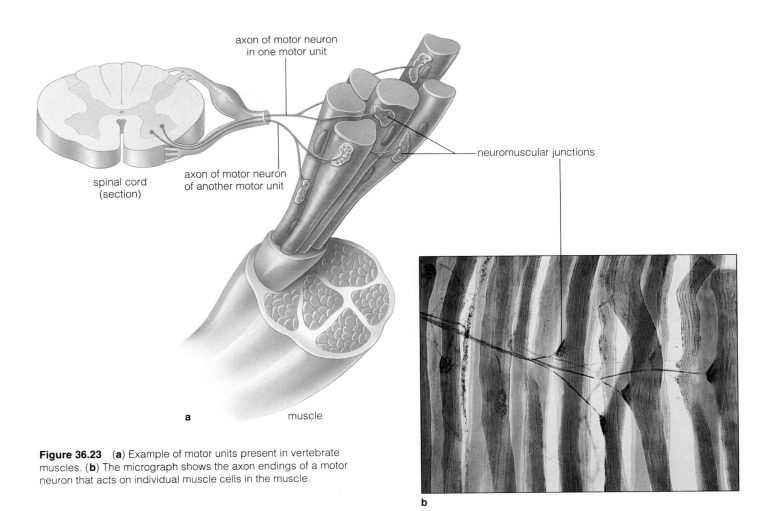

axon of motor neuron
in one motor unit

neuromuscular junctions

spinal cord
(section)

axon of motor neuron
of another motor unit

a

muscle

b

Figure 36.23 (**a**) Example of motor units present in vertebrate muscles. (**b**) The micrograph shows the axon endings of a motor neuron that acts on individual muscle cells in the muscle.

are released, the ions diffuse through the cytoplasm and attach to a protein component (troponin) of actin filaments.

When calcium is attached to troponin, the protein changes shape, as Figure 36.22 shows. The change causes a shift in the position of an adjacent protein (tropomyosin) and a resulting conformational change in the actin filament. Thus, *calcium ions clear the myosin binding sites on actin, allowing cross-bridges to form.*

Muscle contracts when calcium ions are released from the sarcoplasmic reticulum.

Muscle relaxes when calcium ions are actively taken up after contraction and stored in a membrane system around myofibrils (the sarcoplasmic reticulum).

By controlling the action potentials that reach the sarcoplasmic reticulum in the first place, the nervous system controls calcium ion levels in muscle tissue—and so exerts control over muscle contraction.

CONTRACTION OF A SKELETAL MUSCLE

Arnold Schwarzenegger, with his bulging muscles, is one of Hollywood's icons of great physical strength. Ultimately, his strength depends on how forcefully his muscles can contract. That, in turn, depends on the size of a given muscle, how many of its cells are contracting, and the frequency of stimulation by the nervous system. The larger a muscle is in diameter, the more potential it has for strength. By exercising regularly, you can prod your muscle cells to increase in size. (The *Commentary* on the page that follows describes what can happen when this line of reasoning is carried to an extreme.)

A skeletal muscle contains a large number of cells, and these may contract with different degrees of force for different periods of time. To get a sense of what is going on in that muscle, think back on the neuromuscular junction, where a motor neuron makes functional connections with one or more muscle cells (page 556). Together, a motor neuron and the muscle cells under its control are called a **motor unit**. Figure 36.23 shows an example of this.

Physiologists artificially stimulate a motor neuron with electrical impulses and make recordings of the changes in muscle contraction. When a single, brief stimulus activates a motor unit, the muscle contracts briefly, then relaxes. This muscle response is a **muscle twitch** (Figure 36.24). When a motor unit is stimulated again before a twitch response is completed, it twitches again. The strength of the contraction depends on how far the twitch response has proceeded by the time the second signal arrives. A motor unit stimulated repeatedly does not have time to relax (there is not enough time for all of the calcium ions to be transported back into the sarcoplasmic reticulum). Instead, the motor unit is maintained in a state of contraction called **tetanus**. (Recall that in the disease tetanus, described on page 558, muscles cannot be released from the state of contraction.)

In a weak contraction, the nervous system activates only a small number of motor units. In a stronger one, a larger number are activated, at a high frequency.

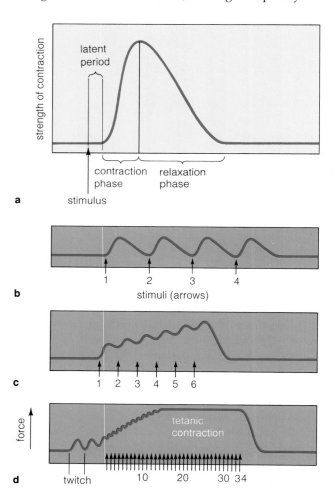

Figure 36.24 Recording of a muscle twitch (**a**). Recordings of a series of muscle twitches caused by about two stimulations per second (**b**); recordings of a summation of twitches resulting from about six stimulations per second (**c**); and a tetanic contraction resulting from about twenty stimulations per second (**d**).

Muscle Mania

Some call it the will-to-win gone bonkers, a consuming desire for muscles that are larger than life. Others say it is a modern requirement for excellence in athletic competition. Either way, they are talking about athletes who illegally use performance-enhancing drugs, mainly anabolic steroids.

Ten athletes were disqualified from the 1988 Olympics—one was even stripped of a gold medal—for using banned drugs, including the anabolic steroid stanazolol. Other competitors dropped out when they heard of the stringent new drug tests.

Cream-of-the-crop amateur athletes account for only a fraction of the surreptitious users of anabolic steroids. By one estimate, 85 percent of all professional football players use or have used the drugs. Adolescent boys as well as their parents sometimes look to the drugs to gain a winning edge in wrestling, football, and weight-lifting tournaments. Each year in the United States alone, perhaps as many as 1 million athletes use anabolic steroids.

What Anabolic Steroids Are

Anabolic steroids are synthetic hormones. They were developed in the 1930s as therapeutic drugs that could mimic the effects of a sex hormone, testosterone. Secondary sexual traits, among other things, depend on testosterone. Under its influence, boys turning into men get a deeper voice; more hair on the skin of their face, underarms, and pubic region; more secretions from sweat glands; and increased muscle mass in the arms, legs, shoulders, and elsewhere. Testosterone also seems to stimulate the more aggressive behavior often associated with maleness. Anabolic steroids can also do these things—but at a significant physical and psychological price.

What Anabolic Steroids Do

The "steroid" part of *anabolic steroid* tells you that molecules of this drug have a backbone of four carbon rings (page 42). The "anabolic" part echoes a name that chemists have for the synthesis of organic compounds (anabolism). The roughly twenty varieties of anabolic steroids stimulate the synthesis of protein molecules—including muscle proteins.

Supposedly, using anabolic steroids while engaged in a weight-training exercise program can lead to rapid gains in lean muscle mass and strength. The claim is disputed, because the results of most studies are based on too few subjects. Even so, testimonials pour in from weight lifters, football players, and athletes who specialize in shotput, discus, hammer throw, and other "brute power" events. Users commonly "stack" their steroid intake, combining daily oral doses with a single hefty injection each month. Much of their self-medication is on the sly, since the nonprescription use of anabolic steroids is illegal.

What, if anything, is bad about anabolic steroids? Physicians, researchers, and athletes themselves report a long list of minor and major side effects.

In men, acne, baldness, shrinking testes, and infertility are the first signs of toxicity. These symptoms are attributable to the fact that high blood levels of anabolic steroids cause the normal production of testosterone to drop precipitously. The drugs may be linked to an early onset of a cardiovascular disease, atherosclerosis. Even brief or occasional use may contribute to kidney damage and to cancer of the liver, testes, and prostate gland.

In women, anabolic steroids trigger the development of a deep voice and pronounced facial hair. Menstrual periods become irregular. Breasts may shrink and the clitoris may become grossly enlarged.

Roid Rage

Not all steroid users have developed severe physical side effects. In fact, studies suggest that severe mental difficulties are more common. Called everything from *'roid rage* to *body-builders' psychosis*, the symptoms range from annoying to frightening. Some men experience irritability and increased aggressiveness. Many competitive athletes look upon the added aggressiveness as a plus. Other men, however, experience uncontrollable aggression, delusions, and wildly manic behavior. In 1988, one steroid-using athlete traveling at 35 miles per hour deliberately drove his car into a tree.

With all of the suspected dangers associated with anabolic steroids, some may wonder why anyone would place his or her body and future in such jeopardy. Possibly not everyone is convinced that the drugs do enough damage to outweigh the "edge" they give in competition. What should a competitor do in a world that accords winning athletes wealth and the status of hero, while relegating others to the pile of also-rans? What would *you* do?

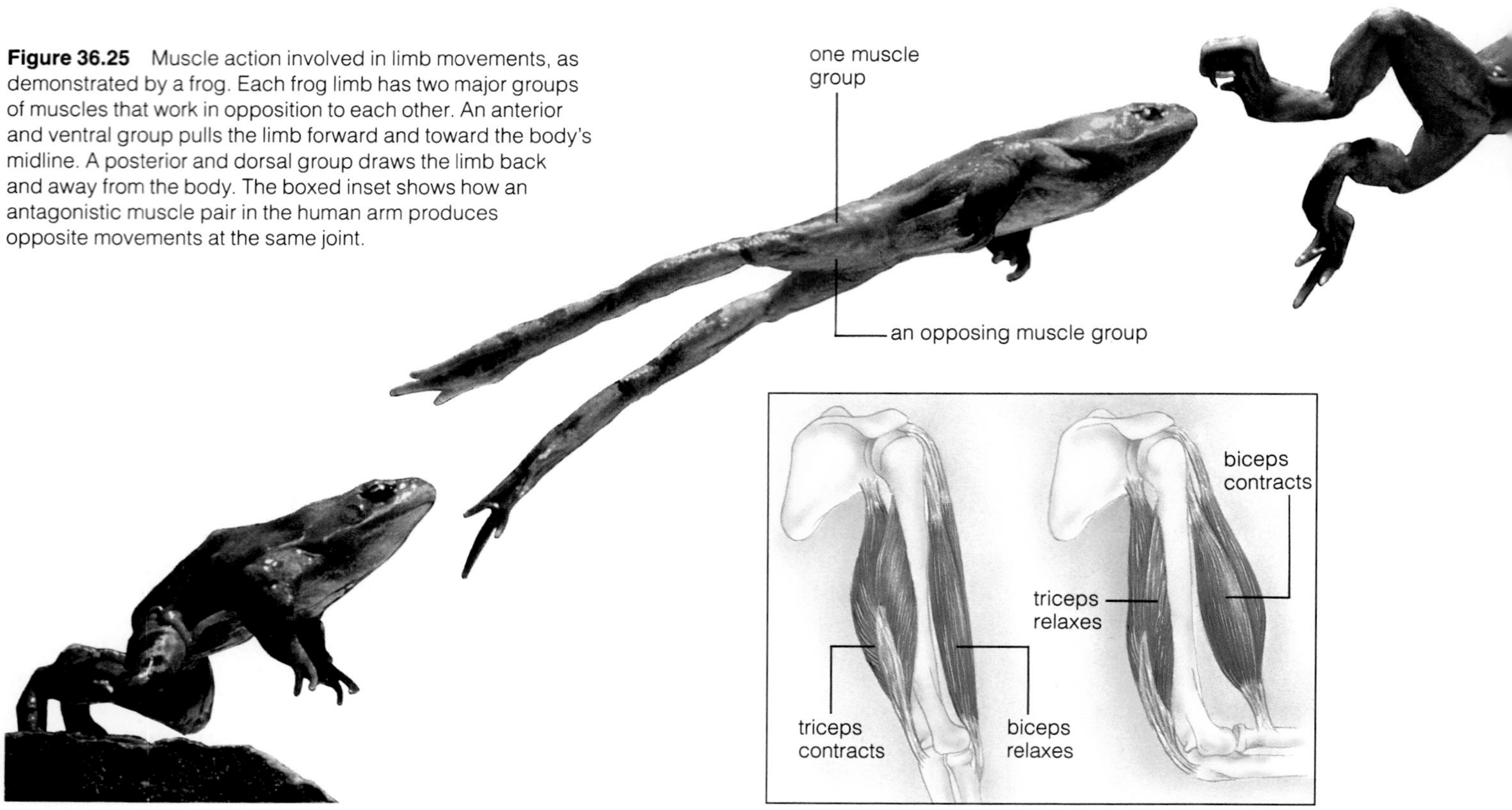

Figure 36.25 Muscle action involved in limb movements, as demonstrated by a frog. Each frog limb has two major groups of muscles that work in opposition to each other. An anterior and ventral group pulls the limb forward and toward the body's midline. A posterior and dorsal group draws the limb back and away from the body. The boxed inset shows how an antagonistic muscle pair in the human arm produces opposite movements at the same joint.

one muscle group

an opposing muscle group

biceps contracts

triceps relaxes

triceps contracts

biceps relaxes

LIMB MOVEMENTS

The human body has more than 600 muscles, arranged as pairs or groups. Some muscles work together (synergistically) to promote the same movement. Others work in opposition (antagonistically), so that the action of one opposes or reverses the action of another. When muscles contract, they transmit force to the bones to which they are attached. Together, the skeleton and the muscles attached to it are like a system of levers in which rigid rods (bones) move about at fixed points (the joints). Most attachments are close to joints. This means a muscle has to contract only a small distance to produce a large movement of some body part.

Figure 36.25 shows an antagonistic pair of muscles, the biceps and triceps of a human arm. When the biceps contracts, the elbow joint flexes (bends). As it relaxes and its partner (the triceps) contracts, the limb extends and straightens. Such coordinated action results partly from **reciprocal innervation** in the spinal cord. By this mechanism, inhibitory signals sent to one muscle's motor neurons prevent that muscle from contracting while the opposing muscle group is being stimulated. In addition, the nervous system uses signals from stretch receptors to coordinate the contractions.

Finally, as a concluding example, Figure 36.25 also shows a few limb movements by one of nature's splendid jumpers.

SUMMARY

1. The vertebrate integumentary system (skin and its derivatives) protects the rest of the body from abrasion, bacterial attack, ultraviolet radiation, and dehydration. It helps control internal temperature, and it serves as a blood reservoir for the rest of the body. Its receptors are essential in detecting environmental stimuli.

2. Bones are the structural elements of vertebrate skeletons. They function in movement (by interacting with skeletal muscles to which they are attached), protection and support of other body parts, mineral storage, and blood cell formation.

3. The human skeleton has an axial portion (skull, backbone, ribs, and breastbone) and an appendicular portion (limb bones, pelvic girdle, and pectoral girdle). Intervertebral disks are shock pads and flex points in the backbone.

4. Smooth, cardiac, and skeletal muscle tissue all show excitability, contraction, and elasticity. Only skeletal muscle interacts with the skeleton to bring about movement of the body through the environment or positional changes of its parts.

5. Each skeletal or cardiac muscle cell contains many threadlike myofibrils, which contain actin and myosin filaments. The filaments are organized in orderly arrays in sarcomeres (the basic units of contraction).

6. Sarcomeres contract when action potentials trigger the release of calcium ions from a membrane system (sarcoplasmic reticulum) in the muscle cell. Calcium binding alters the actin filaments so that the heads of adjacent myosin filaments can bind to them. ATP drives the cross-bridge power strokes that cause actin filaments to slide past the myosin filaments and so shorten the sarcomere.

7. In combination with skeletal muscles, the skeleton works like a system of levers in which rigid rods (bones) move about at fixed points (joints). A limb can be extended and rotated around a joint because of the way pairs or groups of muscles are arranged relative to joints.

Review Questions

1. What are some of the functions of skin? List some derivatives of epidermis. *621, 622*

2. What are some of the functions of bone tissue? *626*

3. Name the three properties that all three muscle tissues (smooth, cardiac, and skeletal) have in common. *630*

4. Look at Figure 36.17. Then, on your own, sketch and label the fine structure of a muscle, down to one of its individual myofibrils. Can you identify the basic unit of contraction in a myofibril? *630*

5. How do actin and myosin interact in a sarcomere to bring about muscle contraction? What role does ATP play? What role does calcium play? *630–635*

Self-Quiz *(Answers in Appendix IV)*

1. The _____ system protects the body from abrasion, ultraviolet radiation, bacterial attack, and other environmental stresses.

2. _____ and _____ systems work together to move the body and specific body parts.

3. The three types of muscle tissue are _____, _____, and _____ .

4. Which of the following is *not* a function of the integumentary system?
   a. protect the body from abrasion
   b. protect the body from dehydration
   c. detect environmental stimuli
   d. bring about body movements
   e. serve as a blood reservoir for the rest of the body

5. Which of the following serve as shock pads and flex points in the human backbone?
   a. vertebrae   c. lumbar bones
   b. cervical bones   d. intervertebral disks

6. Which structure stores calcium ions necessary for muscle contraction?
   a. plasma membrane   c. sarcoplasmic reticulum
   b. motor neuron   d. T tubule

7. The smallest unit of contraction in skeletal muscle is the
   _____ .
   a. myofibril   c. muscle fiber
   b. sarcomere   d. myosin filament

8. Muscle contraction will not occur _____ .
   a. in the absence of calcium ions   c. both a and b
   b. in the absence of ATP   d. neither a nor b

9. Match the terms on muscle structure and function.
   _____ myofibrils   a. contains many myofibrils
   _____ sarcoplasmic   b. the contractile unit of muscle
            reticulum   c. drives the power stroke to slide
   _____ sarcomere   actin filaments past myosin
   _____ muscle cell   filaments
   _____ ATP   d. composed of actin and myosin
                                 filaments
                             e. calcium ion storage site in a muscle

Selected Key Terms

actin *632*
bone *626*
bone tissue *626*
cartilaginous joint *629*
cuticle *620*
dermis *622*
endoskeleton *624*
epidermis *622*
exoskeleton *624*
hydrostatic skeleton *624*
hypodermis *621*
integument *620*
intervertebral disk *628*
joint *628*
keratin *622*

ligament *628*
motor unit *635*
muscle twitch *636*
myofibril *631*
myosin *632*
osteocyte *627*
reciprocal innervation *638*
red marrow *627*
sarcomere *632*
sarcoplasmic reticulum *634*
skeletal muscle *630*
skin *621*
sliding-filament model *632*
tendon *628*
yellow marrow *627*

Readings

Alexander, R. M. July–August 1984. "Walking and Running." *American Scientist* 72(4):348–354. The biomechanics of traveling on foot.

Huxley, H. E. December 1965. "The Mechanism of Muscular Contraction." *Scientific American* 213(6):18–27. Old article, great illustrations.

Vander, A., J. Sherman, D. Luciano. 1990. *Human Physiology.* Fifth edition. New York: McGraw-Hill.

Sorry, Have to Eat and Run

For the pronghorn antelope (*Antilocapra americanus*), home is where the food is. Populations of this medium-sized mammal range from central Canada, down through the American Southwest, and on into northern Mexico. Late winter and fall, you may find herds on windblown mountain ridges, where wild sage grows. In spring you may find them moving to open grasslands and deserts, wherever low grasses and tasty shrubs are sprouting.

Young antelope are vulnerable and tasty to coyotes, bobcats, and golden eagles, so while the herds are browsing, they keep a constant eye out for danger. They can do this even while their head is bent low in the grasses, given how far back their eye sockets are positioned in the skull (Figure 37.1a). And can those animals eat and run! If danger appears imminent, they leave the table with bursts of speed that have been clocked at 95 kilometers per hour.

Just as you can do for other mammals, you can look at a pronghorn antelope's teeth and gain insight into its life-style. Think about your own cheek teeth, for instance, with their flattened crown that serves as a grinding platform. The crown of the antelope's cheek teeth dwarfs them (Figure 37.1b). You probably do not have your mouth pressed close to the ground while eating, but the antelope does. Because the antelope ends up with abrasive soil particles as well as tough plant material in its mouth, its teeth wear down far more rapidly than yours do. For them, natural selection apparently has favored more crown to wear down.

Like other ruminants, pronghorn antelopes spend nearly all their lives alternately eating and then bedding down to chew their cud. Pound for pound, it takes far more plant tissue to provide energy than the same amount of animal tissue provides for predators. To be sure, predators face energy shortages if they have to

a

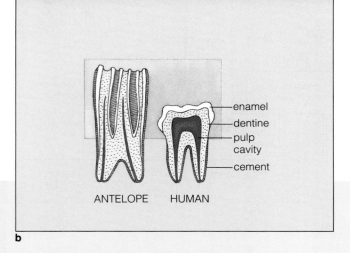

enamel
dentine
pulp
cavity
cement

ANTELOPE   HUMAN

b

Figure 37.1 (**a**) Pronghorn antelopes (*Antilocapra americanus*) busy at work, taking in nutrients. (**b**) A comparison of the general structure of cheek teeth of herbivorous mammals, including antelopes, and humans. Each tooth's crown is positioned over the green background; its root is beneath this.

wait a long time between meals, and the energy deficit increases when they have to run down dinner. But ruminants have to spend far more time *digesting* meals. The plants they eat consist largely of cellulose, the structure of which was shown on page 39. Cellulose digestion requires specific enzymes and a rather long processing time. Not surprisingly, pronghorn antelopes have one of the world's most elaborate stomachs. Not one stomach sac for them—their stomach is partitioned into four interconnected sacs!

The first two sacs of the stomach house vast microbial populations. Among these are symbiotic bacteria. The bacteria synthesize digestive enzymes that act specifically on cellulose. By degrading cellulose, they make its component nutrients available to their host as well as to themselves. While bacterial enzymes are attacking cellulose, the antelope is regurgitating the contents of the first two stomach sacs and rechewing the stuff before swallowing again. (This is what "chewing the cud" means.) Thus, plant material is mixed and pummeled more than once—so more of the cellulose fibers are exposed to agents of digestion before continuing on through the antelope gut.

We humans don't chew cud, but our nutrient-acquiring strategies are just as amazing. An Eskimo might eat only raw whale blubber in a given day, a Nepalese might eat only rice, and an American might partake of pepperoni pizza, chocolate, kiwi fruit, couscous, snake meat, or dandelion wine. Yet through its metabolic magic, the human body converts these and a dizzying variety of other substances into usable energy and tissues of its own.

With this chapter we start our tour of nutrition, which will take us through processes by which food is ingested, digested, absorbed, and later converted to the body's own carbohydrates, lipids, and proteins.

KEY CONCEPTS

1. Interactions among the digestive, circulatory, respiratory, and urinary systems supply the body's cells with raw materials, dispose of wastes, and maintain the volume and composition of extracellular fluid.

2. Most digestive systems have specialized regions for food transport, processing, and storage. Different regions are concerned with mechanical and chemical breakdown of food, absorption of the breakdown products, and elimination of unabsorbed residues.

3. To maintain an acceptable body weight and overall health, energy intake must balance energy output (by way of metabolic activity, physical exertion, and so on). Complex carbohydrates are the main energy source.

4. Nutrition requires the intake of vitamins, minerals, and certain amino acids and fatty acids that the body cannot produce itself.

TYPES OF DIGESTIVE SYSTEMS AND THEIR FUNCTIONS

Generally speaking, a **digestive system** is some form of body cavity or tube in which food is first reduced to particles, then to small molecules. A layer of cells lines the body cavity or tube, and nutrients cross this lining and so enter the internal environment.

Chapter 25 described some invertebrates that are equipped with a saclike gut. Such animals are said to have an **incomplete digestive system** because the gut has only one opening. What goes in but cannot be digested goes out the same way. Recall that planarians, a type of flatworm, have this system. A muscular pharynx opens into a highly branched cavity that serves both digestive and circulatory functions. Food is partly digested and transported to cells even as residues are being sent back out through the pharynx. During the course of flatworm evolution, the two-way traffic must have worked against modification of the gut into specialized regions for food transport, processing, and storage.

641

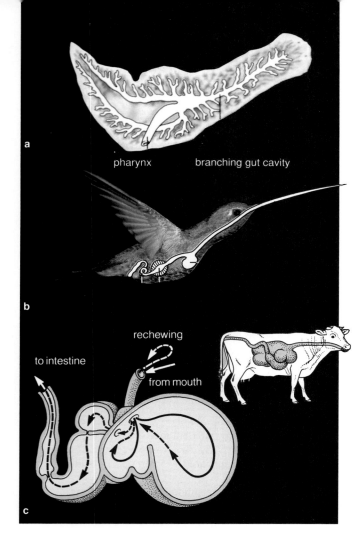

Figure 37.2 (**a**) Incomplete digestive system of a flatworm, with two-way traffic of food and undigested material through one opening (the pharynx). The branched gut cavity serves both digestive and circulatory functions. (**b**) Complete digestive system of a bird—basically a tube with regional specializations and an opening at each end. (**c**) Complete digestive system of a cow. Cattle, antelopes, and other ruminants have a stomach with multiple chambers in which cellulose is digested before being sent on to the small intestine, where nutrients are absorbed.

Recall also from Chapter 25 that chordates, including antelopes, have a **complete digestive system**. So do annelids, mollusks, arthropods, and echinoderms. All of these animals have an internal tube with an opening at one end for taking in food and an opening at the other end for eliminating unabsorbed residues (Figure 37.2). Between the two openings, food generally moves in one direction through the lumen. (*Lumen* refers to the space inside a tube.) The tube itself is subdivided into specialized regions for food transport, processing, and storage. For instance, one part of the digestive tube of birds is modified into a crop, a food storage organ. Another part is modified into a gizzard, a muscular organ that grinds food into smaller bits.

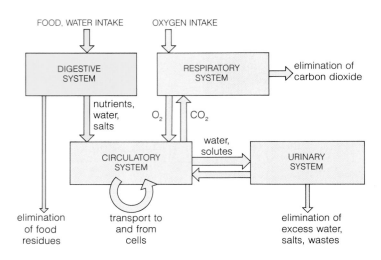

Figure 37.3 Links between the digestive, respiratory, circulatory, and urinary systems. These organ systems work together to supply the body's cells with raw materials and eliminate wastes. This chapter focuses on the digestive system; subsequent chapters will address the other systems shown here.

Specialized regions of complete digestive systems can be correlated with feeding behavior. Predators and scavengers (such as vultures) have *discontinuous* feeding habits. They generally gorge themselves with food when it is available, then may go for long periods without eating at all. Parts of their digestive system store food that is taken in faster than it can be digested and absorbed. Other, accessory parts help maintain an adequate distribution of nutrients between meals. By contrast, antelope, deer, goats, cattle, and other ruminants eat almost continuously when they are not bedding down. (A *ruminant* is any hoofed mammal having multiple stomach chambers in which cellulose can be digested.) Figure 37.2c is a generalized diagram of their digestive system.

We can summarize the overall functions of complete digestive systems in the following way:

1. Motility Muscular movement of the gut wall, leading to the mechanical breakdown, mixing, and passage of ingested nutrients, then elimination of undigested and unabsorbed residues.

2. Secretion Release into the lumen of enzymes, fluids, and other substances required for the functions of the digestive tract.

3. Digestion Breakdown of nutrients into particles, then into molecules small enough to be absorbed.

4. Absorption Passage of digested nutrients, fluid, and ions across the tube wall and into the blood or lymph, which will distribute them through the body.

Table 37.1	Components of the Human Digestive System	
Organ	**Main Functions**
Mouth	Mechanically break down food, mix it with saliva
Salivary glands	Moisten food; start polysaccharide breakdown; buffer acidic foods in mouth
Stomach	Store, mix, dissolve food; kill many microorganisms; start protein breakdown; empty contents in a controlled way
Small intestine	Digest and absorb most nutrients
Pancreas	Enzymatically break down all major food molecules; buffer hydrochloric acid from stomach
Liver	Secrete bile for fat absorption; secrete bicarbonate, which buffers hydrochloric acid from stomach
Gallbladder	Store, concentrate bile from liver
Large intestine	Store, concentrate undigested matter by absorbing water and salts (mineral ions)
Rectum	Control over elimination of undigested and unabsorbed residues

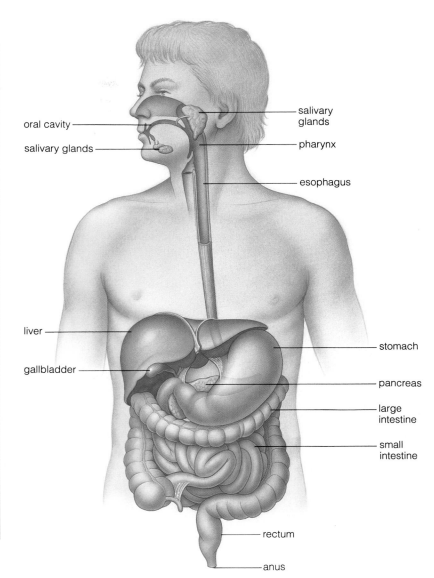

Figure 37.4 Simplified picture of the human digestive system.

The remainder of this chapter will focus on the digestive system and nutritional requirements of humans. As Figure 37.3 suggests, this system does not act alone to meet the body's metabolic needs. Digested nutrients enter the internal environment—that is, the body's extracellular fluid. A circulatory system typically distributes the nutrients to cells throughout the body. A respiratory system helps cells use the nutrients by supplying them with oxygen (for aerobic respiration) and relieving them of carbon dioxide wastes. And even though the kinds and amounts of nutrients being absorbed can vary, depending on the diet, a urinary system helps maintain the volume and composition of extracellular fluid.

Keep these vital interactions in mind as we proceed with our nutritional tour in this chapter and the ones to follow.

HUMAN DIGESTIVE SYSTEM: AN OVERVIEW

Components

Figure 37.4 shows the human digestive system, and Table 37.1 lists the functions of its components. Humans have discontinuous feeding habits, and they ingest a variety of foods. From this you might deduce, correctly, that the human digestive system is a tube with many regional specializations. Stretched out, the tube would be 6.5 to 9 meters (21 to 30 feet) long in adults. Its specialized regions are the mouth (oral cavity), pharynx, esophagus, and the *gut*, or gastrointestinal tract. The gut itself is subdivided into a stomach, small intestine, large intestine (colon), rectum, and anus. Enzymes and other substances from the salivary glands, liver, gallbladder, pancreas, and the gut wall are secreted into different parts of the tube. The secretions assist in digestion and absorption.

Gut Structure and Motility

Figure 37.5 shows the structure of the gut wall. The *mucosa* (an epithelium and underlying layer of connective tissue) faces the gut lumen. It is surrounded by the *submucosa*, a connective tissue layer with blood and lymph vessels and nerve plexuses (local networks of neurons). Next is *smooth muscle*—usually two sublayers, one circular and the other longitudinal in orientation. The outer layer of connective tissue (*serosa*) is almost as thin as Saran wrap. *Sphincters* (rings of muscle in the wall) occur at the beginning and end of the stomach and other specialized regions. They help control the forward movement of food and prevent backflow.

The muscle layers engage in mixing and wavelike contractions (Figure 37.6). During *peristalsis*, rings of circular muscles contract behind food and relax in front of it. The food distends the tube wall, peristaltic movement forces the food onward and expands the next wall region, and so on. During *segmentation*, rings of smooth muscle in the gut wall repeatedly contract and relax, creating an oscillating (back-and-forth) movement in the same place. This movement constantly mixes and forces the contents of the lumen against the absorptive surface of the intestinal wall.

Control of the Digestive System

Recall that homeostatic control mechanisms operate when physical and chemical conditions change in the *internal* environment. By contrast, controls over the stomach and intestines respond to the volume and composition of material in the gut lumen. The nervous system, local nerve plexuses in the gut wall, and the endocrine system interact to exert control.

After a meal, for instance, food distends the gut wall. Signals from mechanoreceptors travel on short reflex pathways that are confined to nerve plexuses. They also may travel on long reflex pathways to the central nervous system. Signals along one or both types of pathways can lead to muscle contractions in the gut wall or secretion of enzymes and other substances into the gut lumen. The hypothalamus and other parts of the brain monitor such activities and coordinate them with other events (Figure 37.7).

Four gastrointestinal hormones are known. *Gastrin* is secreted by endocrine cells in the stomach's lining when amino acids and peptides are in the stomach. It mainly stimulates the secretion of acid into the stomach. Endocrine cells in the lining of the small intestine secrete the other three hormones. *Secretin*, a peptide hormone, stimulates the pancreas to secrete bicarbonate (page 581). *CCK* (cholecystokinin) enhances the actions of secretin and stimulates gallbladder contractions. *GIP*

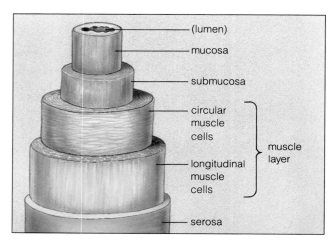

Figure 37.5 Generalized sketch of the wall of the gastrointestinal tract. (The layers are not drawn to scale.)

Figure 37.6 (**a**) Peristaltic wave down the stomach, produced by alternating contraction and relaxation of muscles in the stomach wall. (**b**) Segmentation, or oscillating movement, in the intestines.

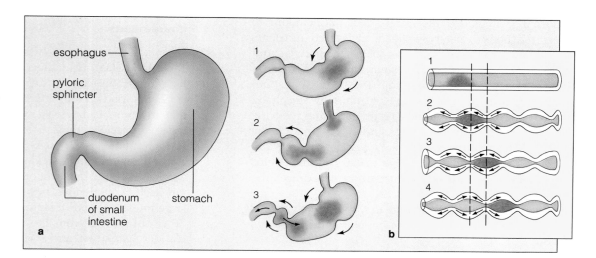

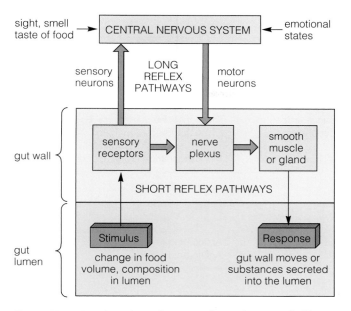

Figure 37.7 Local and long-distance reflex pathways called into action when food is in the digestive tract.

(glucose insulinotropic peptide) is released in response to the presence of glucose and fat in the small intestine. It stimulates insulin secretion.

The nervous system, endocrine system, and local nerve plexuses control conditions in the digestive system in response to the volume and composition of material in the gut lumen.

In this respect, controls over the digestive system differ from homeostatic control mechanisms (which maintain conditions in the internal environment).

INTO THE MOUTH, DOWN THE TUBE

Mouth and Salivary Glands

Food starts getting pummeled and polysaccharide digestion begins in the mouth (oral cavity). Only humans and other mammals *chew* food. Adult humans normally have thirty-two teeth to do this. Each tooth is an engineering marvel, able to withstand many years of chemical insults and mechanical stress. It has an enamel coat (hardened calcium deposits), dentine (a thick bonelike layer), and an inner pulp (with nerves and blood vessels). The chisel-shaped incisors bite off chunks of food, the cone-shaped cuspids tear it, and the flat-topped molars grind it (page 332 and Figure 37.1).

Food in the mouth becomes mixed with saliva, a fluid secreted from **salivary glands**. Saliva includes a starch-degrading enzyme (salivary amylase), bicarbonate (HCO_3^-), and mucins. The buffering action of bicarbonate ions keeps the pH of your mouth between 6.5

and 7.5, even when you eat tomatoes and other acidic foods. Mucins (modified proteins) bind bits of food into a softened, lubricated ball called a bolus.

Muscle contractions of the tongue force the softened ball of food into the **pharynx**. This muscular tube connects with the **esophagus**, which leads to the stomach. Swallowing is initiated when an individual voluntarily pushes a bolus into the pharynx. Sensory receptors in the wall of the pharynx are stimulated, and they trigger contractions (an involuntary response). The pharynx and esophagus do not have roles in digestion; contractions in their walls simply propel food into the stomach.

The pharynx also connects with the trachea, which leads to the lungs. Swallowing opens a sphincter at the start of the esophagus. You normally don't choke on food because a flaplike valve, the epiglottis, closes off the opening into the respiratory tract and keeps you from breathing while food is moving into the esophagus.

The Stomach

The **stomach**, a muscular, stretchable sac, has three main functions. First, it stores and mixes food. Second, its secretions help dissolve and degrade food. Third, it helps control the movement of food into the small intestine.

Stomach Acidity. Each day, cells in the stomach lining secrete about 2 liters of substances, including hydrochloric acid (HCl), pepsinogens, and mucus. The substances make up the fluid in the stomach, the so-called *gastric fluid*. The HCl separates into H^+ and Cl^-, and the increase in acidity helps dissolve bits of food to form a solution called chyme. It also kills most of the microorganisms hitching rides into the body in food.

Stomach secretion begins when your brain responds to the sight, aroma, and taste of food (even to hungry thoughts about it) and fires off signals to the acid-secreting and endocrine cells in the stomach lining. But most of the secretions occur in response to food in the stomach. When food stretches the stomach, it activates receptors in the stomach wall and gives rise to neural signals that call for stepped-up secretions. Also, secretory cells are stimulated directly by certain substances, including partially dismantled proteins as well as the caffeine in coffee, tea, chocolate, and cola drinks.

Protein digestion begins in the stomach. High stomach acidity structurally changes proteins and exposes the peptide bonds. It also converts pepsinogens to active forms (pepsins) that break down proteins. Protein fragments directly stimulate the secretion of gastrin, a hormone that acts on HCl-secreting cells. The more protein you eat, the more gastrin and HCl are released.

What protects the stomach lining itself from HCl and pepsin? Control mechanisms assure that enough mucus

Table 37.2 Major Enzymes of Digestion				
Enzyme	Source	Where Active	Substrate	Main Breakdown Products*
Carbohydrate Digestion:				
Salivary amylase	Salivary glands	Mouth	Polysaccharides	Disaccharides
Pancreatic amylase	Pancreas	Small intestine	Polysaccharides	Disaccharides
Disaccharidases	Small intestine	Small intestine	Disaccharides	Monosaccharides (e.g., glucose)
Protein Digestion:				
Pepsins	Stomach mucosa	Stomach	Proteins	Peptide fragments
Trypsin and chymotrypsin	Pancreas	Small intestine	Proteins, polypeptides	Peptide fragments
Carboxypeptidase	Pancreas	Small intestine	Peptide fragments	Amino acids
Aminopeptidase	Intestinal mucosa	Small intestine	Peptide fragments	Amino acids
Fat Digestion:				
Lipase	Pancreas	Small intestine	Triglycerides	Free fatty acids, monoglycerides
Nucleic Acid Digestion:				
Pancreatic nucleases	Pancreas	Small intestine	DNA, RNA	Nucleotides
Intestinal nucleases	Intestinal mucosa	Small intestine	Nucleotides	Nucleotide bases, monosaccharides

*Yellow parts of table identify breakdown products that can be absorbed into the internal environment.

and buffering molecules (bicarbonate ions especially) are secreted to protect the lining from their destructive effects. Sometimes, however, normal controls are blocked. When the surface of the stomach breaks down, H⁺ diffuses into the lining and triggers the release of a chemical called histamine from tissue cells. Histamine acts on local blood vessels—and it stimulates more HCl secretion. A positive-feedback loop is set up, tissues become further damaged, and bleeding into the stomach and possibly into the abdomen may occur. This outcome is called a *peptic ulcer*.

Stomach Emptying. In the stomach, peristaltic waves mix the chyme and build up force as they approach the sphincter between the stomach and small intestine (Figure 37.6). The arrival of a strong contraction closes the sphincter, so most of the chyme is squeezed back. But a small amount moves into the small intestine.

The volume and composition of chyme affect how fast the stomach empties. For example, large meals activate more receptors in the stomach wall, these call for increases in the force of contraction, and the stomach empties faster. As another example, receptors in the small intestine sense increases in acidity, fat content, and so on, and they call for the release of hormones that slow stomach emptying. Through such slowdowns, food is not moved along faster than it can be processed. Fear, depression, and other emotional upsets also can slow stomach motility.

The Small Intestine

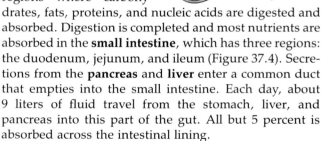

Table 37.2 summarizes the regions where carbohydrates, fats, proteins, and nucleic acids are digested and absorbed. Digestion is completed and most nutrients are absorbed in the **small intestine**, which has three regions: the duodenum, jejunum, and ileum (Figure 37.4). Secretions from the **pancreas** and **liver** enter a common duct that empties into the small intestine. Each day, about 9 liters of fluid travel from the stomach, liver, and pancreas into this part of the gut. All but 5 percent is absorbed across the intestinal lining.

Digestion Processes. Enzymes secreted from the pancreas act on carbohydrates, fats, proteins, and nucleic acids. For example, like pepsin in the stomach, the pancreatic enzymes trypsin and chymotrypsin digest proteins into peptide fragments. The fragments are then degraded to free amino acids by carboxypeptidase (from the pancreas) and by aminopeptidase (present on the surface of the intestinal mucosa). The pancreas also secretes bicarbonate. The buffering action of bicarbonate helps neutralize the HCl arriving from the stomach. Two hormones secreted from the pancreas (insulin and glucagon) do not function in digestion, but they still have roles in nutrition, as described on page 590.

Bile, a secretion from the liver, has a key role in digestion. This secretion contains bile salts, bile pigments, cholesterol, and lecithin (a phospholipid). The

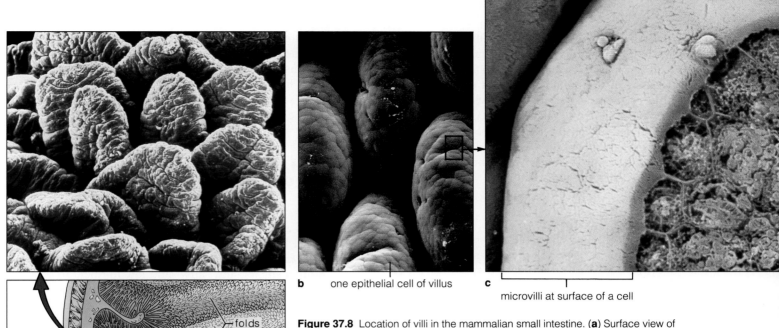

b one epithelial cell of villus

c microvilli at surface of a cell

a location of intestinal villi

folds

Figure 37.8 Location of villi in the mammalian small intestine. (**a**) Surface view of the deep, permanent folds of the inner layer of the intestinal tube. (**b**) Some of the fingerlike projections (villi) that cover the inner layer. The villi are so dense and numerous, they give the surface a velvety appearance. Individual epithelial cells are visible. (**c**) The dense crown of microvilli at the surface of a single cell.

bile salts assist in fat breakdown and absorption. Most fats in our diet are triglycerides, clumped into large fat globules. The globules are mechanically broken apart into smaller droplets in the small intestine. Bile salts keep the droplets from clumping back together into globules, a process called *emulsification*. Through the emulsifying effect of bile salts, fat-degrading enzymes have greater access to more triglycerides, so fat digestion is enhanced.

Between meals, bile is stored and concentrated in the **gallbladder** (Figure 37.4).

Absorption Processes. By the time proteins, lipids, and carbohydrates are halfway through the small intestine, mechanical action and enzymes have broken down most of them to smaller molecules. The breakdown products include glucose and other monosaccharides, amino acids, fatty acids, and monoglycerides, all of which can move across the intestinal lining. The lining is densely folded into **villi** (singular, villus). Villi are absorptive structures that increase the surface area available for interactions with chyme. Epithelial cells at their surface have a crown of **microvilli**. Each microvillus is a threadlike projection of the plasma membrane. Collectively, microvilli greatly increase the surface area available for absorption (Figure 37.8).

At each villus, glucose and most amino acids cross the gut lining by active transport mechanisms, which move them across the plasma membranes of epithelial

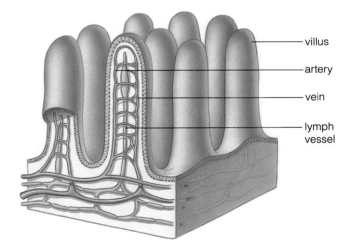

villus

artery

vein

lymph vessel

Figure 37.9 Location of blood vessels and lymph vessels in intestinal villi. Monosaccharides and most amino acids moving across the intestinal lining enter the blood vessels; fats enter the lymph vessels, which drain into the general circulation.

cells. Then they diffuse through extracellular fluid and enter small blood vessels inside the villus (Figure 37.9).

The fatty acids and monoglycerides diffuse across the lipid bilayer of the membranes. (Bile salts help maintain the required concentration gradients for those free lipids. They combine with a number of the lipid molecules, forming small aggregates called *micelles*. The micelles themselves don't diffuse across the membrane. They are more like holding stations *at* the membrane. The bound lipids readily depart from micelles when their concentrations decrease in the lumen as a result of absorption.) Inside the epithelial cells, the fatty acids and monoglycerides recombine into fats. The fats cluster together as small droplets. Then the droplets leave the cells by exocytosis and enter lymph vessels, which drain into the blood circulation system. Water and mineral ions also are absorbed at the intestinal villi.

The Large Intestine

Material not absorbed in the small intestine moves into the **large intestine**, or **colon**. The colon stores and concentrates *feces*, a mixture of undigested and unabsorbed material, water, and bacteria. The concentrating mechanism involves the active transport of sodium ions across the lining of the colon; water follows passively as a result.

The colon is about 1.2 meters long and starts out as a blind pouch (Figure 37.10). The *appendix*, a narrow projection from the pouch, has no known digestive functions. (It may have roles in defense against infectious agents.) The colon ascends on the right side of the abdominal cavity, continues across to the other side, then descends and connects with a small tube, the **rectum** (Figure 37.4). Distension of the rectal wall triggers the expulsion of fecal matter from the body. Expulsion is controlled by the nervous system, which can stimulate or inhibit contractions of a muscle sphincter at the **anus**, the terminal opening of the gut.

The average American diet does not include enough bulk. *Bulk* is the volume of fiber (mainly cellulose) and other undigested food material that cannot be decreased by absorption in the colon. Because of the insufficient volume, it takes longer for feces to move through the colon, with irritating and perhaps carcinogenic effects.

Most people of rural Africa and India cannot afford to eat much more than whole grains—which are high in fiber content—and they rarely suffer colon cancer or appendicitis. (In *appendicitis*, the appendix becomes infected and may rupture; bacteria normally living in the colon may spread into the abdominal cavity and cause serious infection.) When those people move to wealthier nations and leave behind their fiber-rich diet, they are more likely to suffer colon cancer and appendicitis. And this example leads us into a closer look at what constitutes good and bad nutrition.

HUMAN NUTRITIONAL REQUIREMENTS

The earliest human ancestors, it seems, dined on fresh fruits and other fibrous plant material. From this nutritional beginning, humans in many parts of the world have moved to diets rich in saturated fats, cholesterol, refined sugars, and salts—and low in fiber. To be sure, our life span is much longer than that of our earliest ancestors. But many of us are probably suffering more than they did from colon cancer, kidney stones, breast cancer, and circulatory disorders—all of which may be correlated with the long-term shift in diet.

Energy Needs and Body Weight

The body grows and maintains itself when kept supplied with energy and materials from foods of certain types, in certain amounts. Nutritionists measure energy in units called "calories," which unfortunately is supposed to mean the same thing as "kilocalories." A *kilocalorie* is 1,000 calories of energy, and that is the term we will use here.

To maintain an acceptable weight and keep the body functioning normally, caloric intake must be balanced with energy output. The output varies from one person to the next because of differences in physical activity, basic rate of metabolism, age, sex, hormone activity, and emotional state. Some of these factors are influenced by a person's social environment. But others have a genetic basis. For example, long-term studies were made of many identical twins (with identical genes) who were separated at birth and raised apart, in different households. At adulthood, the body weights of the separated twins were remarkably similar.

In most adults, energy input balances the output, so body weight remains much the same over long periods. As any dieter knows, it is as if the body has a set point for what that weight is going to be and works to counteract deviations from its set point.

How many kilocalories should you take in each day to maintain what you consider to be an acceptable body weight? One way to answer this question is to estimate your body's energy requirements. First, multiply the desired weight (in pounds) by 10 if you are not very active physically, by 15 if you are moderately active, and by 20 if you are quite active. Then, depending on your

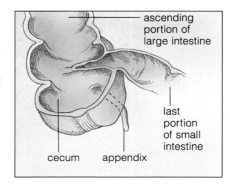

Figure 37.10
Location of the appendix, a narrow projection from the cecum (a cup-shaped pouch at the start of the large intestine).

ascending portion of large intestine

last portion of small intestine

cecum   appendix

Man's Height	Size of Frame		
	Small	Medium	Large
5' 2"	128–134	131–141	138–150
5' 3"	130–136	133–143	140–153
5' 4"	132–138	135–145	142–156
5' 5"	134–140	137–148	144–160
5' 6"	136–142	139–151	146–164
5' 7"	138–145	142–154	149–168
5' 8"	140–148	145–157	152–172
5' 9"	142–151	148–160	155–176
5'10"	144–154	151–163	158–180
5'11"	146–157	154–166	161–184
6' 0"	149–160	157–170	164–188
6' 1"	152–164	160–174	168–192
6' 2"	155–168	164–178	172–197
6' 3"	158–172	167–182	176–202
6' 4"	162–176	171–187	181–207

a

Woman's Height	Size of Frame		
	Small	Medium	Large
4'10"	102–111	109–121	118–131
4'11"	103–113	111–123	120–134
5' 0"	104–115	113–126	122–137
5' 1"	106–118	115–129	125–140
5' 2"	108–121	118–132	128–143
5' 3"	111–124	121–135	131–147
5' 4"	114–127	124–138	134–151
5' 5"	117–130	127–141	137–155
5' 6"	120–133	130–144	140–159
5' 7"	123–136	133–147	143–163
5' 8"	126–139	136–150	146–167
5' 9"	129–142	139–153	149–170
5'10"	132–145	142–158	152–173
5'11"	135–148	145–159	155–176
6' 0"	138–151	148–162	158–179

b

Figure 37.11 (**a**) "Ideal" weights for adults according to one insurance company in 1983. Values shown are for people 25 to 59 years old wearing shoes with 1-inch heels and 3 pounds of clothing (for women) or 5 pounds (for men). (**b**) Extreme obesity puts severe strain on the circulatory system. The body produces many more capillaries to service the increased tissue masses, so blood volume drops. The heart becomes more stressed. It must pump harder to keep blood circulating.

age, subtract the following amount from the value obtained from the first step:

Age	Subtract
25–34	0
35–44	100
45–54	200
55–64	300
Over 65	400

For example, if you want to weigh 120 pounds and are highly active, 120 × 20 = 2,400 kilocalories. If you are thirty-five years old, then you should take in a total of (2,400 − 100), or 2,300 kilocalories a day. Such calculations provide a rough estimate, but other factors, such as height, must be considered also. (An active person who is 5 feet, 2 inches tall doesn't need as many kilocalories as an active person who weighs the same but is 6 feet tall.)

By definition, **obesity** is an excess of fat in the body's adipose tissues, caused by imbalances between caloric intake and energy output. Yet, what is too fat or too thin? What is a person's "ideal weight"? Many charts have been developed (Figure 37.11), mostly by insurance companies that want to identify overweight people who are considered to be insurance risks. Such charts factor in height. People who are 25 percent heavier than the "ideal" are viewed as obese.

Some researchers who study causes of death suspect that the "ideal" actually may be 10 to 15 pounds heavier than the charts indicate. Some nutritionists are convinced the chart values should be less. Whatever the ideal range may be, serious disorders do arise with extremes at either end of that range (see the *Commentary* on page 650).

To maintain an acceptable body weight, energy input (caloric intake) must be balanced with energy output (as through metabolic activity and exercise).

Carbohydrates

Complex carbohydrates are the body's main sources of energy. As we have seen, they can be readily broken down into glucose units (Chapter 6). Glucose is the primary energy source for your brain, muscles, and other body tissues. According to many nutritionists, the fleshy fruits, cereal grains, legumes (including beans and peas), and other fibrous carbohydrates should make up at least 50 to 60 percent of the daily caloric intake.

Each year, the average American eats as much as 128 pounds of refined sugar, or sucrose. That's more than 2 pounds a week! You may think this a far-fetched statement, but take a look at the ingredients listed on your packages of cereal, frozen dinners, soft drinks, and other prepared foods. A common ingredient is sucrose. This simple sugar adds calories to the diet but does so without the fiber of complex carbohydrates.

Proteins

Proteins should make up about 12 percent of the total diet. When proteins are digested and absorbed, their amino acids become available for the body's own protein-building programs. Of the twenty common amino acids, eight are **essential amino acids**. Our cells cannot build these amino acids; they must obtain them indirectly, from food. The eight are cysteine (or methionine), isoleucine, leucine, lysine, phenylalanine (or tyrosine), threonine, tryptophan, and valine.

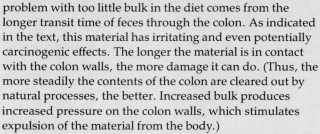

Human Nutrition Gone Awry

Eating Disorders

Anorexia Nervosa. Millions of Americans are dieting in any given day. Unfortunately, in a growing number of cases, obsessive dieting leads to a potentially fatal eating disorder called *anorexia nervosa* (Figure *a*). The disorder occurs primarily in women in their teens and early twenties.

Individuals with anorexia nervosa have a skewed perception of their body weight. They have an overwhelming fear of being fat and being hungry. They embark on a course of self-induced starvation and, frequently, overexercising. Emotional factors contribute to the disorder. Some individuals fear growing up in general and maturing sexually in particular; others have irrational expectations of what they can accomplish. Severe cases require psychiatric treatment.

Bulimia. Another eating disorder on the rise is *bulimia* ("an oxlike appetite"). At least 20 percent of college-age women are now suffering to varying degrees from this disorder. Those afflicted may look outwardly healthy, but their food intake is out of control. During an hour-long eating binge, a bulimic may take in more than 50,000 kilocalories. This is followed by vomiting or purging the body with laxatives, sometimes in doses of 200 tablets or more. The binge–purge routine may occur once a month; it may occur several times a day.

Some women start doing this because it seems like a simple way to lose weight. Others have emotional problems. Often they are well-educated, accomplished individuals, but they strive for perfection and may have problems with control exerted by other family members. According to one view, eating may actually be an unpleasant event for them, but the purging (which they themselves control) relieves them of anger and frustration.

Repeated purgings, however, can damage the gastrointestinal tract. Repeated vomiting, which brings stomach acids into the mouth, can erode teeth to stubs. At its most extreme, bulimia can lead to death through heart failure, stomach rupturing, or kidney failure. Psychiatric treatment and hospitalization may be required in severe cases.

Digestive Disorders

Generally, Americans are among the best-fed people in the world—yet at the same time, they suffer a high incidence of digestive disorders. Along with affluence, it appears that other bad eating habits besides anorexia nervosa and bulimia are rampant. Americans skip meals, eat too much and too fast when they do sit down at the table, and generally give their gut erratic workouts. Worse yet, their diet tends to be rich in sugar, cholesterol, and salt—and low in bulk. The problem with too little bulk in the diet comes from the longer transit time of feces through the colon. As indicated in the text, this material has irritating and even potentially carcinogenic effects. The longer the material is in contact with the colon walls, the more damage it can do. (Thus, the more steadily the contents of the colon are cleared out by natural processes, the better. Increased bulk produces increased pressure on the colon walls, which stimulates expulsion of the material from the body.)

In addition, diet affects the distribution and diversity of bacterial populations living in the gut. Do changes in those populations contribute to digestive disorders? That is not known.

The emotional stress of living in complex societies seems to compound the nutritional problem. Urban populations seem to be more susceptible to the irritable colon syndrome (colitis). Its symptoms include abdominal pain, diarrhea (excretion of watery feces), and constipation. Diarrhea can be brought on by emotional stress. Although the development of ulcers may have a genetic basis, emotional stress also is a contributing factor.

Learning to handle stress is one way to ease up on the digestive tract, and learning how to eat properly is another.

Yet what is "eating properly"? As long ago as 1979, the United States Surgeon General released a report representing a medical consensus on how to promote health and avoid such afflictions as high blood pressure, heart disorders, cancer of the colon, and bad teeth. The report advised Americans to eat "less saturated fat and cholesterol; less salt; less sugar; relatively more complex carbohydrates such as whole grains, cereals, fruits, and vegetables; and relatively more fish, poultry, legumes (for example, peas, beans, and peanuts); and less red meat."

The controversies over what constitutes proper nutrition are still raging today. In the meantime, it might not be a bad idea to think about your own eating habits and how moderation in some things might help you hedge your bets. Put the question to yourself: Do you look upon a bowl of bran cereal with the same passion as you look upon, say, french fries and ice cream, prime rib, and chocolate mousse? Now put the same question to your colon.

No Limiting Amino Acid	Low in Lysine	Low in Methionine, Other Sulfur-Containing Amino Acids	Low in Tryptophan

legumes:
  soybean
  tofu
  soy milk
 	
cereal grains:
  wheat germ
 	
nuts:
 	
milk
cheeses (except
  cream cheese)
yogurt
eggs
meats
	legumes:
	  peanuts
	
	cereal grains:
	  barley
	  buckwheat
	  corn meal
	  oats
	  rice
	  rye
	  wheat
	
	nuts, seeds:
	  almonds
	  cashews
	  coconut
	  English walnuts
	  hazelnuts
	  pecans
	  pumpkin seeds
	  sunflower seeds
		legumes:
		  beans (dried)
		  black-eyed peas
		  garbanzos
		  lentils
		  lima beans
		  mung beans
		  peanuts
		
		nuts:
		  hazelnuts
		
		fresh vegetables:
		  asparagus
		  broccoli
		
		  green peas
		  mushrooms
		  parsley
		  potatoes
		  soybeans
		  Swiss chard
			legumes:
			  beans (dried)
			  garbanzos
			  lima beans
			  mung beans
			  peanuts
			
			cereal grains:
			  corn meal
			
			nuts:
			  almonds
			  English walnuts
			
			fresh vegetables:
			  corn
			  green peas
			  mushrooms
			  Swiss chard

total protein intake

isoleucine
leucine
lysine
methionine
phenylalanine
threonine
tryptophan
valine

a

b

Most animal proteins are "complete," meaning they contain high amounts of all essential amino acids. Plant proteins are "incomplete." They have all the essential amino acids, but not in the proportions required for proper human nutrition. To get enough protein, vegetarians must eat certain combinations of different plants (Figure 37.12). Nutritionists use a measure called *net protein utilization*, or NPU, to compare proteins from different sources (Table 37.3). NPU values range from 100 (all essential amino acids present in ideal proportions) to 0 (one or more absent; the protein is useless when eaten alone).

You know that enzymes and other proteins are vital for the body's structure and function, so it should be readily apparent that protein-deficient diets are no joking matter. Protein deficiency is most damaging among the young, for rapid brain growth and development occur early in life. Unless enough protein is taken in just before and just after birth, irreversible mental retardation occurs. Even mild protein starvation can retard growth and affect mental and physical performance.

Lipids

Fats and other lipids have important roles in the body. For example, phospholipids (such as lecithin) and cholesterol are components of animal cell membranes. Besides being used as energy reserves, fat deposits serve as cushions for many organs, including the eyes and kidneys, and they provide insulation beneath the skin. Fats from the diet also help the body absorb fat-soluble vitamins.

Figure 37.12 (**a**) Essential amino acids—a small portion of the total protein intake. All must be available at the same time, in certain amounts, if cells are to build their own proteins. Milk and eggs have high amounts of all eight essential amino acids in required proportions; they are among the complete proteins.

Nearly all plant proteins are incomplete, so vegetarians should plan their meals carefully to avoid protein deficiency. For example, they can combine *different* foods from any two of the columns shown in (**b**). Also, vegetarians who avoid dairy products and eggs should take vitamin B_{12} and B_2 (riboflavin) supplements. Animal protein is a luxury in most traditional societies, yet good combinations of plant proteins are worked into their cuisines—including rice/beans, chili/cornbread, tofu/rice, lentils/wheat bread, and macaroni/cheese.

Table 37.3	Efficiency of Some Single Protein Sources in Meeting Minimum Daily Requirements			
Source	Protein Content (%)	Net Protein Utilization (NPU)	Amount Needed to Satisfy Minimum Daily Requirement	
			(grams)	(ounces)
Eggs	11	97	403	14.1
Milk	4	82	1,311***	45.9***
Fish*	22	80	244	8.5
Cheese*	27	70	227	7.2
Red meat*	25	68	253	8.8
Soybeans	34	60	210**	7.3**
Kidney beans	23	40	468**	16.4**
Corn	10	50	860**	30.0**

*Average values.
**Dry weight values.
***Equivalent of 6 cups. The figure is somewhat misleading, for most of the volume of milk is water. Milk is actually a rich source of high-quality protein.

Table 37.4 Vitamins Required for Normal Cell Functioning

Vitamin	RDA* (milligrams) Females	RDA* (milligrams) Males	Common Sources	Some Known Functions
Water-Soluble Vitamins:				
B_1 (Thiamin)	1.1	1.5	Lean meats, liver, eggs, whole grains, green leafy vegetables, legumes	Connective tissue formation; iron, folic acid utilization
B_2 (Riboflavin)	1.3	1.7	Milk, egg white, yeast, whole grains, poultry, fish, meat	Coenzyme action (FAD, FMN)
Niacin	15	19	Meat, poultry, fish; also peanuts, potatoes, green leafy vegetables, liver	Coenzyme action (NAD^+, $NADP^+$)
B_6	1.6	2	Meats, potatoes, tomatoes, spinach	Coenzyme role in amino acid metabolism
Pantothenic acid	**	**	In many foods, especially meat, yeast, egg yolk	Coenzyme role in glucose metabolism; fatty acid and steroid synthesis
Folic acid (folate)	0.18	0.2	Dark green vegetables, eggs, liver, yeast, lean meat, whole grains; produced by bacteria in gut	Coenzyme role in nucleic acid and amino acid metabolism
B_{12}	0.002	0.002	Meat, poultry, fish, eggs, dairy foods (not butter)	Coenzyme role in nucleic acid metabolism
Biotin	**	**	Legumes, nuts, liver, egg yolk; some produced by bacteria in gut	Coenzyme action in fat and glycogen formation; amino acid metabolism
Choline	**	**	Whole grains, legumes, egg yolk, liver	Component of phospholipids, acetylcholine
C (Ascorbic acid)	60	60	Citrus, papaya, cantaloupe, berries, tomatoes, potatoes, green leafy vegetables	Structural role in bone, cartilage, teeth; roles in collagen formation, carbohydrate metabolism
Fat-Soluble Vitamins:				
A (Retinol)	0.8	1	Formed from carotene in deep-yellow, deep-green leafy vegetables; already present in fish liver oil, liver, egg yolk, fortified milk	Role in synthesis of visual pigments; required for bone, tooth development; maintains epithelial tissues
D	0.01/0.005	0.01/0.005	Vitamin D_3 formed in skin cells (also in fish liver oils, egg yolk, fortified milk); converted to active form in other body regions	Promotes bone growth, mineralization; increases calcium absorption
E	8	10	Vegetable oils, margarine, whole grains, dark-green vegetables	Prevents breakdown of vitamins A, C in gut; helps maintain cell membranes
K	0.06/0.065	0.07/0.08	Most formed by bacteria in colon; also in green leafy vegetables, cauliflower, cabbage	Role in clot formation; electron transport role in ATP formation

*1989 recommended daily allowance for two age groups: 19–24/25–50.
**Not established.

Today, lipids make up 40 percent of the average diet in the United States. Most of the medical community agrees it should be less than 30 percent. The body can synthesize most of its own fats, including cholesterol, from protein and carbohydrates. (That is exactly what it does when you eat too much protein and carbohydrates.) You only need to take in *one tablespoon a day* of a polyunsaturated fat, such as corn oil or olive oil. These oils contain linoleic acid. It is one of the **essential fatty acids**. This means the body cannot produce this fatty acid; it must be provided by the diet.

Butter and other animal fats are saturated fats, which tend to raise the level of cholesterol in the blood. Cho-lesterol is necessary in the synthesis of bile acids and steroid hormones. However, too much cholesterol may have devastating effects on the circulatory system (page 671).

Vitamins and Minerals

Normal metabolic activity depends on small amounts of more than a dozen organic substances called **vitamins**. Most plant cells synthesize all of these substances. In general, animal cells have lost the ability to do so, so animals must obtain vitamins from food. Human cells need at least thirteen different vitamins, each with specific metabolic roles. Many metabolic reactions depend on

Table 37.5 Minerals Required for Normal Cell Functioning

Mineral	RDA* (milligrams) Females	RDA* (milligrams) Males	Common Sources	Some Known Functions
Calcium	1200/800	1200/800	Dairy products, dark-green vegetables, dried legumes	Bone, tooth formation; clotting, neural signals
Chlorine	**	**	Table salt; usually too much in diet	HCl formation by stomach; helps maintain body pH
Copper	**	**	Meats, legumes, drinking water	Used in synthesis of hemoglobin, melanin, transport chain components
Fluorine	**	**	Fluoridated water, seafood	Bone, tooth maintenance
Iodine	0.15	0.15	Marine fish, shellfish; dairy products	Thyroid hormone formation
Iron	10	15	Liver, lean meats, yolk, shellfish, nuts, molasses, legumes, dried fruit	Hemoglobin, cytochrome formation
Magnesium	280	350	Dairy products, nuts, whole grains, legumes	Coenzyme role in ATP–ADP cycle; role in muscle, nerve function
Phosphorus	1200/800	1200/800	Dairy products, red meat, poultry, whole grains	Component of bone, teeth, nucleic acids, proteins, ATP, phospholipids
Potassium	**	**	Diet provides ample amounts	Muscle, nerve function; role in protein synthesis; acid-base balance
Sodium	**	**	Table salt; diet provides adequate to excess amounts	Key salt in solute-water balance, muscle and nerve function
Sulfur	**	**	Dietary proteins	Component of body proteins
Zinc	12	15	Seafood, meat, cereals, legumes, nuts, yeast	Component of digestive enzymes; roles in normal growth, wound healing, taste and smell, sperm formation

*1981 recommended daily allowance for two age groups: 18–24/25–50.
**Not established.

several vitamins, and the absence of one vitamin can affect the functions of others (Table 37.4).

Metabolic activity also depends on inorganic substances called **minerals** (Table 37.5). For example, most cells use calcium and magnesium in many different reactions. All cells use potassium during muscle activity and nerve function. All cells require iron for cytochrome molecules, these being components of electron transport chains. Red blood cells contain iron in hemoglobin, the oxygen-carrying pigment in blood.

The sensible way to supply cells with essential vitamins and minerals is to eat a well-balanced selection of carbohydrates, lipids, and proteins. About 250–500 grams of carbohydrates, 66–83 grams of lipids, and 32–42 grams of protein should do the trick. Some people claim the body will benefit from massive doses of certain vitamins and minerals. To date, no clear evidence exists that vitamin intake above recommended daily amounts leads to better health. To the contrary, excessive vitamin doses are often merely wasted and may cause chemical imbalances.

For example, the body simply will not hold more vitamin C than it needs for normal functioning. Vitamin C is not fat-soluble, and the excess is eliminated in urine. In fact, any amount above the recommended daily allowance ends up in the urine almost immediately after it is absorbed from the gut! Abnormal intake of at least two other vitamins (A and D) can cause serious disorders. Like all fat-soluble vitamins, vitamins A and D can accumulate in tissues and interfere with normal metabolic function.

Similarly, sodium is present in plant and animal tissues, and it is a component of table salt. Sodium has roles in the body's salt–water balance, muscle activity, and nerve function. Yet, prolonged, excessive intake of sodium is thought to be a cause of high blood pressure.

Severe shortage or massive excess of vitamins and minerals can disturb the delicate balances that promote health.

Figure 37.13 Summary of major pathways of organic metabolism. Urea formation occurs primarily in the liver. Carbohydrates, fats, and proteins are continually being broken down and resynthesized.

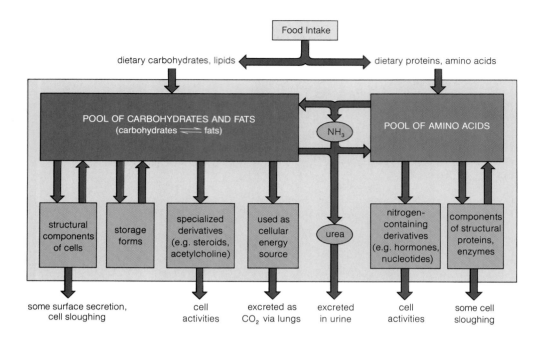

Table 37.6	**Some Activities That Depend on Liver Functioning**

1. Carbohydrate metabolism
2. Control over some aspects of plasma protein synthesis
3. Assembly and disassembly of certain proteins
4. Urea formation from nitrogen-containing wastes
5. Assembly and storage of some fats
6. Fat digestion (bile is formed by the liver)
7. Inactivation of many chemicals (such as hormones and some drugs)
8. Detoxification of many poisons
9. Degradation of worn-out red blood cells
10. Immune response (removal of some foreign particles)
11. Red blood cell formation (liver absorbs, stores factors needed for red blood cell maturation)

NUTRITION AND METABOLISM

Storage and Interconversion of Nutrients

Figure 37.13 summarizes the main routes by which nutrient molecules are shuffled and reshuffled once they have been absorbed into the body. With few exceptions (such as DNA), most carbohydrates, lipids, and proteins are broken down continually, with their component parts picked up and used again in new molecules. At the molecular level, your body undergoes massive and sometimes rapid turnovers.

When you eat, the body builds up its pools of organic molecules. Excess carbohydrates and other dietary molecules are transformed mostly into fats, which are stored in adipose tissue. Some are also converted to glycogen in the liver and in muscle tissue. Most cells use glucose as their main energy source at this time; there is no net breakdown of protein in muscle or other tissues.

Between meals, there is a notable shift in the type of food molecules used to support cell activities. A key factor in this shift is the need to provide brain cells with glucose, the major nutrient they use for energy.

When glucose is being absorbed, its blood levels are readily maintained. How does the body maintain blood glucose concentrations between meals? First, glycogen, stored mainly in the liver, is rapidly broken down to glucose, which is released into blood. Second, body proteins are broken down to amino acids, which are sent to the liver for conversion to glucose that is released into the blood.

Most cells use fats as the main energy source between meals. Fats stored in adipose tissue are broken down into glycerol and fatty acids, which are released into blood. The glycerol can be converted to glucose in the liver; the circulating fatty acids can be used in ATP production.

During a meal, glucose moves into cells, where it can be used for energy and where the excess can be stored.

Between meals, most cells use fat as the main energy source. Stored fats are mobilized, and brain cells are kept supplied with glucose (their major energy source).

Commentary

Case Study: Feasting, Fasting, and Systems Integration

With the possible exception of gastroenterologists, most of us probably would not use the control of digestion as a riveting topic of conversation at a party. Yet every day of your life, you depend absolutely on the integrated functions of those controls.

Suppose, this morning, you are vacationing in the mountains and decide on impulse to follow a forested trail. You fail to notice a wooden trail marker that bears the intriguing name, "Fat Man's Misery." As you walk down the tree-lined corridor, you are enjoying one of the benefits of discontinuous feeding. Having eaten a large breakfast, you have assured your cells of ongoing nourishment; you do not have to forage constantly amongst the ferns as, say, a roundworm must do. Food partly digested in the stomach has already entered the small intestine. Right now, amino acids, simple sugars, and fatty acids are moving across the intestinal wall, then into the bloodstream.

With the surge of nutrients, glucose molecules are entering the bloodstream faster than your cells can use them. The level of blood glucose begins to rise slightly. This is no problem, for your body has a homeostatic program to convert glucose into storage form when it is flooding in, then to release some of the stores when glucose is scarce.

With the rise in blood glucose, pancreatic beta cells are called upon to secrete insulin. Blood concentrations of

insulin rise—and the hormonal targets (liver, fat, and muscle cells) quickly begin using or storing the glucose molecules (Figure 34.14). At the same time, alpha cells are prevented from secreting glucagon—which slows the liver's conversion of stored glycogen into glucose.

As you can see, the liver is central to the storage and interconversion of absorbed carbohydrates, lipids, and proteins. Keep in mind that it has other vital roles (Table 37.6). For example, it helps maintain the concentrations of blood's organic substances and removes many toxic substances from it. The liver inactivates most hormone molecules and sends them on to the kidneys for excretion (in urine). Also, ammonia (NH_3) is produced when cells break down amino acids, and it can be toxic to cells. The circulatory system carries ammonia to the liver, and there it is converted to urea. This is a much less toxic-waste product, and it leaves the body by way of the kidneys, in urine.

Controls Over Metabolism

Both endocrine and neural controls govern metabolism during and between meals. As we saw on page 590, the most important control agents are hormones secreted by the pancreas. One of the hormones, **glucagon**, is secreted in response to a drop in the glucose level in blood between meals. (Between meals, cells use the glucose circulated to them by the bloodstream.) Among other things, glucagon stimulates the conversion of glycogen into glucose and so *increases* the blood glucose level. Another hormone is **insulin**, which is secreted in response to a rise in the blood glucose level after a meal. Among other things, insulin stimulates glucose uptake by cells—and so *decreases* the blood glucose level.

Figure 34.14 illustrates some of the interplays among hormones that help keep the blood glucose level relatively constant, despite great variation in when (and how much) we eat. A discussion of the control mechanisms themselves would be beyond the scope of this book. However, the *Commentary* above will give you a sense of the splendid nature of their interactions.

What is the outcome? High levels of glucose that have entered the circulation from your gut move out of the blood and into cells, where it can be burned as fuel or stored for later use.

Even though you are no longer feeding your body, your brain cells have not lessened their high demands for glucose. Neither have your muscle cells, which are getting a strenuous workout. Little by little, blood glucose levels drop. Now endocrine activities shift in the pancreas. With less glucose binding to them, beta cells decrease their insulin output. With less glucose to inhibit them, alpha cells increase their glucagon output. When glucagon reaches your liver, it causes the conversion of glycogen back to glucose—which is returned to your blood. This prevents blood glucose from falling below levels required to maintain brain function.

Yet the best-laid balance of internal conditions can go astray when external conditions change. In your case, the "miserable" part of the trail has begun. You find yourself scrambling higher and higher on steep inclines. Suddenly you stop, surprised, in pain. You forgot to reckon with the lower oxygen pressure of mountain air, and your leg muscles cramped. Your body has already detected its deficiency of oxygen-carrying red blood cells at this altitude, but it will take days before enough additional red blood cells are available.

In the meantime, your muscle cells are not being supplied with enough oxygen for the prolonged climb. They are relying on an anaerobic pathway in which lactate is the end product (page 129).

Again, systems interact to return conditions to a homeostatic state. Your body detects the reduced oxygen pressure and an accompanying increase in hydrogen ion concentrations in cerebrospinal fluid. Nerve impulses course toward the respiratory center in the medulla. The result: The diaphragm and other muscles associated with inflating and deflating your lungs contract more rapidly. You breathe faster now, and more deeply. In the liver, lactate is converted to glucose—which is returned to the blood.

On checking the sun's position, you see it is well past noon. And guess what: you forgot about lunch. When you start the long walk back, the drop in blood glucose levels triggers new homeostatic control mechanisms.

Under hypothalamic commands, your adrenal medulla begins secreting epinephrine and norepinephrine. Its main targets: the liver, adipose tissue, and muscles. In the liver, glycogen synthesis stops. In body tissues generally, glucose uptake is blocked. In fat cells, fats are converted to fatty acids, which are routed to the liver, muscles, and other tissues as alternative energy sources. For every fatty acid molecule sent down metabolic pathways in those tissues, several glucose molecules are held in reserve for the brain.

You do get back to the start of the trail by sundown. However, your body had enough stored energy to sustain you for many more days, so the situation was never really desperate. The balance of blood sugar and fat is constantly monitored by the liver and controlled by hormones. Glucose levels only drop beyond the set point to stimulate glycogen conversion and fat conversion, and vice versa. It takes several days of fasting before blood sugar levels are markedly reduced.

Even after several days of fasting, your energy supplies would not have run out. Another command from the hypothalamus would have prodded your anterior pituitary into secreting ACTH (page 584). The ACTH would have signaled adrenal cortex cells to secrete glucocorticoid hormones, which have a potent effect on the synthesis of carbohydrates from proteins and on the further breakdown of fat. Slowly, in muscles and other tissues, your body's proteins would have been disassembled. Amino acids from these structural tissues would have been used in the liver to build new glucose molecules—and once more your brain would have been kept active.

As extreme as this last pathway might be, it would be a small price to pay for keeping your brain functional enough to figure out how to take in more nutrients and bring you back to a homeostatic state.

SUMMARY

1. Nutrition includes all the processes by which the body takes in, digests, absorbs, and uses food.

2. A digestive system breaks down food molecules by mechanical, enzymatic, and hormone-assisted means. It also enhances absorption of the breakdown products into the internal environment, and it eliminates the unabsorbed residues.

3. The human digestive system includes the mouth, pharynx, esophagus, stomach, small intestine, large intestine (colon), rectum, and anus. Glands associated with digestion are the salivary glands, liver, gallbladder, and pancreas.

4. Controls over the digestive system operate in response to the volume and composition of food passing through. The response can be a change in muscle activity, the secretion rate of hormones or enzymes, or both.

5. Starch digestion begins in the mouth; protein digestion begins in the stomach. Digestion is completed and most nutrients are absorbed in the small intestine. Following absorption, monosaccharides (including glucose) and most amino acids are sent directly to the liver. Fatty acids and monoglycerides enter lymph vessels and then enter the general circulation.

6. To maintain acceptable weight and overall health, caloric intake must balance energy output. Complex carbohydrates are the body's main energy source. The body produces fats as storage forms of carbohydrates and proteins. Eight essential amino acids, a few essential fatty acids, vitamins, and minerals must be provided by the diet.

Review Questions

1. Study Figure 37.3. Then, on your own, diagram the connections between metabolism and the digestive, circulatory, and respiratory systems. *642*

2. What are the main functions of the stomach? The small intestine? The large intestine? *645, 646, 648*

3. Name four kinds of breakdown products that are actually small enough to be absorbed across the intestinal lining and into the internal environment. *646–648*

4. A glass of milk contains lactose, protein, butterfat, vitamins, and minerals. Explain what happens to each component when it passes through your digestive tract. *645–648, 653*

5. Describe some of the reasons why each of the following is nutritionally important: carbohydrates, proteins, fats, vitamins, and minerals. *648–653*

Self-Quiz *(Answers in Appendix IV)*

1. The _____, _____, _____, and _____, interact in supplying body cells with raw materials, disposing of wastes, and maintaining the volume and composition of extracellular fluid.

2. Different specialized regions of the digestive system function in _____ and _____ food and in _____ unabsorbed food residues.

3. Maintaining good health and normal body weight requires that _____ intake be balanced by _____ output.

4. The main energy sources for the body are complex _____.

5. The human body cannot produce its own vitamins or minerals, and it also cannot produce certain _____ and _____.

6. Which glands are *not* associated with digestion?
   a. salivary glands   d. gallbladder
   b. thymus gland   e. pancreas
   c. liver

7. Digestion is completed and breakdown products are absorbed in the _____.
   a. mouth   c. small intestine
   b. stomach   d. large intestine

8. After absorption, fatty acids and monoglycerides move into the _____.
   a. bloodstream   c. liver
   b. intestinal cells   d. lymph vessels

9. _____ are storage forms of excess carbohydrates and proteins.
   a. Amino acids   c. Fats
   b. Starches   d. Monosaccharides

10. Match each digestive system component with its description.
   _____ liver
   _____ small intestine
   _____ human digestive system
   _____ nutrition
   _____ digestive system controls

   a. begins at mouth, ends at anus
   b. operate in response to food volume and composition
   c. functions are digestion, absorption, use of food
   d. where most digestion is completed
   e. receives monosaccharides and amino acids

Selected Key Terms

absorption *642*
bile *646*
bulk *648*
digestion *642*
digestive system *641*
emulsification *647*
essential amino acid *649*
essential fatty acid *652*
gallbladder *647*
gastrointestinal tract (gut), *643*
large intestine (colon) *648*
liver *655*
micelle *647*
microvillus *647*

mineral *653*
motility *642*
net protein utilization *651*
pancreas *646*
peristalsis *644*
pharynx *645*
ruminant *642*
salivary gland *645*
secretion *642*
segmentation *644*
small intestine *646*
stomach *645*
villus *647*
vitamin *652*

Readings

Campbell-Platt, G. May 1988. "The Food We Eat." *New Scientist*, 19:1–4.

Cohen, L. 1987. "Diet and Cancer." *Scientific American* 257(5): 42–68.

Hamilton, W. 1985. *Nutrition: Concepts and Controversies*. Third edition. Menlo Park, California: West.

Krause, M., and L. Mahan. 1984. *Food, Nutrition, and Diet Therapy*. Seventh edition. Philadelphia: Saunders.

38 CIRCULATION

Heartworks

For Augustus Waller, Jimmie the bulldog was no ordinary pooch. Connected to wires and soaked to his ankles in buckets of salty water, Jimmie was a four-footed window into the workings of the heart.

Feel the pulse of blood that is coursing through the artery at your wrists or the repeated thumpings of your heart at the chest wall. Those same rhythms fascinated Waller and other physiologists of the nineteenth century. They suspected that every heartbeat might produce a characteristic pattern of electrical currents—a pattern that could be recorded painlessly at the body surface. That is where Jimmie and the buckets came in. Saltwater is an efficient conductor of electricity—so efficient that it carried faint signals from Jimmie's beating heart, through the skin of his legs, to a crude monitoring device. With this device, Waller made one of the world's first recordings of a beating heart—an electrocardiogram (Figure 38.1).

Look at the series of peaks of that simple graph. Taken together, they resemble the pattern that shows up in recordings of the electrical activity of your own heart (or of any other vertebrate). In fact, your heart started beating this way within weeks after you began to develop from a fertilized egg. Inside your tiny embryonic body, patches of newly formed cardiac muscle started contracting until one patch took the lead. Ever since that moment, it has been the pacemaker for your heart's activity—and normally it will remain so until the day you die.

Figure 38.1 History in the making—Dr. Augustus Waller's pet bulldog, Jimmie, taking part in a painless experiment (**a**) that yielded one of the world's first electrocardiograms (**b**).

It is the pacemaker that faithfully sets, adjusts, and resets the rate at which blood is pumped from your heart, through a vast network of blood vessels, then back to the heart. Do nothing more than stretch out in a meadow and stare mindlessly at the sky and your heart rate will be moderate, somewhere around 70 beats a minute. Find yourself suddenly confronted by a mean-tempered bull in the meadow, and the demands by your muscles for blood-borne oxygen and glucose will escalate. Then, your heart may start pounding 150 times a minute to deliver sufficient blood to them.

Sometimes injury or disease interrupts the heart's programmed cadence. The heart may race wildly, slow ominously, or alternate sporadically between the two extremes. Blood flow to the body's tissues becomes skewed, so nutrients cannot be delivered and metabolic wastes cannot be removed properly. Each year in the

a

BUCKET

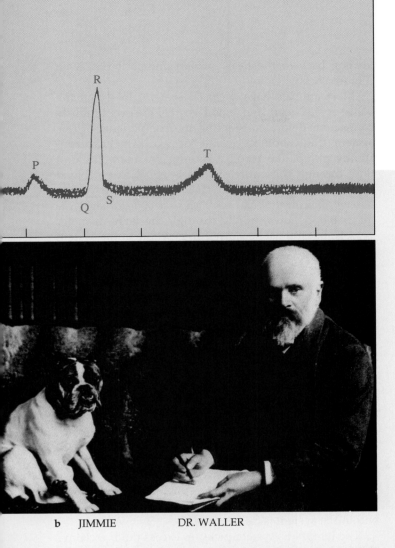

b   JIMMIE   DR. WALLER

United States alone, more than 400,000 individuals face the consequences of disrupted heart function.

We have come a long way from Jimmie and the buckets in our understanding of the patterns and changes in the heart's tempo. Sophisticated external sensors can pick up the faintest of signals. Minute variations that might signal an impending heart attack can now be pinpointed. Hospital computers analyze a patient's beating heart and instantaneously build color images of it on a video screen. Heart surgeons implant artificial pacemakers—small, battery-powered heart regulators that pinch-hit for a malfunctioning one. Even physical fitness buffs attach sensors to their wrists, earlobes, or fingertips and monitor their heart rate as they exercise.

With this chapter, we turn to the circulatory system, the means by which substances are rapidly moved to and from all living cells in animals ranging from worms to bulldogs and other mammals. As you will see, the system is absolutely central to the body's ability to maintain stable operating conditions in the internal environment—a state we call homeostasis.

KEY CONCEPTS

1. Cells survive by exchanging substances with their surroundings. In complex animals, a closed circulatory system allows rapid movement of substances to and from all living cells. Most circulatory systems consist of a heart, blood, and blood vessels, which are supplemented by a lymphatic system.

2. The human body has two circuits for blood flow. In the pulmonary circuit, the heart pumps oxygen-poor blood to the lungs (where it picks up oxygen), then the oxygen-enriched blood flows back to the heart. In the systemic circuit, the heart pumps the oxygen-enriched blood to all body regions. After giving up oxygen in those regions, the blood flows back to the heart. Carbon dioxide, plasma proteins, vitamins, hormones, lipids, and other solutes also make the circuits.

3. Arteries and veins are large-diameter transport tubes. Capillaries and venules are fine-diameter tubes for diffusion. Arterioles, with adjustable diameters, serve as control points for the distribution of different volumes of blood flow to different regions of the body. Metabolically active regions get more of the total volume at a given time.

CIRCULATORY SYSTEM: AN OVERVIEW

Imagine what would happen if an earthquake or flood closed off the highways around your neighborhood. Grocery trucks couldn't enter and waste-disposal trucks couldn't leave—so food supplies would dwindle and garbage would pile up. Every living cell in your body would face similar predicaments if your body's highways were disrupted. Those highways are part of the **circulatory system**, which functions in the rapid internal transport of substances to and from cells. Together with other important organ systems in the body, the circulatory system helps maintain favorable neighborhood conditions, so to speak.

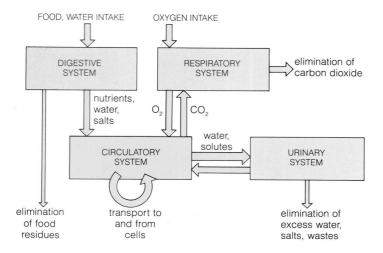

FOOD, WATER INTAKE   OXYGEN INTAKE

DIGESTIVE SYSTEM   RESPIRATORY SYSTEM   elimination of carbon dioxide

nutrients, water, salts

O_2 CO_2

water, solutes

CIRCULATORY SYSTEM   URINARY SYSTEM

elimination of food residues   transport to and from cells   elimination of excess water, salts, wastes

Figure 38.2 The central role of the circulatory system in transporting substances to and from the body's living cells. Together with the other systems shown, it helps maintain favorable operating conditions in the internal environment.

Throughout this unit of the book, we've seen that differentiated cells abound in complex animals. Differentiated cells perform specific tasks with great efficiency, but not without cost. Such cells cannot really fend for themselves. They have few mechanisms for adjusting to drastic changes in the composition, volume, and temperature of the tissue fluid surrounding them, the *interstitial fluid*. To stay alive, they depend on circulating blood to maintain stable operating conditions in their internal environment. Figure 38.2 diagrams the central role of the circulatory system in this task.

Components of the System

In most invertebrates and all vertebrates, the circulatory system has these components:

blood	*a fluid connective tissue composed of water, solutes, and formed elements (blood cells and platelets)*
heart	*a muscular pump that generates the pressure required to help keep blood flowing through the body*
blood vessels	*tubes of different diameters through which blood is transported (for example, arteries and fine capillaries)*

Circulatory systems can be open or closed. Arthropods and most mollusks have an open system, with blood (or bloodlike fluid) pumped from the heart into tubes that dump it into a space or cavity in body tissues. There, blood mingles with tissue fluids. The fluid has nowhere to go except through open-ended tubes

leading back to the heart, which pumps it out again (Figure 38.3a).

Most animals have a closed system, in which the walls of the heart and blood vessels are continuously connected. Before considering the components of this system, think about its overall "design." The heart constantly pumps the blood it receives, so the volume of flow through the entire system is equal to the blood returned to the heart. *Yet the rate and volume of flow through individual blood vessels must be adjusted along the route.* Blood flows rapidly through the large-diameter vessels, but it must be slowed somewhere in the system to allow enough time for substances to diffuse to and from cells. As you will see, blood flow is rather leisurely in *capillary beds*, where it is funneled through vast numbers of small-diameter tubes. As a result, there is sufficient time for diffusion (Figure 38.3b).

Functional Links with the Lymphatic System

Whether open or closed, a circulatory system unfortunately is an ideal highway by which bacteria and other agents of disease can spread through the body. Fortunately, the body can detour the invaders through supplementary highways that are part of the **lymphatic system**. This system consists of a network of tubes (lymph vessels) as well as structures and organs that house vast numbers of infection-fighting cells. The fluid within the system is called *lymph*.

The lymphatic system picks up excess fluid, reclaimable solutes, *and disease agents* from interstitial fluid—and runs them past its armies of infection-fighting cells. The system then returns the cleansed fluid to the circulatory system.

The lymphatic system works to prevent the spread of disease agents through the circulatory system. It also reclaims water and solutes from interstitial fluid.

CHARACTERISTICS OF BLOOD

Functions of Blood

Blood is classified as a connective tissue, and it serves multiple functions. It carries oxygen as well as nutrients to cells, and it carries away secretions (including hormones) and metabolic wastes. Phagocytic cells travel the blood highways as mobile scavengers and infection fighters. Blood helps stabilize internal pH. In birds and mammals, blood also helps equalize body temperature

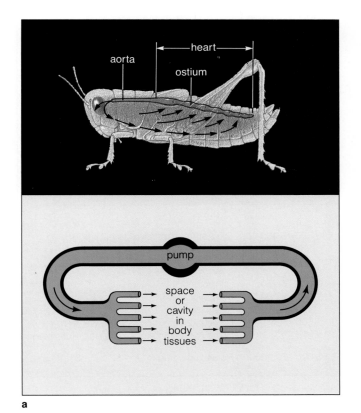

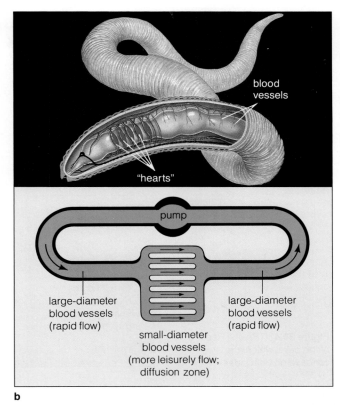

Figure 38.3 (a) Fluid flow through an open circulatory system. A grasshopper, for example, has a "heart" that pumps blood through a vessel (aorta), which dumps the blood into body tissues. Blood diffuses through the tissues, then back into the heart through openings (ostia). (b) Fluid flow through a closed circulatory system. An example is the earthworm, with blood vessels leading away from and back to several muscular "hearts" near its head end. The walls of the hearts and blood vessels interconnect.

by carrying excess heat from regions of high metabolic activity (such as skeletal muscles) to the skin, where the heat can be dissipated.

In most animals, blood is a circulating fluid that carries raw materials to cells, carries products and wastes from them, and helps maintain an internal environment that is favorable for cell activities.

Blood Volume and Composition

The volume of blood in a human individual depends on body size and on the concentrations of water and solutes. For average-sized adults, blood volume generally is about 6 to 8 percent of the body weight. That amounts to about 4 to 5 quarts.

Blood is a rather sticky, viscous fluid, thicker than water and more slow-flowing. As for all vertebrates, human blood consists of plasma, red blood cells, white blood cells, and platelets. When you place a sample of blood in a test tube and prevent it from clotting, it separates into a layer of straw-colored liquid (the plasma) that floats over the red-colored cellular portion of blood. Normally, the plasma portion accounts for 50 to 60 percent of the total blood volume.

Plasma. The portion of blood called plasma is mostly water. Besides serving as a transport medium for the cellular components of blood, the water functions as a solvent for various molecules. Among these molecules are hundreds of different *plasma proteins*, including albumins, globulins, and fibrinogen. The concentration of plasma proteins influences the distribution of water between blood and interstitial fluid. Therefore, it also influences blood's fluid volume. Albumin is important in this water-balancing act, for it represents as much as 60 percent of the total amount of plasma proteins. Some alpha and beta globulins transport lipids and fat-soluble vitamins. As you will see later in this chapter and the next, gamma globulins function in immune responses, and fibrinogen serves in blood clotting.

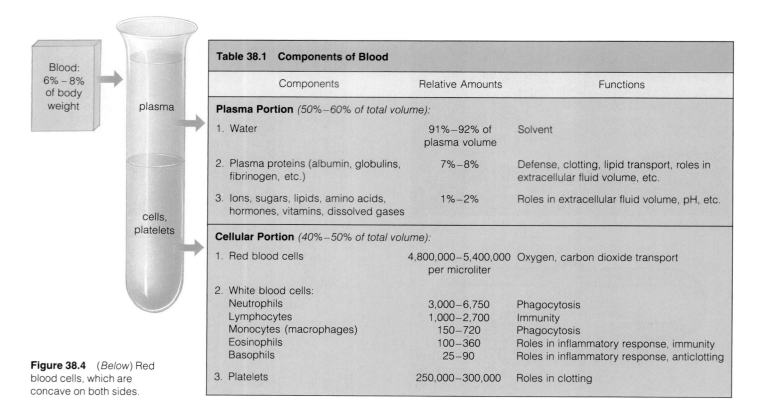

Blood:
6% – 8%
of body
weight

plasma

cells,
platelets

Figure 38.4 (*Below*) Red blood cells, which are concave on both sides.

Table 38.1	Components of Blood		
	Components	Relative Amounts	Functions
Plasma Portion *(50%–60% of total volume):*			
1. Water		91%–92% of plasma volume	Solvent
2. Plasma proteins (albumin, globulins, fibrinogen, etc.)		7%–8%	Defense, clotting, lipid transport, roles in extracellular fluid volume, etc.
3. Ions, sugars, lipids, amino acids, hormones, vitamins, dissolved gases		1%–2%	Roles in extracellular fluid volume, pH, etc.
Cellular Portion *(40%–50% of total volume):*			
1. Red blood cells		4,800,000–5,400,000 per microliter	Oxygen, carbon dioxide transport
2. White blood cells:			
Neutrophils		3,000–6,750	Phagocytosis
Lymphocytes		1,000–2,700	Immunity
Monocytes (macrophages)		150–720	Phagocytosis
Eosinophils		100–360	Roles in inflammatory response, immunity
Basophils		25–90	Roles in inflammatory response, anticlotting
3. Platelets		250,000–300,000	Roles in clotting

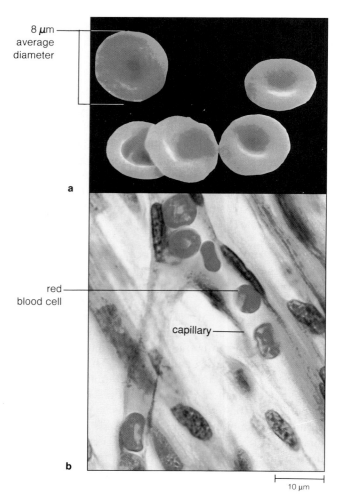

8 μm
average
diameter

a

red
blood cell

capillary

b

10 μm

Plasma also contains ions, glucose and other simple sugars, lipids, amino acids, vitamins, hormones, and dissolved gases—mostly oxygen, carbon dioxide, and nitrogen (Table 38.1). The ions help maintain extracellular pH and fluid volume. The lipids include fats, phospholipids, and cholesterol. Lipids being transported from the liver to different regions generally are bound with proteins, forming lipoproteins.

Red Blood Cells. Erythrocytes, or **red blood cells**, transport oxygen to cells. Your own red blood cells are shaped like doughnuts without the hole (Figure 38.4). Their red color comes from hemoglobin, the iron-containing protein described on page 44. When oxygen from your lungs diffuses into the bloodstream, it binds with hemoglobin. Oxygenated blood is bright red. Blood somewhat depleted of oxygen is darker red but appears blue when observed through blood vessel walls. (That is why veins close to the body surface, as near our wrists, look "blue.") Hemoglobin also transports some of the carbon dioxide wastes of aerobic metabolism.

As Figure 38.5 shows, red blood cells form in red bone marrow. Like white blood cells and platelets, they are derived from stem cells. (By definition, *stem cells* retain the capacity to divide and give rise to different populations of cells.) Mature red blood cells no longer have their nucleus, but they also no longer require its protein-synthesizing instructions. They have enough enzymes and other proteins to remain functional for about 120 days.

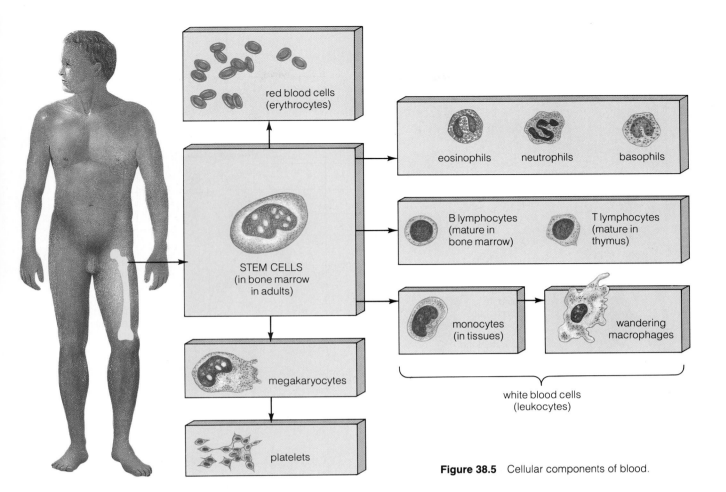

red blood cells
(erythrocytes)

eosinophils   neutrophils   basophils

B lymphocytes
(mature in
bone marrow)

T lymphocytes
(mature in
thymus)

STEM CELLS
(in bone marrow
in adults)

monocytes
(in tissues)

wandering
macrophages

megakaryocytes

white blood cells
(leukocytes)

platelets

Figure 38.5 Cellular components of blood.

Phagocytic cells continually remove the oldest red blood cells from the bloodstream, but ongoing replacements keep the red blood cell count fairly stable. A *cell count* is the number of cells of a given type in a microliter of blood. The red blood cell count is 5.4 million in males and 4.8 million in females, on the average.

Feedback mechanisms help stabilize the red blood cell count. Suppose you have just started vacationing in the Swiss Alps. Because air is "thinner" at high altitudes, your body must work harder to obtain the required oxygen. First your kidneys secrete an enzyme that converts a plasma protein into a hormone (erythropoietin). Then the hormone stimulates an increase in red blood cell production in red bone marrow. New oxygen-carrying red blood cells enter the bloodstream, and within a few days there is a rise in the oxygen level in your tissues. Information about the increase is fed back to the kidneys, production of the hormone dwindles, and the production of red blood cells drops accordingly.

White Blood Cells. Leukocytes, or **white blood cells**, function in day-to-day housekeeping and defense. These are the cells that scavenge dead or worn-out cells in the body and that respond to tissue damage or invasion by bacteria, viruses, and other foreign agents. All white blood cells arise from stem cells in bone marrow (Figure 38.5). They travel the circulation highways, but they perform most housekeeping and defense functions after they squeeze out of blood capillaries and enter tissues.

There are five types of white blood cells, based on differences in size, nuclear shape, and staining traits. They are lymphocytes, monocytes, neutrophils, eosinophils, and basophils (Table 38.1 and Figure 38.5). Two major classes of lymphocytes, the "B cells" and "T cells," are central to immune responses, which are described in the next chapter.

Neutrophils and mature monocytes are "search-and-destroy" cells of the body's immune system. Monocytes follow chemical trails to inflamed tissues. There, they differentiate into wandering macrophages ("big eaters") that engulf invaders and cellular debris.

White blood cell counts vary, depending on whether the body is highly active, healthy, or under siege. Table 38.1 shows the general range of counts for each type of white blood cell. Their vast numbers and continual replacement testify to the fact that the human body is an inviting environment for a great variety of bacteria, viruses, fungi, and protozoans. At any moment, some of those foreign agents can penetrate the body by way of the mouth, nose, skin pores, or injuries. A phago-

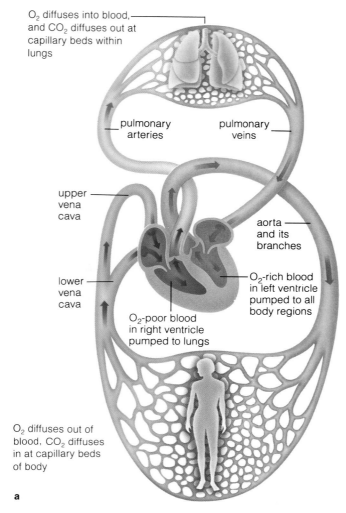

O$_2$ diffuses into blood, and CO$_2$ diffuses out at capillary beds within lungs

pulmonary arteries

pulmonary veins

upper vena cava

lower vena cava

O$_2$-poor blood in right ventricle pumped to lungs

aorta and its branches

O$_2$-rich blood in left ventricle pumped to all body regions

O$_2$ diffuses out of blood, CO$_2$ diffuses in at capillary beds of body

a

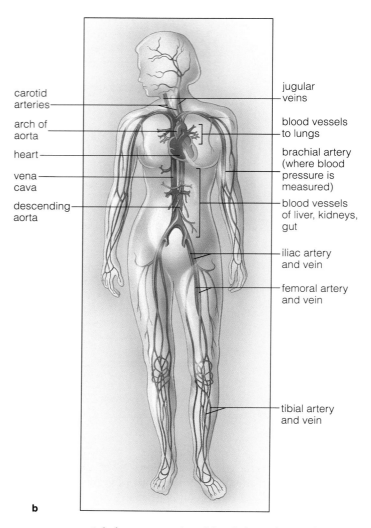

carotid arteries

arch of aorta

heart

vena cava

descending aorta

jugular veins

blood vessels to lungs

brachial artery (where blood pressure is measured)

blood vessels of liver, kidneys, gut

iliac artery and vein

femoral artery and vein

tibial artery and vein

b

Figure 38.6 (**a**) Diagram of the pulmonary and systemic circuits. The right half of the heart pumps blood through the pulmonary circuit; the left half, through the systemic circuit. (**b**) Human circulatory system, showing the locations of the heart and some major blood vessels.

cytic white blood cell may engulf so much foreign material that its own metabolism becomes disrupted. Life for most white blood cells is challenging and short, typically measured in days or, during a major battle, a few hours.

Platelets. In bone marrow, some stem cells develop into "giant" cells (megakaryocytes). Those cells shed fragments of cytoplasm, which become enclosed in a bit of their plasma membrane. The fragments, called **platelets**, are oval or rounded disks about 2 to 4 micrometers across. They have no nucleus.

Each platelet lasts only five to nine days, but hundreds of thousands are always circulating in blood. They

are essential for preventing blood loss from slightly damaged blood vessels. Substances released from platelets initiate a chain of reactions leading to blood clotting, as will be described shortly.

CARDIOVASCULAR SYSTEM OF VERTEBRATES

"Cardiovascular" comes from the Greek *kardia*, meaning heart, and the Latin *vasculum*, meaning vessel. In the cardiovascular system of all vertebrates, a heart pumps blood into large-diameter arteries. From there the blood flows into small, muscular arterioles, which branch into tiny capillaries. Blood flows from capillaries into small venules, then into large veins. The veins return blood to the heart. Figure 26.00 is a general picture of the system in fishes, amphibians, birds, and mammals. Figure 38.6 shows where some of its major components are located in the human body.

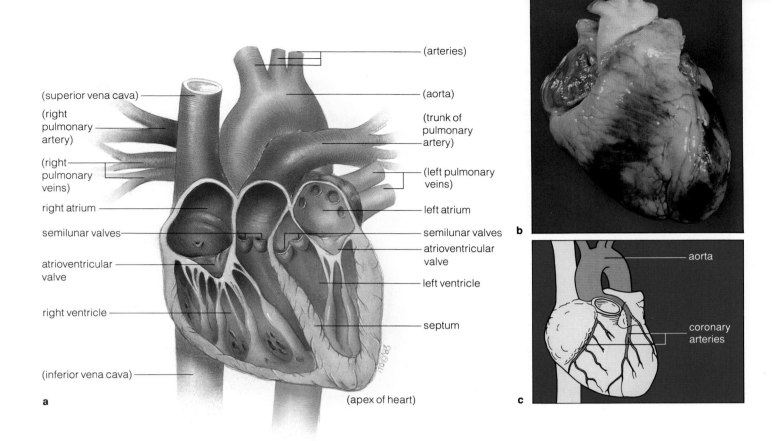

Labels on figure (a):
(superior vena cava)
(right pulmonary artery)
(right pulmonary veins)
right atrium
semilunar valves
atrioventricular valve
right ventricle
(inferior vena cava)
a
(arteries)
(aorta)
(trunk of pulmonary artery)
(left pulmonary veins)
left atrium
semilunar valves
atrioventricular valve
left ventricle
septum
(apex of heart)

Labels on figure (b): b

Labels on figure (c):
aorta
coronary arteries
c

Figure 38.7 The human heart, partial view of the interior (**a**) and external view (**b**). Location of the coronary arteries (**c**).

Blood Circulation Routes in Humans

The human heart is divided into two halves, which are the basis of two cardiovascular circuits through the body. These are the pulmonary and systemic circuits.

In the **pulmonary circuit**, blood from the right half of the heart is pumped to the lungs, where it picks up oxygen and gives up carbon dioxide. From the lungs, the freshly oxygenated blood flows to the left half of the heart. In the **systemic circuit**, the oxygen-enriched blood is pumped through the rest of the body. After passing through tissue regions (where oxygen is used and carbon dioxide is produced), the blood flows to the right half of the heart. In both circuits, blood travels through arteries, arterioles, capillaries, venules, and finally veins.

As Figure 38.6a suggests, a given volume of blood making either circuit generally passes through only one capillary bed. A notable exception is the blood passing through capillary beds in the digestive tract, where it picks up glucose and other substances absorbed from food. That blood moves on through another capillary bed, in the liver—an organ with a key role in nutrition. The slow flow of blood through this second bed gives the liver time to metabolize absorbed substances.

The Human Heart

Heart Structure. During a seventy-year life span, the human heart beats some 2.5 billion times, and it rests only briefly between heartbeats. The heart's structure reflects its role as a durable pump. The heart is mostly cardiac muscle tissue surrounded and protected by a tough, fibrous sac (pericardium). Its inner chambers have a smooth lining (endocardium) composed of connective tissue and a single layer of epithelial cells. The epithelial cell layer is known as *endothelium*. It lines the inside of blood vessels as well as the heart.

Each half of the heart has two chambers—an **atrium** (plural, atria) and a **ventricle**. The flaps of membrane separating the two chambers serve as a one-way valve. As Figure 38.7 shows, the flaps are called an *AV valve* (short for "atrioventricular"). Each half of the heart also has a *semilunar valve*, located between the ventricle and the arteries leading away from it. Each time the heart beats, its valves open and close in ways that prevent backflow and keep blood moving in one direction.

Most cardiac muscle cells are not serviced by the blood moving inside the heart's chambers. The heart has its own "coronary circulation," with two main arteries leading into a large capillary bed (Figure 38.7c). Coronary arteries are the first to branch off the *aorta*, the major artery carrying oxygen-enriched blood away from the heart. Given their small diameter, coronary arteries can become clogged during some cardiovascular disorders (see the *Commentary* on page 671).

Cardiac Cycle. Each time the heart beats, its four chambers go through phases of contraction (systole) and relaxation (diastole). This sequence of muscle contraction and relaxation is a **cardiac cycle** (Figure 38.8).

When the relaxed atria are filling, fluid pressure inside them increases and forces the AV valves to open. Blood flows into the ventricles, which become completely filled when the atria contract. As the filled ventricles begin to contract, fluid pressure inside them increases,

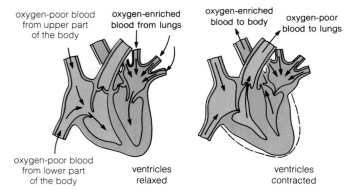

oxygen-poor blood from upper part of the body

oxygen-enriched blood from lungs

oxygen-enriched blood to body

oxygen-poor blood to lungs

oxygen-poor blood from lower part of the body

ventricles relaxed

ventricles contracted

Figure 38.8 Blood flow through the heart during part of a cardiac cycle.

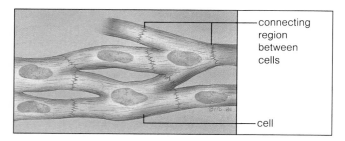

connecting region between cells

cell

Figure 38.9 End-to-end regions between cardiac muscle cells. Communication junctions at these regions permit rapid signaling between cells that causes them to contract nearly in unison.

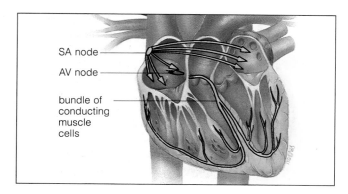

SA node

AV node

bundle of conducting muscle cells

Figure 38.10 Location of cardiac muscle cells that conduct signals for contraction through the heart.

and the AV valves snap shut. As the ventricles continue contracting, ventricular pressure rises sharply above that in blood vessels leading away from the heart. The semilunar valves are forced open by the increased pressure and blood flows out of the heart, into the aorta and pulmonary artery. After blood has been ejected, the ventricles relax, the semilunar valves close—and the already filling atria are ready to repeat the cycle.

Look back at the recording of Jimmie the bulldog's heartbeat (Figure 38.1). The atria of his heart were contracting between times P and Q. The ventricles began contracting forcefully through time QRS and were recovering from contraction at time T.

The blood and heart movements during the cardiac cycle generate vibrations that produce a "lub-dup" sound. You can hear the sound through a stethoscope positioned against the chest wall. At each "lub," the AV valves are closing as the ventricles contract. At each "dup" the semilunar valves are closing as the ventricles relax.

During a cardiac cycle, atrial contraction simply helps fill the ventricles. *Ventricular contraction* is the driving force for blood circulation.

Mechanisms of Contraction. In skeletal muscle tissue, the ends of individual cells are attached to bones. In cardiac muscle tissue, however, the cells branch and then connect with one another at their endings (Figure 38.9). Communication junctions occur where the plasma membranes of abutting cells are joined together. With each heartbeat, signals calling for contraction spread so rapidly across the junctions that cardiac muscle cells contract together as if they were a single unit.

Cardiac and skeletal muscle cells differ in another way. Signals from the nervous system bring about skeletal muscle contraction. But the nervous system can only *adjust* the rate and strength of cardiac muscle contraction. Even if all nerves leading to the heart are severed, the heart will keep on beating! How? Some cardiac muscle cells are self-excitatory; they produce and conduct the action potentials that initiate contraction. These cells are the basis of the **cardiac conduction system** shown in Figure 38.10.

Excitation begins with conducting cells in the *SA node* (short for "sinoatrial"), where major veins enter the right atrium. The SA node generates one wave of excitation after another, usually seventy or eighty times a minute. Each wave spreads over both atria, causes them to contract, then reaches the *AV node* (again, for "atrioventricular"). The wave spreads more slowly here. The delay gives the atria time to finish contracting before the wave of excitation spreads over the ventricles.

Although all cells of the system are self-excitatory, the SA node fires off action potentials at the highest frequency and comes to threshold first in each cardiac cycle. Thus, the SA node is the *cardiac pacemaker*, mentioned at the start of this chapter. Its rhythmic firing is the basis for the normal rate of heartbeat.

The SA node is the cardiac pacemaker. Its spontaneous, repetitive excitation spreads along a system of muscle cells that stimulate contractile tissue in the atria, then the ventricles, in a rhythmic cycle.

Blood Pressure in the Vascular System

Blood pressure, the fluid pressure generated by heart contractions, is not the same throughout the circulatory system. Pressure normally is high to begin with in the aorta, then drops along the circuit away from and back to the heart. The pressure drops result from the loss of energy used to overcome resistance to flow as blood moves through blood vessels of the sort shown in Figure 38.11.

Arterial Blood Pressure.   As we have seen, **arteries** conduct oxygen-poor blood into the lungs and oxygen-enriched blood to all body tissues. Arteries are pressure reservoirs that can "smooth out" the pulsations in blood pressure that are generated during each cardiac cycle. The thick, muscular wall of arteries bulges somewhat under the pressure surge caused by ventricular contraction, then the wall recoils and forces blood onward. With their relatively large diameters, arteries present little resistance to flow, so pressure does not drop much in the arterial portion of the blood circuits. Figure 38.12 shows how blood pressure is measured at large arteries of the upper arms.

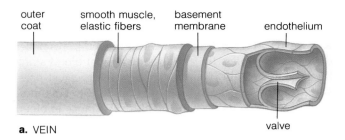

a. VEIN

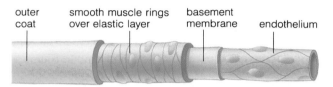

b. ARTERY

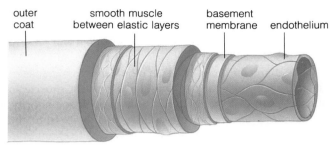

c. ARTERIOLE

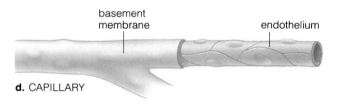

d. CAPILLARY

Figure 38.11   Structure of blood vessels. The basement membrane is collagen-containing connective tissue.

Figure 38.12   Measuring blood pressure with a device called a sphygmomanometer. A hollow cuff, attached to a pressure gauge, is wrapped around the upper arm and inflated with air to a pressure above the highest pressure of the cardiac cycle (at systole, when the ventricles contract). Above the systolic pressure, no sounds can be heard through a stethoscope positioned above the artery (because no blood is flowing through it).

Air in the cuff is slowly released, allowing some blood to flow into the artery. The turbulent flow causes soft tapping sounds, and when this first occurs, the value on the gauge is the systolic pressure—about 120mm Hg in young adults at rest. (This means the measured pressure would make a column of mercury rise a distance of 120 millimeters.)

More air is released until the sounds become dull and muffled. Just after this occurrence, blood flow is continuous; the turbulence and tapping sounds stop. The silence corresponds to the diastolic pressure (at the end of a cardiac cycle, just before the heart pumps out blood again). Generally the reading is about 80mm Hg. In this example, the *pulse pressure* (the difference between the highest and lowest pressure readings) is 120 − 80, or 40mm Hg.

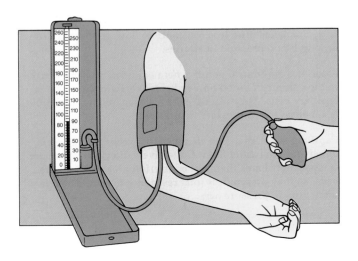

Resistance at Arterioles. Arteries branch into smaller diameter **arterioles**. By tracking the flow of blood along the systemic circuit, you can see that the greatest pressure drop occurs at arterioles (Figure 38.13). This indicates that arterioles offer the greatest resistance to blood flow in the circulation. By analogy, suppose you turn two open bottles of ketchup upside down. Both hold the same amount of ketchup, but one bottle has a very wide neck and the other has a narrow neck. Guess which bottle will be slowest to drain—that is, which will present more resistance to flow.

The diameter of arterioles can be made to increase or decrease by mechanisms that cause smooth muscle in their wall to contract or relax. As you will see shortly, adjustments are made in response to signals from the nervous and endocrine systems, as well as to changes in local chemical conditions. Blood is directed to a region of great metabolic activity when the diameters of arterioles in those regions enlarge. Blood is directed away from less active regions when the diameters of arterioles in those regions constrict. Thus, the more active the cells of a given region, the greater the blood flow to them.

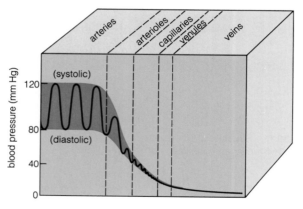

Figure 38.13 Drops in blood pressure in the systemic circulation.

Capillary Function. Capillary beds are *diffusion zones* for exchanges between blood and interstitial fluid. Except for venules, all blood vessels are transport tubes. A **capillary** has the thinnest wall of any blood vessel. Its wall consists of a layer of flat endothelial cells, separated from one another by narrow clefts. The layer rests on a basement membrane (Figure 38.11).

Capillaries have such a small diameter, red blood cells must squeeze through them single file (Figure 38.3). Each capillary presents high resistance to blood flow. Yet there are so many of them in a capillary bed, their combined diameters are greater than the diameters of arterioles leading into them. Hence they present less *total* resistance to flow than arterioles. As a result, the drop in blood pressure here is not as great.

Capillaries thread through nearly every tissue in the body, coming within 0.01 centimeter of every living cell. Most solutes, including oxygen and carbon dioxide, move across the capillary wall by diffusion. Some proteins cross it by endocytosis or exocytosis (page 86), and certain ions probably pass through the clefts between endothelial cells. The clefts are wider in some capillary beds than in others, and those beds are functionally more "leaky" to solutes.

Finally, fluid also moves across the wall by bulk flow, with its water molecules being forced under pressure to move in the same direction. Such movements help maintain the proper fluid balance between the bloodstream and the surrounding tissues. This fluid distribution is important because blood pressure is maintained only when there is an adequate blood volume. Interstitial fluid is a reservoir that is tapped when blood volume drops to the point where there is a decrease in blood pressure, as during hemorrhage.

Figure 38.14 describes the two opposing forces (filtration and absorption) that bring about the fluid movements in capillary beds.

Arteries are pressure reservoirs that keep blood flowing away from the heart while the ventricles are relaxing. Their large diameters offer low resistance to flow, so there is little drop in blood pressure in arteries.

Arterioles are control points where adjustments can be made in the volume of blood flow to be delivered to different capillary beds. They offer great resistance to flow, so there is a major drop in pressure in arterioles.

Capillary beds are diffusion zones for exchanges between blood and interstitial fluid. Collectively, they have a greater cross-sectional area than that of arterioles leading into the beds, so they present less total resistance to flow. There is some drop in pressure here.

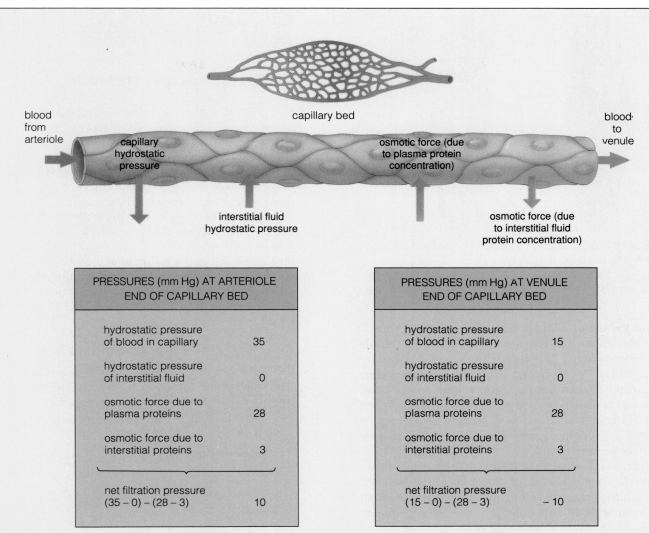

Figure 38.14 Fluid movements in an idealized capillary bed. The movements play no significant role in diffusion. But they are important in maintaining the distribution of extracellular fluid between the bloodstream and interstitial fluid. The movements result from two opposing forces, called *filtration* and *absorption*.

At the arteriole end of a capillary, the difference between capillary blood pressure and interstitial fluid pressure leads to filtration. Because of the difference, some plasma (but very few plasma proteins) leaves the capillary. "Filtration" refers to this fluid movement out of the capillary.

In contrast, "absorption" refers to the movement of some interstitial fluid into the capillary. The difference in water concentration between plasma and interstitial fluid brings it about. Plasma has a greater solute concentration, with its protein components, and therefore a lower water concentration.

Fluid filtration at the arteriole end of a capillary bed tends to be balanced by absorption at the venule end. Normally, there is only a small *net* filtration of fluid, which the lymphatic system returns to the blood.

Edema is a condition in which excess fluid accumulates in interstitial spaces. This happens to some extent during exercise. As arterioles dilate in local tissue regions, capillary pressure increases and triggers increased filtration. Edema also results from an obstructed vein or from heart failure. Here again, capillary pressure and then filtration increase. Edema reaches its extreme during elephantiasis, which is brought on by a roundworm infection and subsequent obstruction of lymphatic vessels (page 417).

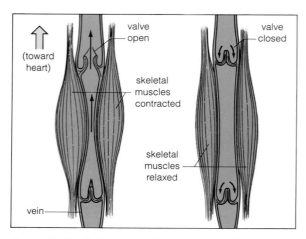

Figure 38.15 Role of skeletal muscle contractions and venous valves in returning blood to the heart.

Venous Pressure.   Capillaries merge into "little veins," or **venules**. Some diffusion also occurs across the venule wall, which is only a little thicker than that of a capillary (Figure 38.11). Venules merge into large-diameter **veins**, the transport tubes leading back to the heart. Blood movement is assisted by valves inside the veins. When blood starts moving backward because of gravity, it pushes the valves into a closed position. In this manner the valves prevent backflow.

Veins also serve as blood volume reservoirs, for they contain 50 to 60 percent of the total blood volume. Although the vein wall is thin and can bulge more than an arterial wall, it does contain some smooth muscle (Figure 38.11). When body activities increase, blood must circulate faster. The smooth muscle cells in vein walls contract; the walls stiffen and don't bulge as much. Venous pressure rises and drives more blood to the heart. The increased blood volume returned to the heart is now ejected and flows to active tissue regions.

Venous pressure rises when limbs move and skeletal muscles bulge against adjacent veins (Figure 38.15). It also is influenced by how rapidly you are breathing. When inhaled air pushes down on internal organs, it changes the pressure gradient between the heart and veins.

Venules overlap with capillaries in function; they afford some control over capillary pressure.

Veins are highly distensible blood volume reservoirs and help adjust flow volume back to the heart. They offer only low resistance to flow, so there is little drop in blood pressure here.

Commentary

On Cardiovascular Disorders

More than 40 million Americans have cardiovascular disorders, which claim about 750,000 lives every year. The most common cardiovascular disorders are *hypertension* (sustained high blood pressure) and *atherosclerosis* (a progressive narrowing of the arterial lumen). They are the major causes of most *heart attacks*—that is, the damage or death of heart muscle due to an interruption of its blood supply. (They also can cause *stroke*, or damage to the brain due to an interruption of blood circulation to it.)

Most heart attacks bring a "crushing" pain behind the breastbone that lasts a half-hour or more. Frequently, the pain radiates into the left arm, shoulder, or neck. The pain can be mild but usually is excruciating. Often it is accompanied by sweating, nausea, vomiting, and dizziness or loss of consciousness.

Risk Factors in Cardiovascular Disorders

Cardiovascular disorders are the leading cause of death in the United States. Curiously, many factors associated with those disorders have been identified *and are controllable.* These are the known risk factors:

1. High level of cholesterol in the blood
2. High blood pressure
3. Obesity (page 649)
4. Lack of regular exercise
5. Smoking (page 710)
6. Diabetes mellitus (page 590)
7. Genetic predisposition to heart disorders
8. Age (the older you get, the greater the risk)
9. Gender (until age fifty, males are at much greater risk than are females)

The last four factors obviously cannot be avoided; but in most people, the first five can be. The risk associated with all five can be minimized simply by watching your diet, exercising, and not smoking.

For example, the fatter you become, the more your body develops additional blood capillaries to service the increased number of cells, and the harder the heart has to work to pump blood through the increasingly divided vascular circuit. As another example, the nicotine in tobacco stimulates the adrenal glands to secrete epinephrine, which constricts blood vessels and so triggers an accelerated heartbeat and a rise in blood pressure. The carbon monoxide present in cigarette smoke has a greater affinity for binding sites on hemoglobin than does carbon dioxide—and its action means that the heart has to pump harder to rid the body of carbon dioxide wastes. In short, smoking can destroy not only your lungs but also your heart.

Some examples of the tissue destruction resulting from cardiovascular disorders will now be described.

Hypertension

Hypertension arises through a gradual increase in resistance to blood flow through the small arteries; eventually, blood pressure is sustained at elevated levels even when the person is at rest. Heredity may be a factor here (the disorder tends to run in families). Diet also is a factor; for example, high salt intake can raise the blood pressure in persons predisposed to the disorder. High blood pressure increases the workload of the heart, which in time can become enlarged and fail to pump blood effectively. High blood pressure also can cause arterial walls to "harden" and so influence the delivery of oxygen to the brain, heart, and other vital organs.

Hypertension has been called the silent killer because affected persons may show no outward symptoms; they often believe they are in the best of health. Even when their high blood pressure has been detected, some hypertensive persons tend to resist medication, corrective changes in diet, and regular exercise. Of 23 million Americans who are hypertensive, most are not undergoing treatment. About 180,000 will die each year.

Atherosclerosis

"Arteriosclerosis" refers to a condition in which arteries thicken and lose their elasticity. In atherosclerosis, conditions worsen because lipid deposits also build up in the arterial walls and shrink the diameter of the arterial lumen. How does this occur?

Recall that lipids such as fats and cholesterol are insoluble in water (page 40). Lipids absorbed from the digestive tract are picked up by lymph vessels that empty into the bloodstream. There, the lipids become bound to protein carriers that keep them suspended in the blood plasma. In atherosclerosis, abnormal smooth muscle cells have multiplied and connective tissue components have increased in arterial walls. Lipids have been deposited within cells and extracellular spaces of the wall's endothelial lining. Calcium salts have been deposited on top of the lipids, and a fibrous net has formed over the whole mass. This *atherosclerotic plaque* sticks out into the lumen of the artery (Figures *a,b*).

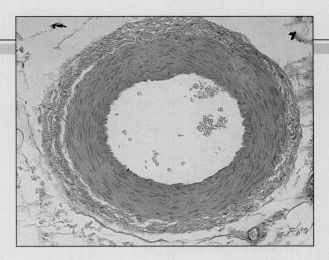

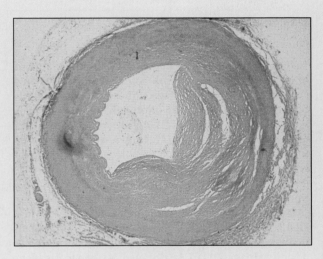

a Cross-section of a normal artery (above) and a partially obstructed one (below).

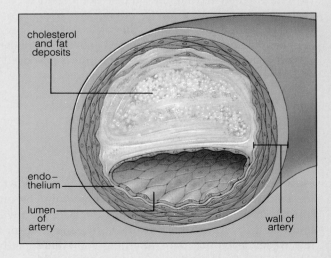

cholesterol and fat deposits

endo- thelium

lumen of artery

wall of artery

b Diagram of an atherosclerotic plaque.

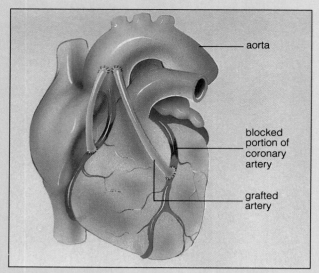

c Two coronary bypasses (green).

With their narrow diameter, the coronary arteries and their branches (Figure *c*) are extremely susceptible to clogging through plaque formation or occlusion by a clot. When such an artery becomes narrowed to one-quarter of its former diameter, the resulting symptoms can range from mild chest pain (angina pectoris) to a full-scale heart attack.

Atherosclerosis can be diagnosed on the basis of several procedures. These include stress electrocardiograms, or EKGs (recording the electrical activity of the cardiac cycle while a person is exercising on a treadmill) and *angiography* (injecting a dye that will stain plaques and then taking x-rays of the arteries; see page 42). Treatments of serious blockages include *coronary bypass surgery*. During this operation, a section of a vein taken from the arm or leg is stitched to the aorta and to the coronary artery below the narrowed or blocked region (Figure *c*). In another technique, called *laser angioplasty*, highly focused laser beams are used to vaporize the atherosclerotic plaques. *Balloon angioplasty* is more common. Here, a small balloon is inflated within a blocked artery to flatten a plaque and thereby increase the arterial diameter. All such procedures do not cure the underlying cardiovascular problem. They only buy time for the individual.

Arrhythmia

Arrhythmias are irregular or abnormal heart rhythms. They can be detected by an EKG (Figure *d*). Some arrhythmias are normal. For example, the resting cardiac rate of many athletes who are trained for endurance is lower than average, a condition called *bradycardia*. Inhibition of their cardiac pacemaker by parasympathetic signals has increased as an adaptive response to ongoing strenuous exercise. A cardiac rate above 100 beats per minute (*tachycardia*) also occurs normally during exercise or stressful situations.

Serious tachycardia can be triggered by drugs (including caffeine, nicotine, and alcohol), hyperthyroidism, and other factors. Coronary artery disease also can cause arrhythmias.

A coronary occlusion and certain other disorders also may cause abnormal rhythms that can degenerate rapidly into a dangerous condition called *ventricular fibrillation*.

Plaque formation is related to cholesterol intake, but other factors are also at work here. For example, when cholesterol is transported through the bloodstream, it is bound to one of two kinds of protein carrier molecules: high-density lipoproteins (HDL) and low-density lipoproteins (LDL). High levels of LDL are related to a tendency toward heart trouble. LDLs, with their cholesterol cargo, have a penchant for infiltrating arterial walls. In contrast, the HDLs seem to attract cholesterol out of the walls and transport it to the liver, where it can be metabolized. (Atherosclerosis is uncommon in rats; rats have mostly HDLs. It is common in humans, who have mostly LDLs.) In addition, it appears that unsaturated fats, including olive oil and fish oil, can reduce the level of LDLs in the blood.

Sometimes platelets become caught on the rough edges of plaque and are stimulated into secreting some of their chemicals. When they do, they initiate clot formation. As the clot and plaque grow, the artery can become narrowed or blocked. Blood flow to the tissue that the artery supplies diminishes or may be blocked entirely. A clot that stays in place is called a *thrombus*. If it becomes dislodged and travels the bloodstream, it is called an *embolus*.

Controls Over Blood Flow

Maintaining Blood Pressure. Suppose you decide to measure your blood pressure every day while you are resting, in the manner shown in Figure 38.12. You will find that the resting value remains fairly constant over a few weeks, even months. Assuming you are a healthy adult, it will be somewhere around 120/80mm Hg.

This raises an interesting question. What mechanisms work to maintain the level of blood pressure over time and so ensure adequate blood flow to all regions of the body?

In the medulla oblongata of the brain are integrating centers that control blood pressure. The centers integrate information coming in from sensory receptors in cardiac muscle tissue and in certain arteries, such as the aorta and the carotid arteries in the neck. They use this information to coordinate the rate and strength of heartbeats with changes in the diameter of arterioles and, to some extent, of veins.

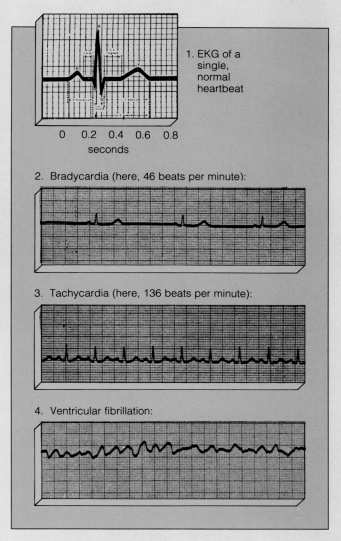

1. EKG of a single, normal heartbeat

0 0.2 0.4 0.6 0.8
seconds

2. Bradycardia (here, 46 beats per minute):

3. Tachycardia (here, 136 beats per minute):

4. Ventricular fibrillation:

d Examples of EKG readings.

Here, cardiac muscle in different parts of the ventricles contracts haphazardly, and the ventricles are unable to pump blood. Loss of consciousness occurs within a few seconds and may signify impending death. Sometimes a strong electric shock delivered to the chest can stop the fibrillation and may restore normal cardiac function.

When an *increase* in blood pressure is detected, the medulla commands the heart to beat more slowly and contract less forcefully. It also sends signals to smooth muscle cells in the wall of arterioles. The cells relax, the outcome being **vasodilation**—an enlargement (dilation) of arteriole diameter.

Conversely, when a *decrease* in blood pressure is detected, the medulla commands the heart to beat faster and contract more forcefully. It also stimulates the smooth muscle cells of arterioles to contract. In this case, the outcome is **vasoconstriction**—a decrease in arteriole diameter.

Hormones assist in maintaining blood pressure. Arterioles in various regions have different receptors that can be activated by epinephrine, angiotensin, and other hormones. (Epinephrine triggers vasoconstriction or vasodilation. Angiotensin triggers widespread vasoconstriction.)

Integrating centers in the medulla oblongata of the brain function to keep the resting level of blood pressure fairly constant over time.

Control of Blood Distribution. The nervous system, endocrine system, and changes in local chemical conditions interact to assure that blood circulation meets the metabolic demands of various tissues. Think about what happens after you have eaten a large meal. More blood is diverted to your digestive system, which swings into full gear as other systems more or less idle. When your body is exposed to cold wind or snow for an extended time, blood is diverted away from the skin to deeper tissue regions, so that the metabolically generated heat that warmed the blood in the first place can be conserved.

Chapter 33 indicated that the heart is serviced by parasympathetic nerves (which signal for decreases in heart rate) and sympathetic nerves (which signal for increases in heart rate). Adjustments in the signals traveling along those nerves lead to adjustments in blood circulation. Moreover, sympathetic nerves as well as hormones can stimulate vasoconstriction in many tissues, including the kidneys and muscle tissues that are not being called upon to contract. Thus they can cause blood to be diverted away from areas requiring less blood flow.

Finally, think about what happens when you swim or run. In your skeletal muscle tissue, the oxygen level decreases and the levels of carbon dioxide, hydrogen ions, potassium ions, and other substances increase. The changes in local chemical conditions cause smooth muscle cells of arterioles to relax, so arteriole diameter enlarges. Now more blood flows past the active muscles, delivering more raw materials and carrying away cell products and wastes. While this is occurring, arteriole diameter is decreasing in tissues of the digestive tract and kidneys.

Adjustments in the distribution of blood flow are made in response to signals from the nervous system, endocrine system, and changes in local chemical conditions.

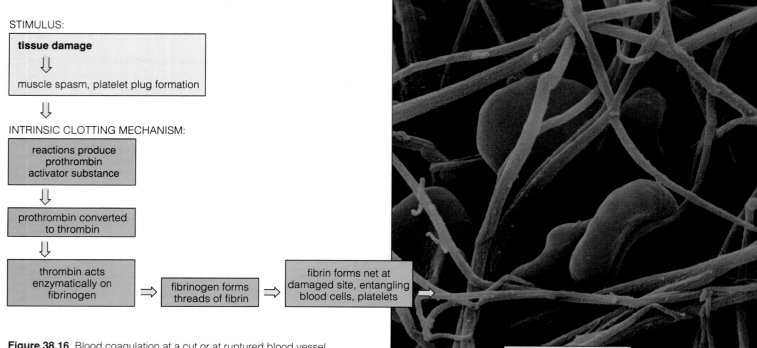

STIMULUS:

tissue damage
⇩
muscle spasm, platelet plug formation
⇩
INTRINSIC CLOTTING MECHANISM:

reactions produce
prothrombin
activator substance
⇩
prothrombin converted
to thrombin
⇩
thrombin acts
enzymatically on
fibrinogen ⇒ fibrinogen forms
threads of fibrin ⇒ fibrin forms net at
damaged site, entangling
blood cells, platelets ⇒

clot formation

Figure 38.16 Blood coagulation at a cut or at ruptured blood vessel tissue. The micrograph shows red blood cells trapped in a fibrin net.

Hemostasis

Don't even think about what would happen if the body could not repair breaks or cuts even in its small blood vessels. In **hemostasis**, blood vessel spasm, platelet plug formation, blood coagulation, and other mechanisms can stop bleeding.

First, smooth muscle in the wall of a damaged blood vessel contracts in a reflex response called a spasm. The blood vessel constricts, and the flow of blood is temporarily stopped. Second, platelets clump together, temporarily plugging the rupture. They also release substances that help prolong the spasm and attract more platelets. Third, blood *coagulates* (converts to a gel) and forms a clot. Finally, the clot retracts into a compact mass, drawing the walls of the vessel together.

Blood coagulates when damage exposes collagen fibers in blood vessel walls. The response is called the *intrinsic clotting mechanism*. A plasma protein becomes activated and triggers reactions that lead to the formation of an enzyme (thrombin), which acts on a large, rod-shaped plasma protein (fibrinogen). The rods adhere to one another, forming long, insoluble threads that stick to exposed collagen. The result is a net in which blood cells and platelets become entangled (Figure 38.16). The entire mass is a blood clot.

Blood also can coagulate through an *extrinsic clotting mechanism*. "Extrinsic" means that the series of reactions leading to blood clotting is triggered by the release of enzymes and other substances *outside* of the blood itself (that is, from damaged blood vessels or from the surrounding tissues). The substances lead to thrombin formation, and the remaining steps parallel those shown in

Figure 38.16. Overall, fewer steps are involved than in the intrinsic clotting mechanism, and the reactions occur much more rapidly.

The contribution of the extrinsic clotting mechanism to hemostasis is unclear. However, it is definitely involved in walling off bacteria and preventing the spread of bacterial infection from invaded tissue regions.

Blood Typing

All of your cells carry membrane proteins at their surface that serve as "self" markers; they identify the cells as being part of your own body. Your body also has proteins called **antibodies**, which can recognize markers on *foreign* cells (page 683). When the blood of two people mixes during transfusions, antibodies will act against any cells bearing the "wrong" marker. They can do the same thing during pregnancy, if antibodies diffuse from the mother's circulation system to that of her unborn child.

ABO Blood Typing. As we have seen, *ABO blood typing* is based on some of the surface markers on red blood cells (page 172). Type A blood has A markers on those cells, type B blood has B markers, type AB has both, and type O has neither one.

If you are type A, you do not carry antibodies against A markers—but you have antibodies against B markers. If you are type B, you have antibodies against A but not B markers. If you are type AB, you have no antibodies against A or B markers, so your body will tolerate donations of type A, B, or AB blood. If you are type O, however, you have antibodies against A and B mark-

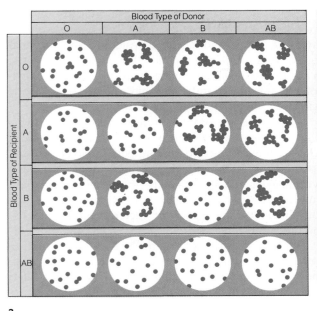

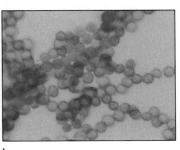

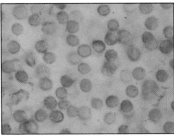

Figure 38.17 (**a**) Agglutination responses in blood types O, A, B, and AB when mixed with blood samples of the same and different types. (**b**) Micrographs showing the absence of agglutination in a mixture of two different but compatible types (above) and agglutination in a mixture of incompatible blood types (below).

ers—and those antibodies will act against cells bearing one or both types.

Figure 38.17 shows what happens when blood from different types of donors and recipients is mixed together. In a response called **agglutination**, antibodies act against the "foreign" cells and cause them to clump. Such clumps can clog small blood vessels. They may lead to tissue damage and death. In looking at Figure 38.17, can you say what the agglutination responses will be to type AB blood? To type O blood?

Rh Blood Typing. Other surface markers on red blood cells also can cause agglutination responses. For example, *Rh blood typing* is based on the presence or absence of an Rh marker (so named because it was first identified in the blood of *rh*esus monkeys). Rh+ individuals have blood cells with this marker; Rh− individuals do not. Ordinarily, people do not have antibodies that act against Rh markers. However, if someone has been given a transfusion of Rh+ blood, antibodies will be produced against it and will continue circulating in the bloodstream.

If an Rh− female becomes pregnant by an Rh+ male, there is a chance the fetus will be Rh+. During pregnancy or childbirth, some red blood cells of the fetus may leak into the mother's bloodstream (Chapter 43). If they do, they will stimulate her body into producing antibodies against the Rh markers (Figure 38.18). If the woman becomes pregnant *again*, Rh antibodies will enter the fetal bloodstream. If this second fetus happens to have Rh+ blood, the antibodies will cause red blood cells to swell and then rupture, releasing hemoglobin into the bloodstream.

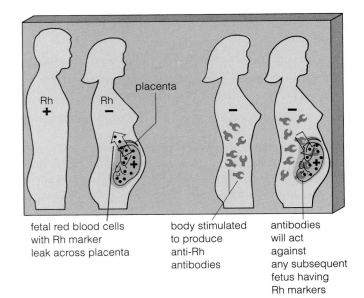

fetal red blood cells with Rh marker leak across placenta

body stimulated to produce anti-Rh antibodies

antibodies will act against any subsequent fetus having Rh markers

Figure 38.18 Development of antibodies in response to Rh+ blood.

In extreme cases of this disorder, called *erythroblastosis fetalis*, too many cells are destroyed and the fetus dies before birth. If it is born alive, all the newborn's blood can be slowly replaced with blood free of the Rh antibodies. Currently, a known Rh− female can be treated right after her first pregnancy with a drug, called Rho-Gam, that will protect her next fetus. The drug will inactivate any Rh+ fetal blood cells circulating in the mother's bloodstream before she can become sensitized and begin producing anti-Rh+ antibodies.

LYMPHATIC SYSTEM

We conclude this chapter with a brief section on the lymphatic system, which supplements the circulatory system by returning excess tissue fluid to the bloodstream. But think of this section as a bridge to the next chapter, on immunity, for the lymphatic system is also vital to the body's defenses against injury and attack. As Figure 38.19 shows, the system's components include transport vessels and lymphoid organs. Tissue fluid that has moved into the transport vessels is called **lymph**.

Lymph Vascular System

The **lymph vascular system** includes lymph capillaries, lymph vessels, and ducts. Collectively, these transport vessels serve the following functions:

1. Return of excess filtered fluid to the blood

2. Return of small amounts of proteins that leave the capillaries

3. Transport of fats absorbed from the digestive tract

4. Transport of foreign particles and cellular debris to disposal centers—that is, the lymph nodes

At one end of the lymph vascular system are *lymph capillaries*, no larger in diameter than blood capillaries. They occur in the tissues of almost all organs and serve as "blind-end" tubes. They have no entrance at the end located in tissues; their only "opening" merges with larger lymph vessels, as shown by Figure 38.20. Extracellular fluid simply diffuses into them through gaps in the capillary wall.

Like veins, *lymph vessels* have smooth muscle in their walls and flaplike valves that prevent backflow. When you breathe, movements of the rib cage and skeletal muscle adjacent to the lymph vessels help move fluid through lymph vessels, just as they do for veins. Lymph vessels converge into collecting ducts, which drain into veins in the lower neck. In this way, the lymph fluid is returned to the circulation.

Lymphoid Organs

The **lymphoid organs** include the lymph nodes, spleen, thymus, tonsils, adenoids, and patches of tissue in the small intestine and appendix. These organs and tissue patches are production centers for infection-fighting cells, including lymphocytes. They also are sites for some defense responses.

Like all white blood cells, lymphocytes are derived from stem cells in bone marrow. The derivative cells enter the blood and take up residence in lymphoid organs. With proper stimulation, they divide by mitosis. In fact, most new lymphocytes are produced by divisions in the blood and lymphoid organs, not in bone marrow.

Lymph nodes are located at intervals along lymph vessels, as suggested by Figure 38.19. All lymph trickles through at least one node before being delivered to the bloodstream. Each node has several inner chambers. Lymphocytes and plasma cells (the progeny of certain lymphocytes) pack each chamber. Macrophages in the node help clear the lymph of bacteria, cellular debris, and other substances.

The largest lymphoid organ, the *spleen*, is a filtering station for blood and a holding station for lymphocytes. The spleen also has inner chambers, but these are filled with red and white "pulp." The red pulp contains large stores of red blood cells and macrophages. Red blood cells are produced here in developing human embryos.

The *thymus* secretes hormones concerned with the activity of lymphocytes. It also is a major organ where lymphocytes multiply, differentiate, and mature into fighters of specific types of disease agents. The thymus is central to immunity, the focus of the chapter to follow.

SUMMARY

1. Animals ranging from worms to humans have a circulatory system consisting of a muscular pump (heart or heartlike structure), blood, and blood vessels.

    a. In closed circulatory systems, blood flows continuously inside the walls of these components; it exchanges substances with interstitial fluid only in diffusion zones.

    b. In open systems, a heart or heartlike structure pumps blood into tissues. The blood diffuses through tissue fluid and returns to the heart.

2. Blood is a transport fluid that carries oxygen and other substances to cells, and products and wastes (including carbon dioxide) from them. It helps maintain an internal environment favorable for cell activities.

3. Blood consists of red and white blood cells, platelets, and plasma.

    a. Plasma contains water, ions, nutrients, hormones, vitamins, dissolved gases, and the plasma proteins.

    b. Red blood cells transport oxygen (bound to hemoglobin) between the lungs and cells. They also transport some carbon dioxide.

    c. In all vertebrates, some white blood cells are scavengers of dead or worn-out cells and other debris. Others serve in the defense of the body against bacteria, viruses, and other foreign agents.

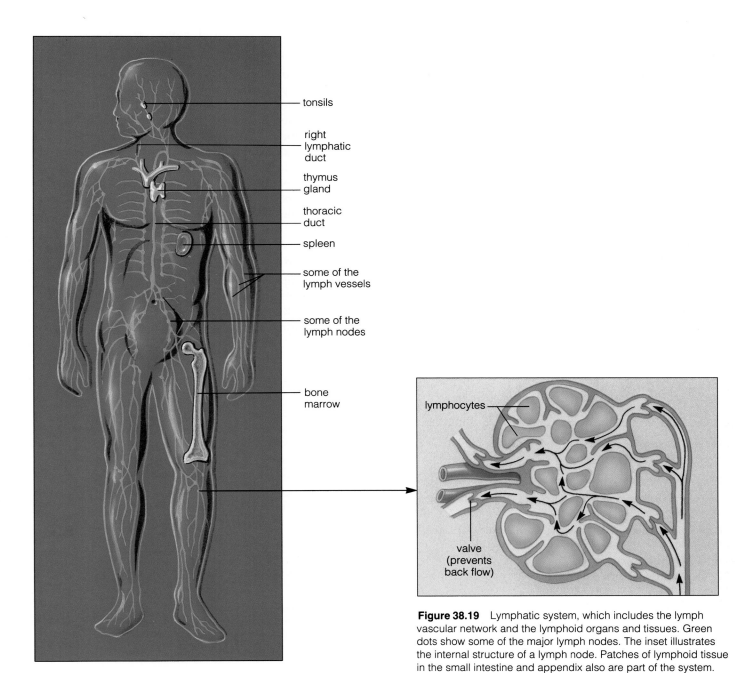

— tonsils

— right
lymphatic
duct

— thymus
gland

— thoracic
duct

— spleen

— some of the
lymph vessels

— some of the
lymph nodes

— bone
marrow

lymphocytes

valve
(prevents
back flow)

Figure 38.19 Lymphatic system, which includes the lymph vascular network and the lymphoid organs and tissues. Green dots show some of the major lymph nodes. The inset illustrates the internal structure of a lymph node. Patches of lymphoid tissue in the small intestine and appendix also are part of the system.

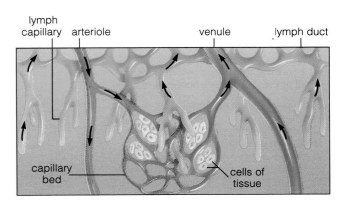

lymph
capillary   arteriole   venule   lymph duct

capillary
bed   cells of
                                  tissue

Figure 38.20 Lymph vessels near a capillary bed.

4. In all vertebrates, the heart pumps blood into arteries. From there it flows into arterioles, capillaries, venules, veins, and back to the heart.

5. The human heart is divided into two halves, each with two chambers (an atrium and a ventricle). The division is the basis of two cardiovascular circuits. These are called the pulmonary and systemic circuits.

a. In the pulmonary circuit, the right half of the heart pumps oxygen-poor blood to capillary beds inside the lungs, then oxygen-enriched blood flows back to the heart.

b. In the systemic circuit, the left half of the heart pumps oxygen-enriched blood to all body regions, where it nourishes all tissues and organs. Then oxygen-poor blood flows from those regions back to the heart.

6. Heart contractions (specifically, the contracting ventricles) are the driving force for blood circulation. Fluid pressure is high at the start of a circuit, then drops in arteries, arterioles, capillaries, then veins. It is lowest in the relaxed atria.

7. Blood moves through several types of blood vessels in both the pulmonary and systemic circuits.

a. Arteries are elastic pressure reservoirs that smooth out fluid pressure changes caused by heart contraction and relaxation.

b. Arterioles are control points for the distribution of different volumes of blood to different body regions.

c. Beds of capillaries are diffusion zones between the blood, interstitial fluid, and cells.

d. Venules overlap capillaries and veins somewhat in function.

e. Veins are blood volume reservoirs and help adjust volume flow back to the heart.

8. The lymphatic system supplements the circulatory system by returning excess fluid that seeps out of blood vessels back to the circulation. Some of its components have major roles in immune responses.

Review Questions

1. What are some of the functions of blood? *660–661*

2. Describe the cellular components of blood. Describe the plasma portion of blood. *661–664*

3. Define the functions of the following: heart, cardiovascular system, and lymphatic system. *664–665, 676*

4. Distinguish between the following:
   a. open and closed circulation *660*
   b. systemic and pulmonary circuits *665*
   c. lymph vascular system and lymphoid organs *676*

5. Label the component parts of the human heart: *665*

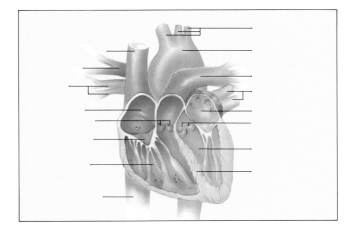

6. Explain how the medulla oblongata of the brain helps regulate blood flow to different body regions. *672*

7. State the main function of blood capillaries. What drives solutes out of and into capillaries in capillary beds? *668–669*

8. State the main function of venules and veins. What forces work together in returning venous blood to the heart? *670*

Self-Quiz *(Answers in Appendix IV)*

1. In large, complex animals, a _____ system functions in the rapid exchange of substances to and from all living cells, and usually it is supplemented by a _____ system.

2. _____ and _____ are large-diameter blood vessels for fluid transport; _____ and _____ are fine-diameter, thin-walled blood vessels for diffusion; and _____ serve as control points over the distribution of different blood volumes to different body regions.

3. Which of the following are *not* components of blood?
   a. red and white blood cells
   b. platelets and plasma
   c. assorted solutes and dissolved gases
   d. all of the above are components of blood

4. Red blood cells are produced in the _____ and function in transporting _____ and some _____.
   a. liver; oxygen; mineral ions
   b. liver; oxygen; carbon dioxide
   c. bone marrow; oxygen; hormones
   d. bone marrow; oxygen; carbon dioxide

5. White blood cells are produced in the _____ and function in both _____ and _____ .
   a. liver; oxygen transport; defense
   b. lymph glands; oxygen transport; pH stabilization
   c. bone marrow; day-to-day housekeeping; defense
   d. bone marrow; pH stabilization; defense

6. In the pulmonary circuit, the _____ half of the heart pumps _____ blood to capillary beds inside the lungs, then _____ blood flows back to the heart.
   a. left; oxygen-poor; oxygen-enriched
   b. right; oxygen-poor; oxygen-enriched
   c. left; oxygen-enriched; oxygen-poor
   d. right; oxygen-enriched; oxygen-poor

7. In the systemic circuit, the _____ half of the heart pumps _____ blood to all body regions, then _____ blood flows back to the heart.
   a. left; oxygen-poor; oxygen-enriched
   b. right; oxygen-poor; oxygen-enriched
   c. left; oxygen-enriched; oxygen-poor
   d. right; oxygen-enriched; oxygen-poor

8. Fluid pressure in the circulatory system is _____ at the beginning of a circuit, then _____ in arteries, arterioles, capillaries, and then veins. It is _____ in the relaxed atria.
   a. low; rises; highest
   b. high; drops; lowest
   c. low; drops; lowest
   d. high; rises; highest

9. Match the type of blood vessel with its major function.
   _____ arteries
   _____ arterioles
   _____ capillaries
   _____ venules
   _____ veins
   a. diffusion
   b. control of blood volume distribution
   c. transport, blood volume reservoirs
   d. overlap of capillary function
   e. transport and pressure reservoirs

10. Match the circulation components with their descriptions.
   _____ capillary beds
   _____ lymph vascular system
   _____ heart chambers
   _____ veins
   _____ heart contractions
   _____ arteries
   a. two atria, two ventricles
   b. pressure reservoirs
   c. driving force for blood
   d. zones of diffusion
   e. interstitial fluid
   f. blood volume reservoirs

Selected Key Terms

ABO blood typing 674
agglutination 675
antibody 674
aorta 665
arteriole 668
artery 667
atherosclerosis 671
atrium 665
AV valve 665
blood 660
blood pressure 667
capillary 668
capillary bed 668
cardiac conduction system 666
cardiac cycle 666
cardiac pacemaker 667
cell count 663
circulatory system 660
endothelium 665
heart 665
hemostasis 674
interstitial fluid 660

lymph 676
lymphatic system 660
lymph node 676
lymphoid organ 676
lymph vessel 676
plasma protein 661
platelet 664
pulmonary circuit 665
pulse pressure 667
red blood cell (erythrocyte) 662
Rh blood typing 675
semilunar valve 665
spleen 676
systemic circuit 665
thymus gland 676
vasoconstriction 673
vasodilation 673
vein 670
ventricle 665
venule 670
white blood cell (lymphocyte) 663

Readings

Eisenberg, M. S., et al. May 1986. "Sudden Cardiac Death." *Scientific American* 254(5):37–43.

Golde, D. W., and J. C. Gasson. July 1988. "Hormones That Stimulate the Growth of Blood Cells." *Scientific American* 259(1):62–70.

Kapff, C. T., and J. H. Jandl. 1981. *Blood: Atlas and Sourcebook of Hematology.* Boston: Little, Brown. Beautiful micrographs of normal and abnormal blood and marrow cells.

Little, R., and W. Little. 1989. *Physiology of the Heart and Circulation.* Fourth edition. Chicago: Year Book Medical Publishers, Inc. Comprehensive coverage of cardiovascular physiology.

Robinson, T. F., et al. June 1986. "The Heart as a Suction Pump." *Scientific American* 254(6):84–91.

39 IMMUNITY

a

Figure 39.1 Immunization past and present. (**a**) Statue honoring Edward Jenner's development of an immunization procedure against smallpox, one of the most dreaded diseases in human history. (**b**) Micrograph of an immune cell (T lymphocyte) being attacked by the virus (blue particles) that causes AIDS. Immunologists are working to develop weapons against this modern-day scourge.

Russian Roulette, Immunological Style

Until about a century ago, smallpox swept repeatedly through the world's cities. Some outbreaks were so severe, only half of those stricken managed to survive. No one emerged unscathed. Even the survivors ended up with permanent scars on the face, neck, shoulders, and arms. Scarring was a small price to pay, however, because survivors seldom contracted the disease again—they were "immune" to smallpox.

No one knew what caused smallpox, but the possibility of acquiring immunity was dreadfully fascinating. In Asia, Africa, and then Europe, many who were in good health gambled with inoculations. They allowed themselves to be intentionally infected with material from the sores of diseased people. Thus Chinese of the twelfth century ground up crusts from smallpox sores and inhaled the powder. By the seventeenth century Mary Montagu, wife of the ambassador to Turkey, was championing inoculation. She went so far as to inject bits of smallpox scabs into the veins of her children; even the Prince of Wales did the same. Others soaked threads in the fluid from smallpox sores, then poked the threads into scratches on the body. The survivors of such practices acquired immunity to smallpox, but many came down with raging infections. As if the odds were not dangerous enough, those who were inoculated by such crude procedures also risked coming down with leprosy, syphilis, or hepatitis.

While this immunological version of Russian roulette was going on, Edward Jenner was growing up in the English countryside. At the time it was known that cowpox, a rather mild disease, could be transmitted from cattle to humans. Yet people who contracted cowpox never became ill with smallpox. No one thought much about this until 1796, when Jenner, by now a physician, took some material from a sore on a cowpox-infected person and injected it into the arm of a young boy (Figure 39.1). Six weeks later, after the reaction to cowpox subsided, Jenner inoculated the boy with fluid from smallpox sores. He hypothesized that the earlier inoculation would provoke immunity to smallpox— and he was right. The boy remained free of infection. The French mocked the procedure, calling it "vaccination" (which translates as "encowment").

Much later a French chemist, Louis Pasteur, devised similar procedures for other diseases. Pasteur also called his procedures vaccinations; only then did the term become respectable.

By Pasteur's time, improved microscopes were revealing a variety of bacteria, fungal spores, and other previously invisible forms of life. As Pasteur discovered, microorganisms abound in ordinary air. Did some of them cause contagious diseases? Probably. Could they settle into food or drink and cause it to spoil? Pasteur proved that they did. Boiling could kill them—Pasteur and others knew this. (Being a wine connoisseur, he also knew that you cannot simply boil wine—or beer or milk, for that matter—and end up with the same beverage. He devised a way to heat food or beverages at a temperature low enough not to ruin them but high enough to kill most of the microorganisms that cause spoilage. We still depend on his partial sterilization methods, which were named pasteurization in his honor.)

But it was a German physician, Robert Koch, who actually proved that a specific microorganism can cause a specific disease—namely, anthrax. In the late 1870s, Koch repeatedly transferred blood from infected animals to uninfected ones. Each time, the blood of the recipient ended up teeming with cells of a rather large bacterium (*Bacillus anthracis*). And each time, the animal developed symptoms of anthrax. More than

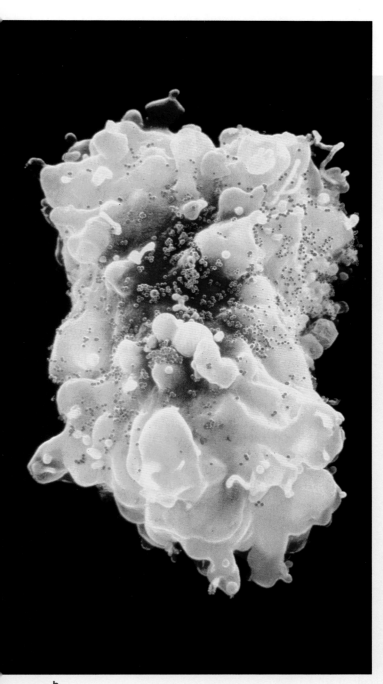

b

this, bacterial cells cultured in nutrients outside the body could, when injected into an animal, cause the same disease!

Thus, by the beginning of the twentieth century, the promise of understanding the basis of immunity loomed large. The battle against infectious disease was about to begin in earnest. Those battles are the focus of this chapter.

KEY CONCEPTS

1. The vertebrate body defends itself against viruses, bacteria, and other foreign agents that enter the internal environment. Some defense responses are nonspecific, in that they occur when any kind of invasion is detected. Other responses are specific, with certain white blood cells being mobilized against a particular invader, not invaders in general.

2. White blood cells responsible for immune responses can distinguish between self-markers on the body's own cells and antigens. An antigen is any large molecule that white blood cells recognize as foreign and that triggers an immune response. Different foreign agents have different antigens on their surface.

3. Antibody-mediated immune responses are made against antigens circulating in the body's tissues or attached to the surface of an invader. Cell-mediated immune responses are made only against body cells already infected and against cancerous or mutated cells. In most cases, both responses proceed simultaneously.

681

When you suffer sneezes, a puffed-up dripping nose, and watery eyes, you have evidence that your body is being attacked by some type of cold virus. Yet you probably are not even aware that it is simultaneously fighting off attacks by many other pathogens—and does so every day of your life. **Pathogens** are a diverse assortment of viruses, bacteria, fungi, and protozoans, each able to infect and cause diseases in humans and other organisms. Examples of their infectious cycles were described earlier, in Chapters 22 and 25. With his pioneering procedure, Jenner was actually mobilizing cells to make an immune response to a specific virus. That type of response is one of the elegant defenses described in this chapter. Before we turn to the specific responses, however, let's start with the body's generalized defenses against attack.

NONSPECIFIC DEFENSE RESPONSES

Barriers to Invasion

When Julie Andrews sang "The hills are alive . . ." in the motion picture *The Sound of Music*, she might well have been referring to the microbial world. Hills, streams, plants, air, the animal body abound with invisible organisms, many harmless but some pathogenic. You and other vertebrates coevolved with most of them, however, so you need not lose sleep over this. Most of the time, the pathogens cannot even get past your body's physical and chemical barriers. Those barriers include the following:

1. Intact skin and mucous membranes. Few microorganisms can penetrate these.

2. Ciliated, mucous membranes in the respiratory tract. Like sticky brooms, the cilia trap and sweep out airborne bacteria.

3. Secretions from exocrine glands in the skin, mouth, and elsewhere. The enzyme lysozyme, for instance, destroys the cell wall of many bacteria. It is present in the fluid (tears) that bathes the eyes.

4. Gastric fluid. The acids in this fluid destroy many pathogens that are present in food when it enters the gut.

5. Normal bacterial inhabitants of the skin, gut, and vagina. They outcompete the pathogens for resources and so help keep them in check.

Phagocytes: The Macrophages and Their Kin

What happens when physical barriers to invasion are breached by microorganisms, as when skin is cut or scraped? Then, the invasion mobilizes phagocytic white blood cells. These cells engulf and destroy foreign agents. As Figure 38.5 shows, these cells arise from stem cells in bone marrow. They include neutrophils, eosinophils, and monocytes that mature into macrophages, the "big eaters." Figure 39.2 shows a macrophage.

Phagocytes are strategically distributed cells. Some circulate within blood vessels, then enter damaged or invaded tissues by squeezing between endothelial cells making up the walls of capillaries. Some take up stations in lymph nodes and the spleen. (You may wish to review Figure 38.19, which shows the lymphatic system.) Other phagocytes are located in the liver, kidneys, lungs, joints, and the brain.

Complement System

When certain microorganisms invade the body, about twenty plasma proteins interact as a system—the **complement system**. Those circulating proteins have roles in both nonspecific and specific defense responses. They are activated one after another in a "cascade" of reactions.

Once activated, each protein molecule helps activate many molecules of a different protein at the next reaction step. Each of these helps activate many molecules of a different protein at the next reaction step, and so on until huge numbers of complement proteins are mobilized. The reactions have these results:

1. Chemical gradients, created by the huge cascades of certain complement proteins, attract phagocytes to the scene (Figure 39.3).

2. Some complement proteins coat the surface of invading cells, and phagocytes zero in on the coat.

3. Other complement proteins help kill the pathogen by promoting lysis of its plasma membrane. (*Lysis* refers to gross induced leakage across the membrane that leads to cell death.)

Inflammation

Many cells, the complement system, and other substances take part in the **inflammatory response**. This response is a series of events that destroy invaders and

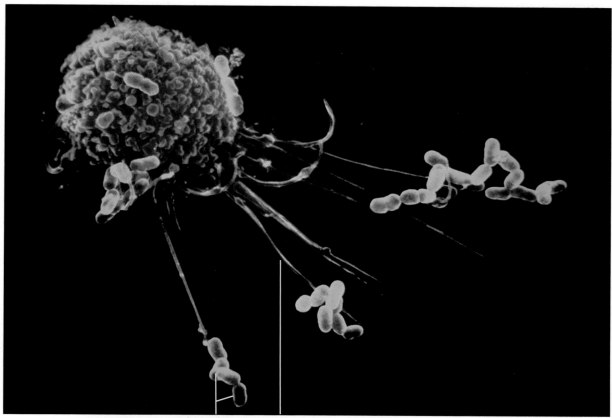

bacterial cells   cytoplasmic extension of macrophage

restore tissues to normal. The events are not limited to nonspecific defense responses. As you will see, they proceed also when the body acts against specific invaders.

For example, when the complement system is activated, circulating basophils (and mast cells, their counterparts in tissues) release *histamine*. This potent substance dilates capillaries and makes them "leaky," so fluid seeps out. The complement proteins and other substances used to fight an invasion are dissolved in this fluid and so gain access to tissues. Also, clotting mechanisms (page 674) are working to keep blood vessels intact and to wall off infected or damaged tissues. In short, the inflammatory response involves these events:

1. Localized warmth and redness occur in damaged or invaded tissues when capillaries dilate and become leaky.

2. Fluid seeping from capillaries causes local swelling and delivers infection-fighting proteins to the tissues.

3. Phagocytes, following chemical gradients to affected tissues, engulf foreign invaders and debris.

4. Clotting mechanisms help wall off the pathogen and help repair tissues.

Figure 39.2 Scanning electron micrograph of a macrophage, probing its surroundings with cytoplasmic extensions. The macrophage engulfs bacterial cells that come in contact with it.

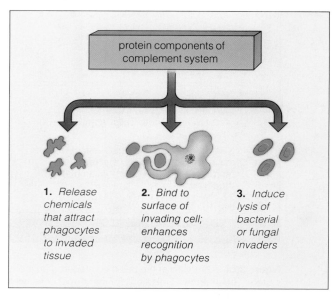

Figure 39.3 Functions of proteins of the complement system. These proteins take part in specific as well as in nonspecific defense responses.

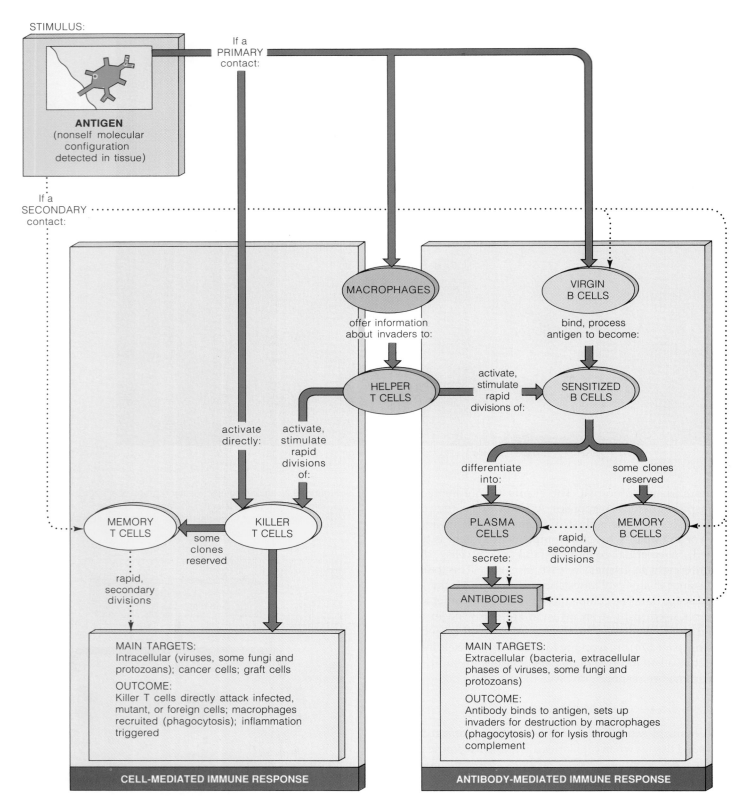

Figure 39.4 Overview of the cell-mediated and antibody-mediated branches of the vertebrate immune system. Green arrows indicate a "primary" response, which follows a first-time encounter with a specific antigen. Dashed arrows indicate a "secondary" response to a subsequent encounter with the same kind of antigen. This illustration can be used as a road map as you make your way through the descriptions in the text. The details of the vertebrate immune system are astonishingly complex; even here, many events have been omitted so the main sequences can be seen clearly.

SPECIFIC DEFENSE RESPONSES: THE IMMUNE SYSTEM

The body's nonspecific defenses—its physical barriers, complement system, and inflammatory response—are effective against many pathogens. Sometimes, however, the general attack responses are not enough to stop the spread of an invader, and illness follows. When that happens, three types of white blood cells—macrophages, T cells, and B cells—make precise counterattacks. Their interactions are the basis of the **immune system**.

The hallmarks of the immune system are *specificity* (its cells zero in on specific invaders) and *memory* (a portion of its cells can mount a rapid attack if the same type of invader returns).

The Defenders: An Overview

Of every 100 cells in your body, one is a lymphocyte—a white blood cell. Here are the names and functions of the white blood cells responsible for immune responses:

1. Macrophages. Besides engulfing anything perceived as foreign, these phagocytic cells alert helper T cells to the presence of *specific* foreign agents.

2. B cells. The B cells and their progeny (plasma cells) produce antibodies. *Antibodies* are molecular weapons that lock onto specific targets and tag them for destruction by phagocytes or the complement system.

3. Cytotoxic T cells. These directly destroy body cells already infected by certain viruses or parasitic fungi.

4. Helper T cells. Helper T cells serve as master switches of the immune system. Among other things, they stimulate the rapid division of B cells and cytotoxic T cells.

5. Suppressor T cells. These "controller cells" slow down or prevent immune responses.

6. Memory cells. Memory cells are a portion of B cell and T cell populations produced during a first encounter with a specific invader but not used in battle. They circulate freely and respond rapidly to any subsequent attacks by the same type of invader.

The white blood cells just listed belong to two fighting branches of the immune system. Both are called into action during most battles. T cells dominate one branch; they carry out a "cell-mediated" response. B cells dominate the other branch; they carry out an "antibody-mediated" response. Figure 39.4 hints at how the two responses are interrelated.

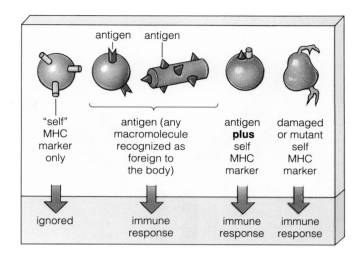

Figure 39.5 Molecular cues that stimulate lymphocytes to make immune responses.

Recognition of Self and Nonself

Before getting into the immunological battles, think about an important question. How do the defenders distinguish *self* (the body's own cells) from *nonself* (harmful foreign agents)? Such recognition is vital, for lymphocytes unleash extremely destructive immune reactions. We know this because on rare occasions the distinction is blurred—such that T and B cells make an autoimmune response. As you will see, this means that the cells turn on the body itself and irritate or damage tissues, sometimes with lethal consequences.

Among the surface proteins on your own cells are **MHC markers** (named after the genes coding for them). Your white blood cells recognize these as "self" markers and normally ignore the cells bearing them. MHC markers are unique to each individual. Except in the case of identical twins, no one has the same kinds.

But viruses, bacteria, fungi, ragweed pollen, bee venom, cells of organ transplants, and just about any other foreign agent have antigens on their surface, which lymphocytes do not ignore. An **antigen** is any large molecule with a distinct configuration that triggers an immune response. Most antigens are protein or oligosaccharide molecules.

B cells and T cells will not attack cells bearing MHC markers only. They will mount an immune response when they encounter an antigen. They will do so regardless of whether antigen is merely present in tissues or associated with MHC markers on cell surfaces (Figure 39.5).

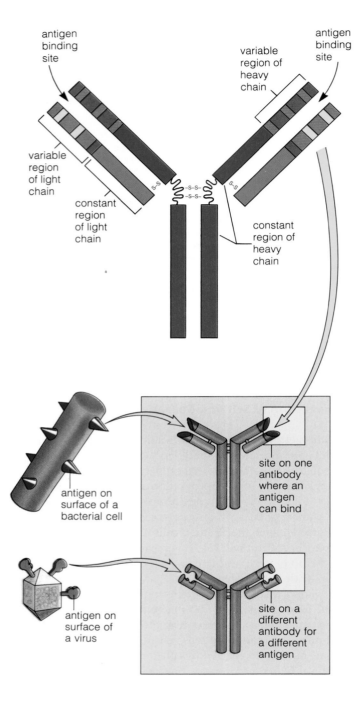

Figure 39.6 Structure of antibodies. An antibody molecule has four polypeptide chains joined into a Y-shaped structure. Some regions are always the same in all antibody molecules. But the molecular configuration varies in one region; this is the antigen-binding site.

antigen binding site

antigen binding site

variable region of heavy chain

variable region of light chain

constant region of light chain

constant region of heavy chain

-s-s-
-s-s-

antigen on surface of a bacterial cell

antigen on surface of a virus

site on one antibody where an antigen can bind

site on a different antibody for a different antigen

Primary Immune Response

A *first-time* encounter with an antigen elicits a **primary immune response** from macrophages, T and B cells, and their products. Here we will consider an antibody-mediated response to such an encounter, then a cell-mediated one.

Antibody-Mediated Immune Response. An **antibody** is a Y-shaped protein molecule with binding sites for a specific antigen. Figure 39.6 is a diagram of its general structure. Only B cells and their progeny, called **plasma cells**, make antibodies.

B cells, recall, mature in bone marrow. While each B cell is differentiating, it makes many copies of just one kind of antibody. Some of these Y-shaped molecules become positioned at the cell surface, where they serve as receptors for a specific antigen. The tail of each "Y" is embedded in the plasma membrane, and the arms stick out above the surface. Now the cell is released into the circulation as a "virgin" B cell. This term signifies it has membrane-bound antibodies but has not yet made contact with antigen.

Suppose bacteria enter the body through a small cut (step *1* in Figure 39.7). The invasion triggers a general inflammatory response, and macrophages manage to engulf a few bacterial cells. The engulfed bacteria move into the cytoplasm inside endocytic vesicles, which fuse with other vesicles containing lysosomal enzymes (page 64). Although the enzymes digest the bacterial cells, they do not completely destroy their antigens. Antigen fragments are transported to the surface of the macrophage's plasma membrane. There, the fragments become bound to MHC markers (step *2* in Figure 39.7). *Each macrophage now displays antigen–MHC complexes at its surface.*

Some bacterial cells escape detection by macrophages, however. They multiply, and for a time they move undetected past a number of virgin B cells. Eventually they encounter the one B cell with antibodies able to bind to antigen on the bacterial cell surface (step *3*). Once the B cell binds with antigen, it becomes sensitive to stimulatory signals from macrophages and helper T cells.

What happens is this: When helper T cells make contact with the battling macrophages, some of their membrane receptors lock onto the antigen–MHC complexes at the macrophage surface (step *4* in Figure 39.7). Once the connection is made, macrophages secrete an *interleukin* that stimulates the helper T cells to secrete their own interleukins. The interleukins are communication signals. In their presence, *any B cell that has become sensitized to the antigen will start dividing.*

Rapid divisions among the stimulated B cell progeny give rise to a clone—a population of identical B cells.

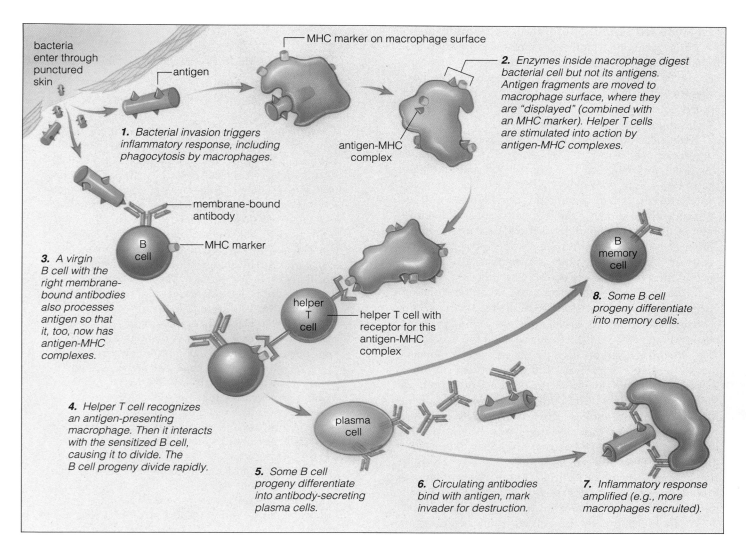

1. Bacterial invasion triggers inflammatory response, including phagocytosis by macrophages.

bacteria enter through punctured skin

antigen

MHC marker on macrophage surface

antigen-MHC complex

2. Enzymes inside macrophage digest bacterial cell but not its antigens. Antigen fragments are moved to macrophage surface, where they are "displayed" (combined with an MHC marker). Helper T cells are stimulated into action by antigen-MHC complexes.

membrane-bound antibody

B cell

MHC marker

3. A virgin B cell with the right membrane-bound antibodies also processes antigen so that it, too, now has antigen-MHC complexes.

helper T cell

helper T cell with receptor for this antigen-MHC complex

B memory cell

8. Some B cell progeny differentiate into memory cells.

4. Helper T cell recognizes an antigen-presenting macrophage. Then it interacts with the sensitized B cell, causing it to divide. The B cell progeny divide rapidly.

plasma cell

5. Some B cell progeny differentiate into antibody-secreting plasma cells.

6. Circulating antibodies bind with antigen, mark invader for destruction.

7. Inflammatory response amplified (e.g., more macrophages recruited).

Figure 39.7 Amplification of the inflammatory response by specific immune reactions. This example is of an *antibody-mediated response* to a bacterial invasion. Plasma cells (the progeny of activated B cells) release antibodies, which circulate and mark invaders for destruction by other defense agents, including more macrophages recruited to the battle scene.

Part of the population differentiates into plasma cells. The plasma cells are weapons factories. They make vast numbers of copies of the particular antibody that had been generated in the virgin B cell (step 5). For the next few days they secrete about 2,000 antibody molecules per second into their surroundings! The circulating antibodies do not destroy pathogens directly. They simply tag the invader for disposal by other means.

There are five classes of antibodies that serve different roles in defense. All belong to a group of plasma proteins, the **immunoglobulins** (Ig). When bound to antigen, the ones designated IgM and IgG enlist the aid of macrophages and complement proteins. IgG also can cross the placenta and help defend a developing fetus from pathogens. IgA, which is present in tears, saliva, and mucus, helps repel invaders at the start of the respiratory system, digestive system, urinary tract, and elsewhere. IgE calls histamine-secreting cells into action. Lastly, IgD and IgM work together to help bind antigen to B cells.

The main targets of an antibody-mediated response are bacteria and *extracellular phases* of viruses, some fungal parasites, and some protozoans. In other words, antibodies can't lock onto antigen if the invader has entered the cytoplasm of a host cell. The antigen must be circulating in tissues or at the cell surface.

Cell-Mediated Immune Response. This brings us to the viruses and other pathogens that have already penetrated host cells, where they remain hidden from antibodies. In a cell-mediated immune response, the host cells are killed before the pathogens can replicate and spread to other cells. Cytotoxic T cells serve as the executioners.

Stem cells in bone marrow give rise to the forerunners of cytotoxic T cells (Figure 38.5). These travel the circulatory highways to the thymus gland, where they mature into cytotoxic T cells. Each cell produces protein receptors that become positioned at its surface. (As you will see shortly, gene segments are shuffled into many different combinations that code for many different receptor proteins.) With these surface receptors, the cytotoxic T cell will be able to recognize any antigen encountered once it starts patrolling the body.

Cytotoxic T cells will patrol right on past circulating virus particles without even recognizing them. When a virus particle infects a cell, however, viral proteins become associated with MHC markers on the host's surface. The MHC markers identify the cell as belonging to the body. At the same time, the viral protein (in other words, the antigen) identifies the presence of something foreign. Cytotoxic T cells bind to this combination of MHC marker and antigen. Then they secrete *perforins*, which are proteins that effectively punch holes in the infected cell. It is possible that they also induce the infected cell to self-destruct, although the mechanism is not known. The remarkable scanning electron micrographs in the *Commentary* suggest that cytotoxic T cells have the same deadly effect on cancerous cells.

When the body rejects a tissue graft or an organ transplant, cytotoxic T cells are one of the reasons why. They recognize MHC markers on the grafted cells as being foreign, unless the donor is an identical twin. (Such twins have identical DNA, hence identical MHC markers.) Organ recipients take drugs to destroy cytotoxic T cells, but this compromises their ability to mount immune responses to pathogens. For example, pneumocystis infections are one of the leading causes of death among transplant recipients; the body cannot overcome the invading bacterium responsible for the disease.

Cytotoxic T cells may execute cancerous cells, but only when viruses have induced the cancer. More than 80 percent of all human cancers are not virus-induced. However, there may be other cells in the body that defend against cancer. Macrophages are candidates. *Natural killer cells*, designated NK cells, are others. NK cells, which are somewhat like lymphocytes, kill tumor cells and viral-infected cells. They do so in the absence of antibodies. They do so spontaneously, regardless of what type of cancerous cell they encounter. Like cytotoxic T cells, they punch holes in infected cells or, possibly, induce them to commit suicide.

Commentary

Cancer and the Immune System

Cancer is a disease in which cells have lost controls over cell division. It can arise when viral attack, chemical change, or irradiation induces mutation in genes that are central to the cell division cycle (page 232). The mutated cells start dividing relentlessly, and unless something stops them, they will destroy surrounding tissues and, ultimately, kill the individual.

Cytotoxic T cells and NK cells can destroy cancer cells—when they detect them. Typically, cancerous transformation involves alterations of glycoproteins positioned at the cell surface. The altered molecules are analogous to foreign antigens, in terms of how cytotoxic T cells and NK cells respond to them.

However, it may be that glycoproteins do not undergo alteration in all cases. It may be that they become chemically disguised or masked. Perhaps they are even released from the cell surface and begin circulating through the bloodstream and so lead NK cells down false trails. Whatever the reason, the transformed cells are free to divide uncontrollably and produce a tumor.

At present, surgery, drug treatment (chemotherapy), and irradiation are the only weapons against cancer. Surgery works when a tumor is fully accessible and has not spread, but it offers little hope when cancer cells have begun wandering. When used alone, chemotherapy and irradiation destroy good cells as well as bad.

Immune therapy is a promising prospect. The idea here is to mobilize cytotoxic T cells by deliberately introducing agents that will set off the immune alarm. *Interferons*, a group of small proteins, were early candidates for immune therapy. Most cells produce and release interferon following a viral attack. The interferon binds to the plasma membrane of other cells in the body and induces resistance to many viruses. So far, however, interferon has been useful only against some rare forms of cancer.

Monoclonal antibodies hold promise for immune therapy. It is difficult to get normal, antibody-secreting B cells to

Control of Immune Responses. Antibody-mediated and cell-mediated responses are regulated events. When the tide of battle turns, antibody molecules are "saturating" the binding sites on pathogens that have not yet been disposed of. With fewer exposed antigens, less antibody is secreted. Also, secretions from suppressor T cells call off the counterattack and keep the reactions from spiraling out of control.

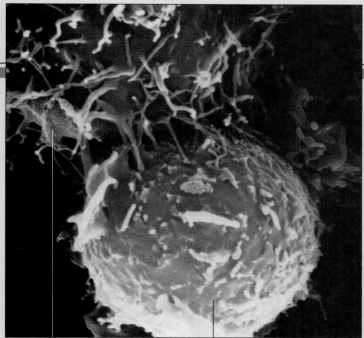

cytotoxic T cell   tumor cell

a A cytotoxic T cell recognizes and binds tightly to a tumor cell, then secretes pore-forming proteins that will destroy the integrity of the target cell membrane.

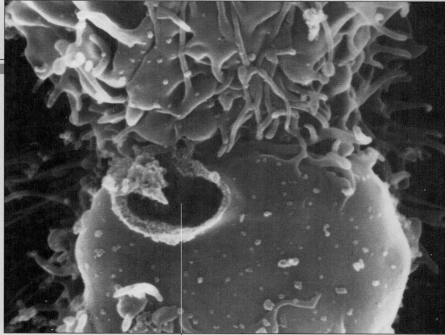

hole "punched" in tumor cell

b The target cell has become grossly leaky and has ballooned under an influx of the surrounding fluid; soon there will be nothing left of it.

grow indefinitely and mass-produce pure antibody in useful amounts. But Cesar Milstein and Georges Kohler discovered a way to do this. They immunized a mouse with a specific antigen. (The point was to allow lymphatic tissues in the mouse—the spleen especially—to become enriched with B cells specific for the immunizing antigen.) Later, B cells were extracted from the mouse spleen and were fused with a malignant B cell that showed indefinite growth. Some of the hybrid cells multiplied as rapidly as the malignant parent and produced quantities of the same antibodies as the parent B cells from the immunized mouse. Clones of such hybrid cells can be maintained indefinitely and they continue to make the same antibody. Hence the name "monoclonal antibodies." All the antibody molecules are identical, and all are derived from the same parent cell.

Monoclonal antibodies are being studied for use in passive immunization against malaria, flu viruses, and hepatitis B. They also are candidates for *cancer imaging*. This procedure uses scanning machines along with radio-actively labeled monoclonal antibodies that are specific for certain types of cancer. It allows us to home in on the exact location of cancer in the body (compare page 22). Such scans indicate whether cancer is present, where it is located, and the size of the tumor.

Monoclonal antibodies might also help overcome one of the major drawbacks to chemotherapy. Such treat-ments have severe side effects because the drugs used are highly toxic and cannot discriminate between normal cells and cancerous ones. A current goal is to hook up drug molecules with a monoclonal antibody. As Milstein and Kohler speculated, "Once again the antibodies might be expected to home in on the cancer cells—only this time they would be dragging along with them a depth charge of monumental proportions." Such is the prospect of *targeted drug therapy*.

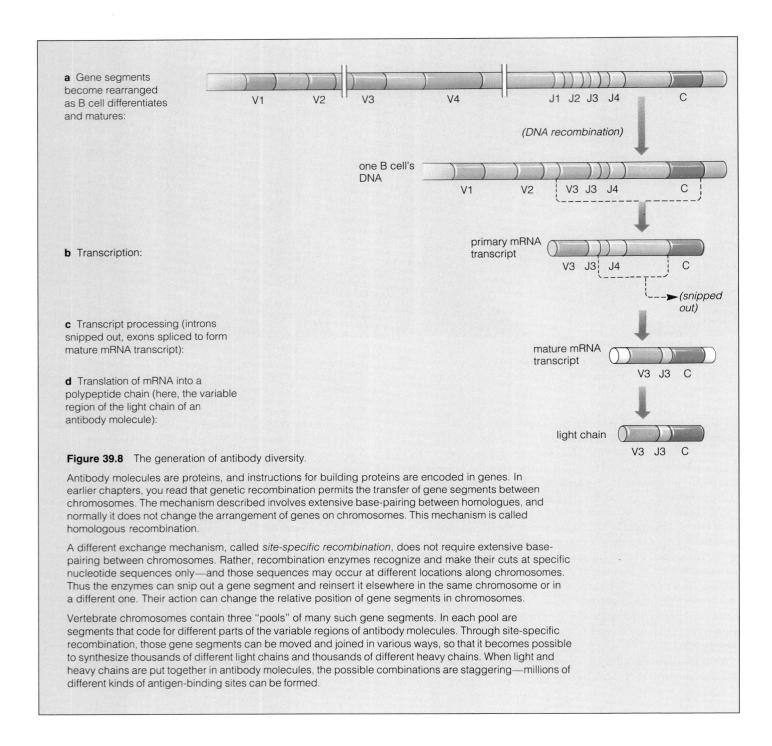

a Gene segments become rearranged as B cell differentiates and matures:

V1   V2   V3   V4   J1 J2 J3 J4   C

(DNA recombination)

one B cell's DNA

V1   V2   V3 J3 J4   C

b Transcription:

primary mRNA transcript

V3 J3 J4   C

(snipped out)

c Transcript processing (introns snipped out, exons spliced to form mature mRNA transcript):

mature mRNA transcript

V3 J3 C

d Translation of mRNA into a polypeptide chain (here, the variable region of the light chain of an antibody molecule):

light chain

V3 J3 C

Figure 39.8 The generation of antibody diversity.

Antibody molecules are proteins, and instructions for building proteins are encoded in genes. In earlier chapters, you read that genetic recombination permits the transfer of gene segments between chromosomes. The mechanism described involves extensive base-pairing between homologues, and normally it does not change the arrangement of genes on chromosomes. This mechanism is called homologous recombination.

A different exchange mechanism, called *site-specific recombination*, does not require extensive base-pairing between chromosomes. Rather, recombination enzymes recognize and make their cuts at specific nucleotide sequences only—and those sequences may occur at different locations along chromosomes. Thus the enzymes can snip out a gene segment and reinsert it elsewhere in the same chromosome or in a different one. Their action can change the relative position of gene segments in chromosomes.

Vertebrate chromosomes contain three "pools" of many such gene segments. In each pool are segments that code for different parts of the variable regions of antibody molecules. Through site-specific recombination, those gene segments can be moved and joined in various ways, so that it becomes possible to synthesize thousands of different light chains and thousands of different heavy chains. When light and heavy chains are put together in antibody molecules, the possible combinations are staggering—millions of different kinds of antigen-binding sites can be formed.

Antibody Diversity and the Clonal Selection Theory

Your body can be assaulted by an enormous variety of pathogens, each with a unique antigen. How do B cells produce the millions of different receptors (antibodies) required to detect all the potential threats? The answer lies with DNA recombinations occurring in the antibody genes of each B cell as it matures in bone marrow. Part of each arm of an antibody is a polypeptide chain, folded into a groove or cavity. All B cells have the same genes coding for the chain—but each shuffles the genes into one of millions of possible combinations, so they can give rise to virtually unlimited chain configurations (Figure 39.8).

Thus, it is not that you or any other individual inherited a limited genetic war chest from your ancestors, useful only against pathogens that were successfully

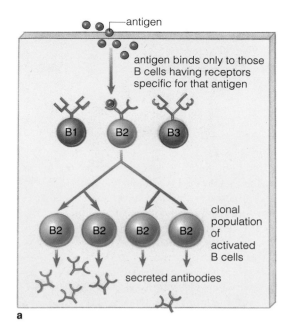

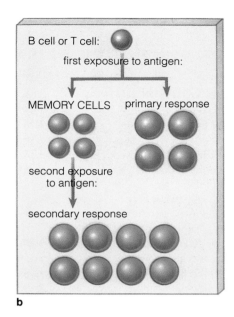

a

b

fought off in the past. Even if you encounter an entirely new antigen (as might occur when an influenza virus has mutated), your body may not be helpless against attack. It may be that DNA recombinations in one of your maturing B cells produced the exact chain configuration that can lock onto the invader. By happy accident, you have the precise weapon needed.

According to the **clonal selection theory**, proposed by Macfarlane Burnet, an activated B cell (or T cell) multiplies rapidly by mitotic cell division, and all of its descendants will retain specificity against the antigen causing the activation. They constitute a *clone* of cells, immunologically identical for the antigen that "selected" them (Figure 39.9).

Figure 39.9 (**a**) Clonal selection of lymphocytes having receptors for specific antigens. The proteins from which the receptors are constructed are produced through random shufflings of DNA segments while lymphocytes are maturing. Only antigen-selected lymphocytes will become activated and give rise to a population of immunologically identical clones. (**b**) Immunological memory. Not all cells of the activated lymphocyte populations are used in the primary immune response to an antigen. Many continue to circulate as memory lymphocytes, which become activated during a secondary immune response.

Secondary Immune Response

The clonal selection theory also explains how a person has "immunological memory" of a first-time response to an invasion. The term refers to the body's capacity to make a very rapid response to a subsequent invasion by the same type of pathogen. A **secondary immune response** to a previously encountered antigen can occur in two or three days. It is greater in magnitude than the primary response and of longer duration (Figure 39.10).

Why is this so? During a primary immune response, some B and T cells of the clonal populations do not engage in battle. They continue to circulate for years, even decades in some cases, as patrolling battalions of *memory cells*. When a memory cell encounters the same type of antigen, it divides at once. A large clone of active B cells or T cells can be unleashed, and it can be unleashed in a matter of days.

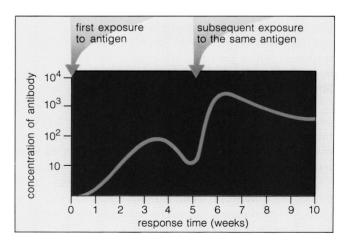

Figure 39.10 Differences in magnitude and duration between a primary and a secondary immune response to the same antigen. (The secondary response starts at week 5.)

AIDS — The Immune System Compromised

AIDS is a constellation of disorders that follow infection by the human immunodeficiency virus, or HIV. The virus cripples the immune system and leaves the body dangerously susceptible to opportunistic infections and some otherwise rare forms of cancer. Currently there is no vaccine against the known forms of this virus, which are called HIV-1, 2, and 3. And currently there is no cure for those already infected.

From 1981 to November of 1991, nearly 200,000 cases of AIDS had been reported in the United States alone. At that time the World Health Organization (WHO) had estimated that nearly half a million people had already died from AIDS and more than 1.5 million were infected with the virus. No one can say how many will be infected in the next decade. The number could be as high as 10 million.

HIV Replication Cycle

HIV compromises the immune system by attacking helper T cells (also called CD4 or T4 lymphocytes) as well as macrophages. Sometimes the virus directly attacks the nervous system, causing mental impairment and loss of motor function.

HIV is a *retrovirus*; its genetic material is RNA rather than DNA. A protein core surrounds the RNA and several copies of an enzyme, a reverse transcriptase. The core itself is wrapped in a lipid envelope derived from the plasma membrane of a host helper T cell. Once inside a host, the enzyme uses the viral RNA as a template for making DNA, which then is inserted into a host chromosome (page 350).

IMMUNIZATION

Jenner didn't know why his cowpox vaccine provided immunity against smallpox. Today we know that the viruses causing the two diseases are related, and they bear similar antigens at their surface. Let's express what goes on in modern terms.

Immunization means deliberately introducing an antigen into the body that can provoke an immune response and the production of memory cells. A **vaccine** (a preparation designed to stimulate the immune response) is injected into the body or taken orally. The first injection elicits a primary immune response. A second injection (the "booster shot") elicits a secondary response, which provokes the production of more antibodies and memory cells to provide long-lasting protection against the disease.

Many vaccines are made from killed or weakened pathogens. Sabin polio vaccine, for example, is a preparation of a weakened polio virus. Other vaccines are made from toxic but inactivated by-products of dangerous organisms, such as the bacteria causing tetanus.

Recently, selected antigen-encoding genes from pathogens were incorporated into the vaccinia virus. The virus was then used successfully to immunize laboratory animals against hepatitis B, influenza, rabies, and other serious diseases. A genetically engineered virus is not as potentially dangerous as a weakened but still-intact pathogen, which could revert to the virulent form.

For people already exposed to diphtheria, tetanus, botulism, and some other bacterial diseases, antibodies purified from some other source are injected directly to confer **passive immunity**. The effects are not lasting because the person's own B cells are not producing the antibodies. The injected antibodies may help counter the immediate attack.

ABNORMAL OR DEFICIENT IMMUNE RESPONSES

Allergies

Many of us suffer from **allergies**, in which the body makes a secondary immune response to a normally harmless substance. Exposure to dust, pollen, insect venom, drugs, certain foods, perfumes, cosmetics, and other substances triggers the abnormal response. Some allergic reactions occur within minutes; others are delayed. In either case, they can cause tissue damage.

Some individuals are genetically predisposed to allergies. But infections, emotional stress, even changes in air temperature can trigger or complicate reactions to dust and other substances that the body perceives as antigens. Every time allergic individuals are exposed to certain antigens, they produce IgE antibodies. The IgE initiates a local inflammatory response; it provokes cells into secreting histamine, prostaglandins, and other potent substances. Besides promoting fluid seepage from

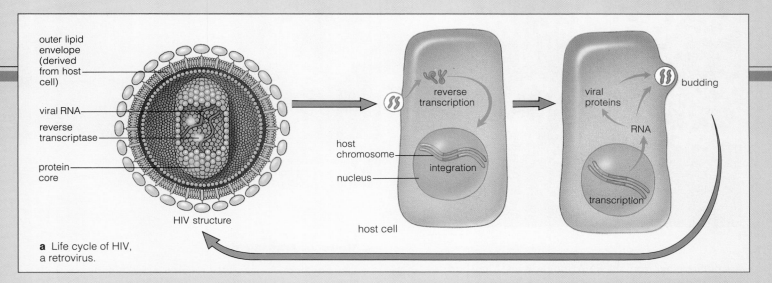

outer lipid envelope (derived from host cell)

viral RNA

reverse transcriptase

protein core

HIV structure

reverse transcription

host chromosome

nucleus

integration

host cell

viral proteins

RNA

budding

transcription

a Life cycle of HIV, a retrovirus.

It may take up to three years after infection before antibodies to several HIV proteins can be detected in the body. The antibodies do not inactivate the circulating virus particles or target infected cells for elimination. Those cells can harbor the foreign DNA for months, even years.

However, when the body is called upon to make an immune response, the infected cell may be activated. When that happens, it transcribes parts of its DNA—including the foreign insert. Transcription yields copies of viral RNA, which are translated into viral proteins. New virus particles

capillaries, histamine also stimulates exocrine glands to secrete mucus. Prostaglandins constrict smooth muscle in different organs and contribute to platelet clumping. In *asthma* and *hay fever*, the resulting symptoms of an allergic response include congestion, sneezing, a drippy nose, and labored breathing.

On rare occasions, inflammatory responses are explosive, and they can trigger a life-threatening condition called *anaphylactic shock*. For example, the few individuals who are hypersensitive to wasp or bee venom can die within minutes following a single sting. Air passages leading to their lungs undergo massive constriction. Fluid escapes too rapidly from grossly permeable capillaries. Blood pressure plummets and can lead to circulatory collapse.

Allergy-producing substances often can be identified by tests, and in some cases the body can be stimulated to make a different type of antibody (IgG) that can block the inflammatory response. Over an extended time, increasingly larger doses of the antigen are administered to allergy patients. Then, circulating IgG antibodies will be produced that bind with and mask molecules of the offending substance before they interact with IgE to produce the abnormal response.

Autoimmune Disorders

In an **autoimmune response**, lymphocytes are unleashed against the body's cells. An example is *rheumatoid arthri-*

tis, in which movable joints especially are inflamed for long periods. Often, affected persons have high levels of an antibody (rheumatoid factor) that locks onto the body's IgG molecules as if they were antigens, then deposits them on membranes at the joints. The deposits trigger the complement cascade and inflammation. The membranes become prime targets for abnormal events, including increased fluid seepage from capillaries. The accumulating fluid separates the membrane from underlying tissues, membrane cells divide repeatedly in response, and the joint thickens. These and other events continue in cycles of inflammation that do not end until the joint is totally destroyed.

Deficient Immune Responses

On rare occasions, cell-mediated immunity is weakened and the body becomes highly vulnerable to infections that might not otherwise be life-threatening. This is what happens in **AIDS** (acquired immune deficiency syndrome).

AIDS is caused by the "human immunodeficiency virus," or HIV. The *Commentary* above describes some immunological aspects of HIV infection—how the virus replicates inside a human host and what the prospects are for treating or curing infected persons. Social implications of the worldwide AIDS epidemic are described on page 780, in a *Commentary* on sexually transmitted diseases.

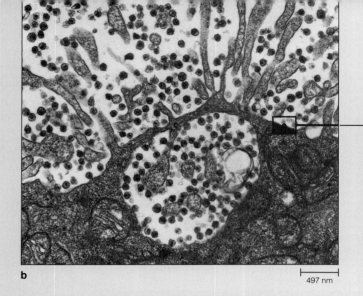

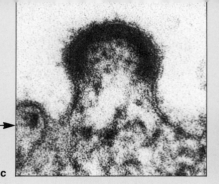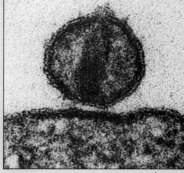

b

⊢————⊣
497 nm

⊢—⊣
45 nm

(b) Transmission electron micrograph of HIV particleś (black specks) escaping from an infected cell. **(c)** Closer views of a virus particle budding from the host cell's plasma membrane.

c

are put together from the RNA and proteins. They bud from the plasma membrane of the host helper T cell or are released when the membrane ruptures (Figures *a–c*). With each new round of infection, more and more helper T cells are destroyed or impaired.

In time, the helper T cell population is depleted, and the body loses its ability to mount immune responses. Initially, infected persons may feel well or have a bout of flulike symptoms. In time, as the population of functional helper T cells undergoes serious depletion, a condition called ARC (AIDS-related complex) may develop. There may be persistent weight loss, joint and muscle pain, fatigue and malaise, nausea, bed-drenching night sweats, enlarged lymph nodes, various minor infections, and other symptoms. Eventually, when the body's ability to mount immune responses is entirely lost, full-blown AIDS appears. Often it is heralded by opportunistic infections, such as a form of pneumonia caused by a fungus (*Pneumocystis carinii*) and tuberculosis. Blue-violet or brown-colored spots may appear on the legs especially. These are signs of Kaposi's sarcoma, a deadly form of cancer that affects blood vessels in the skin and some internal organs.

Modes of Transmission

Like any human virus, HIV requires a medium by which it can leave the body of its host, survive in the environment into which it is released, and enter a susceptible cell that can support its replication.

HIV is transmitted when bodily fluids of an infected person enter another person's tissues. Initially in the United States, transmission occurred most often among male homosexuals and among intravenous drug abusers who shared needles. As many as two-thirds of all drug abusers may now carry the virus. The incidence of HIV carriers also is increasing among heterosexuals in the general population. Besides this, HIV has been transmitted from infected mothers to their infants during pregnancy, birth, and breast-feeding. Contaminated blood supplies accounted for some cases before screening for HIV was implemented in 1985. In 1991 there were four cases of HIV transmission by way of donated tissues. In several developing countries, HIV has spread through contaminated transfusions and through reuse of unsterile needles by health care providers.

HIV generally cannot survive for more than about one or two hours outside the human body. Virus particles

Case Study: The Silent, Unseen Struggles

Let us conclude this chapter with a case study of how the immune system helps *you* survive attack. Suppose on a warm spring day you are walking barefoot to class. Abruptly you stop: A thorn on the ground punctured one of your toes. Even though you remove the thorn at once, the next morning the punctured area is red, tender, and swollen. Yet a few days later, your foot is back to normal and you have forgotten the incident.

All that time your body had been struggling against an unseen enemy. Both the thorn and your bare foot carried some soil bacteria. When the thorn broke through your skin, it carried several thousand bacterial cells with it. Inside, the bacteria found conditions suitable for

growth. They soon doubled in number and were on their way to doubling again. Meanwhile, the products of bacterial metabolism were already starting to interfere with your own cell functions. If unchecked, the invasion would have threatened your life.

Yet as soon as your skin was punctured, defenses were being mobilized. Blood from ruptured blood vessels began to pool and clot around the wound. Histamine and other secretions from basophils and mast cells caused capillaries to dilate and become more permeable to plasma proteins, including complement. Now phagocytic white blood cells crawled through clefts between cells in capillary walls. Like bloodhounds on the trail, they moved toward higher complement concentrations. They began engulfing bacteria, dirt, and damaged cells.

on needles and other objects are readily destroyed by disinfectants, including household bleach. At this time, there is no evidence that HIV can be effectively transmitted by way of food, air, water, or casual contact. The virus *has* been isolated from blood, semen, vaginal secretions, saliva, tears, breast milk, amniotic fluid, cerebrospinal fluid, and urine. However, only infected blood, semen, vaginal secretions, and breast milk contain the virus in concentrations that seem high enough for successful transmission.

Prospects for Treatment

At present, researchers may be close to developing vaccines. Among other things, they have isolated the genes for HIV proteins and are attempting to genetically engineer them in ways that might provoke effective immune responses. The task of developing an effective vaccine is formidable—HIV has the highest mutation rate of any known virus. It may be difficult to produce a vaccine that will work against all its mutated forms. Even if a vaccine can be developed that could coax the body into producing antibodies to HIV, the antibodies may not protect against AIDS. There is laboratory evidence that antibodies do not neutralize the virus.

The drug AZT (azidothymidine) is being used to prolong the life of AIDS patients. In combination with other drugs (such as interferon) or with bone marrow transplants, it may turn out to be useful in developing a cure. The search also is proceeding for compounds that might disrupt the ability of HIV to bind to the receptors by which it gains entry to cells. Other efforts involve looking for ways to interrupt the HIV replication cycle by inactivating a key protein-cutting enzyme that is necessary for viral replication. In the meantime, checking the spread of HIV depends absolutely on implementing behavioral controls through education on a massive scale. We return to this topic in Chapter 43.

If there had been no bacteria on the thorn or if they were unable to multiply rapidly in your tissues, then the inflammatory response would have cleaned things up. This time, bacterial cell divisions outpaced the non-specific defenses—and B and T cells were called up.

If this had been your first exposure to the bacterial species that invaded your body, few B and T cells would have been present to respond to the call. The immune response would have been a primary one, and it would have been a week or more before B cells divided enough times to produce enough antibody. But when you were a child, your body did fight off this bacterial species, and it still carries vestiges of the struggle—memory cells. When the invader showed up again, it encountered an immune trap ready to spring.

As inflammation progressed, B and T cells were also leaving the bloodstream. Most were specific for other antigens and did not take part in the battle. But some memory cells locked onto antigens and became activated. They moved into lymph vessels with their cargo, tumbling along until they reached a lymph node and were filtered from the fluid. For the next few days, memory cells accumulated in the node, secreted communication signals and divided rapidly.

For the first two days the bacteria appeared to be winning; they were reproducing faster than phagocytes, antibody, and complement were destroying them. By the third day, antibody production peaked. The tide of battle turned. For two weeks or more, antibody production will continue until the bacteria are destroyed. After the response draws to a close, memory cells will go on circulating, prepared for some future struggle with this same invader.

SUMMARY

1. The vertebrate body is equipped for these tasks:

   a. *Defense* against many viruses and bacteria, certain fungi, and some protozoans.

   b. *Defense* against mutant or cancerous cells.

   c. *Extracellular housekeeping* that eliminates dead cells and cellular debris from the internal environment.

2. "External" lines of defense against invasion include intact skin, exocrine gland secretions, gastric fluid, normal bacterial inhabitants of the body (which compete effectively against many invaders), and ciliated, mucous membranes of the respiratory tract.

3. The initial "internal" lines of defense include phagocytic cells (which engulf pathogens, dead body cells, and other debris) and the complement system. Complement proteins circulate in inactive form and act against bacteria and certain fungi. The ones generated during cascade reactions attract phagocytes. Some coat the surface of invading cells, enhancing its recognition by phagocytes; others cause the invading cells to lyse.

4. The body makes nonspecific (general) and specific responses to foreign agents that enter the internal environment. During the inflammatory response, phagocytes, complement proteins, and other factors are mobilized to destroy any agents detected as foreign, then to restore tissue conditions. They also are mobilized during immune responses.

5. White blood cells and their products are the basis of the immune system. Some of these cells show specificity (they attack only a particular pathogen, not invaders in general). They also show memory (they make rapid, secondary responses to the same pathogen whenever it is

Table 39.1 Summary of White Blood Cells and Their Roles In Defense

Cell Types	Take Part In	Main Characteristics
Lymphocytes:		
1. Cytotoxic T cell	Cell-mediated immune response	Each cell equipped with membrane receptors specific for one type of antigen; each can directly destroy virus-infected cells (and possibly cancer cells) by punching holes in them
2. Helper T cell	Cell-mediated and antibody-mediated immune responses	Master switch of immune system; stimulates rapid divisions of cytotoxic T cell and B cell populations
3. Suppressor T cell	Same as above	Modulates degree of immune response (slows down or prevents activity by other lymphocytes)
4. Virgin B cell	Antibody-mediated immune response	Not-yet-activated lymphocyte with *membrane-bound* antibodies (serving as antigen-specific receptors)
5. Plasma cell	Same as above	*Antibody-secreting* descendant of an activated B cell
6. Memory cell	Cell-mediated or antibody-mediated immune responses	One of a clonal population of T cells or B cells set aside during a primary immune response that can make a rapid, secondary immune response to another encounter with the same type of invader
Macrophages	Inflammatory, cell-mediated, and antibody-mediated immune responses	Phagocytic (engulfs foreign agents and infected, damaged, or aged cells); develop from circulating monocytes and take up stations in tissues; present antigens to immune cells; secretions trigger T cell and B cell proliferation
Neutrophil	Inflammatory response	Phagocytic; most abundant type of white blood cell; dominates early stage of inflammation
Eosinophil	Inflammatory response	Phagocytic (engulfs antigen-antibody complexes, kills certain parasites); combats effect of histamine in allergic reactions
Basophil and mast cell	Inflammatory response	Release histamine and other substances that contribute to vasodilation and a rapid inflammatory response
Natural killer cell (NK)	Nonspecific response	Directly destroy tumor cells, some virus-infected cells; distinct from T and B cells

encountered again). Table 39.1 summarizes these cells and their functions.

6. Cells of the immune system communicate with one another by chemical secretions (notably interleukins), which stimulate rapid growth and division of certain lymphocytes (B cells, cytotoxic T cells, and helper T cells) into large armies against particular invaders.

7. Lymphocytes mount immune responses against circulating antigen or against cells bearing foreign antigen *in combination with* self-MHC markers.

8. An antigen is any large molecule that lymphocytes perceive as foreign and that triggers an immune response. Antigens occur at the surface of viruses, bacterial cells, fungal cells, and so on.

9. An antibody-mediated immune response is made against antigen circulating in the body's tissues or attached to the surface of an invading pathogen. First, macrophages and virgin B cells become sensitized to antigen (they display antigen–MHC complexes at their surface). When helper T cells encounter antigen–MHC complexes, they stimulate the virgin B cell to divide. Some B cell progeny develop into antibody-secreting plasma cells, others become memory B cells. Antibody molecules bind to specific antigens and mark the invaders for disposal (by macrophages or by complement).

10. A cell-mediated immune response is made against infected body cells and cancerous or mutant cells. Viral-infected cells have viral proteins combined with self-MHC markers on their surface. Cytotoxic T cells have receptors on their surface that may recognize the antigen–MHC complex. They directly destroy infected cells before the virus can replicate and spread to other cells. Natural killer cells, which are less specific about their targets, do the same.

11. After a primary (first-time) immune response, portions of the B and T cell populations produced continue to circulate as memory cells. They are available for a rapid, amplified response to subsequent encounters with the same antigen (a secondary immune response).

Review Questions

1. The vertebrate body has physical and chemical barriers against invading pathogens. Name five such barriers. *682*

2. Which four events characterize an inflammatory response? *683*

3. The vertebrate immune system is characterized by *specificity* and *memory*. Describe what these terms mean. *685, 691*

4. Define the following types of white blood cells: macrophages, helper T cells, B cells, cytotoxic T cells, suppressor T cells, and memory cells. *685*

5. Are phagocytes deployed during nonspecific defense responses, immune responses, or both? *683, 685*

6. Antibodies and interleukins are central to immune responses. Define them and state their functions. *685, 686*

7. What is immunization? What is a vaccine? *692*

8. What is the difference between an allergy and an autoimmune response? What type of disease is AIDS? *692–693*

Self-Quiz *(Answers in Appendix IV)*

1. _____ are any large molecules that white blood cells perceive as foreign and that elicit an immune response.

2. *Antibody-mediated* immune responses are made against _____; *cell-mediated* ones against _____.

3. External barriers to invasion include _____.
   a. unbroken skin
   b. lysozyme
   c. gastric fluid
   d. ciliated mucous membranes
   e. all of the above

4. Inflammatory responses require _____ and _____ as well as other factors to destroy foreign agents.
   a. complement, anticomplement
   b. phagocytes, antigens
   c. red blood cells, antigen
   d. phagocytes, complement

5. _____ are the fighting cells of the immune system.
   a. Red blood cells
   b. White blood cells
   c. Blue blood cells
   d. Antigens

6. The immune system shows _____ in that its cells are stimulated by specific antigens. It also exhibits _____, the ability to recognize the same invader upon subsequent attacks.
   a. communication; perception
   b. specificity; memory
   c. general responses; specific responses
   d. flexibility; recognition

7. The body's own uninfected cells are ignored by the immune response when they bear _____ at their surface.
   a. complement
   b. self-MHC markers
   c. antigen
   d. antigen plus self-MHC markers

8. Which of the following is *not* a molecular cue that triggers a normal immune response?
   a. self-MHC marker alone
   b. antigen
   c. antigen combined with self-MHC marker
   d. damaged or mutant self-MHC marker
   e. all of the above serve as molecular cues

9. An antibody is _____.
   a. an activated plasma cell
   b. a receptor molecule with binding sites for virgin B cells
   c. a receptor molecule with binding sites for antigen
   d. an out-of-body experience

10. Match the immunity concepts.
    _____ cytotoxic T cells
    _____ helper T cells
    _____ macrophages
    _____ some B cell progeny
    _____ portions of B and T cell populations

    a. stimulate virgin B cells and cytotoxic T cells to divide
    b. destroy infected cells by punching holes in them
    c. circulate as memory cells
    d. origin of plasma cells
    e. phagocytosis; stimulate helper T cells

Selected Key Terms

allergy *692*
antibody *686*
antibody-mediated response *686*
antigen *685*
autoimmune response *693*
B cell *685*
clonal selection theory *691*
clone *691*
complement system *682*
cytotoxic T cell *685*
helper T cell *685*
histamine *683*
immune system *685*
immunization *692*
immunoglobulin *687*
inflammatory response *682*
interferon *688*
interleukin *686*
lysis, *682*
macrophage *685*
memory *685*
memory cell *685*
MHC marker *685*
natural killer cell *688*
passive immunity *692*
pathogen *682*
perforin *688*
plasma cell *686*
primary immune response *686*
secondary immune response *691*
suppressor T cell *685*
vaccine *692*

Readings

Golub, E. 1987. *Immunology: A Synthesis*. Second edition. Sunderland, Massachusetts: Sinauer Associates.

Kimball, J. 1990. *Introduction to Immunology*. Third edition. New York: Macmillan.

Leder, P. May 1982. "The Genetics of Antibody Diversity." *Scientific American* 246(5):102–115.

Roitt, I., J. Brostoff, and D. Male. 1989. *Immunology*. St. Louis: Mosby. Second edition. Lavishly illustrated.

Tizard, I. 1988. *Immunology: An Introduction*. Second edition. Philadelphia: Saunders.

40 RESPIRATION

Figure 40.1 A climber approaching the summit of Chomolungma, where oxygen is brutally scarce.

Conquering Chomolungma

To experienced climbers, possibly the ultimate challenge is Chomolungma, a Himalayan mountain that also goes by the name Everest (Figure 40.1). Its summit, 9,700 meters (29,108 feet) above sea level, is the highest place on earth.

To conquer Chomolungma, climbers must be skilled enough to ascend vertical, iced-over rock in driving winds, smart enough to survive blinding blizzards, and lucky enough to escape heart-stopping avalanches. To conquer Chomolungma, they also must come to terms with a severe scarcity of oxygen that can result in long-lasting damage to the brain.

Of the air we breathe, only one molecule in five is oxygen. The earth's gravitational pull keeps oxygen molecules concentrated near sea level—or, as we say for gases, under pressure. We use so much oxygen (for aerobic respiration) that a pressure gradient exists between the outside air and the tissues inside our body. Red blood cells pick up and deliver oxygen along that gradient. Their hemoglobin becomes more or less saturated with oxygen in the lungs (the top of the gradient), then releases it in oxygen-depleted tissues (the bottom of the gradient).

Most of us live at low elevations. When we find ourselves in mountains taller than 3,300 meters (10,000 feet), the breathing game changes. Gravity's pull is less pronounced, gaseous molecules spread out—and the pressure gradient decreases. What would be a normal breath at sea level simply will not deliver enough oxygen into the lungs. The oxygen deficit can produce *altitude sickness*. Symptoms range from shortness of breath, headache, and heart palpitations, to loss of appetite, nausea, and vomiting.

The Chomolungma base camp is 6,300 meters (19,000 feet) above sea level. When climbers reach it, they have left behind more than half of the oxygen in the earth's atmosphere. Higher up, at 7,000 meters (23,000 feet), conditions start to get murderous. Apparently, when air pressure drops dramatically and oxygen becomes extremely scarce, blood vessels become leaky. Tissues in the brain and lungs become swollen with excess fluid, a condition called edema. With severe edema, climbers become comatose and die.

Given the risks, experienced climbers keep themselves in top physical condition. Besides this, climbers live for several weeks at high elevations before their assault on the summit. They know that "thinner air" triggers the formation of billions of additional red blood cells. Thinner air also stimulates the production of more blood capillaries, mitochondria, and myoglobin. (Myoglobin, an oxygen-binding protein, is present in skeletal muscle cells that depend on blood flow for oxygen deliveries.) These and other physiological changes improve the odds of survival. They are transient changes only. After the descent, the capacity

1. Of all organisms, animals are the most active. The energy to drive their activities comes mainly from aerobic metabolism, which uses oxygen and produces carbon dioxide wastes. In a process called respiration, animals move oxygen into their internal environment and give up carbon dioxide to the external environment.

2. All respiratory systems make use of the tendency of any gas to diffuse down its pressure gradient. Such a gradient exists between oxygen in the atmosphere (high pressure) and the metabolically active cells in body tissues (where oxygen is used rapidly; pressure is lowest here). Another gradient exists between carbon dioxide in body tissues (high pressure) and the atmosphere (with its lower amount of carbon dioxide).

3. Respiratory systems are all alike in having a respiratory surface—a thin, moist layer of epithelium that gases can readily diffuse across. In most animals, the oxygen is picked up by the general circulation and transported to body tissues, where carbon dioxide is picked up and transported back to the respiratory surface.

4. Respiratory systems differ in their adaptations for increasing gas exchange efficiency. They differ also in how they match air flow to blood flow.

for oxygen transport and utilization will return to normal.

Even with extensive preparation, climbers still can become incapacitated. One treatment for acute mountain sickness is to inhale bottled oxygen. More recently, a few casualties on Chomolungma were zipped inside an experimental inflatable bag. For the next two hours, oxygen was pumped inside the airtight bag and carbon dioxide removed from it until the internal pressure was the equivalent of descending 6,000 to 9,000 feet. Those climbers were among the lucky; they survived.

Few of us will ever find ourselves on Chomolungma, the roof of the world, pushing our reliance on oxygen to the limits. Here in the lowlands, disease, smoking, and other environmental insults push it in more ordinary ways—although the risks can be just as great, as you will see by this chapter's end.

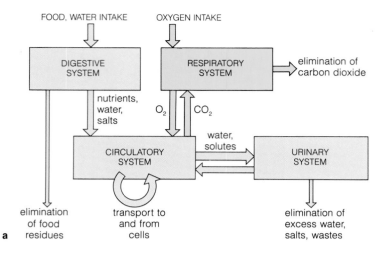

Figure 40.2 (**a**) Roles of the respiratory system in complex animals. Unlike humans (**b**), flatworms (**c**) are small enough that a circulatory system is not required; oxygen can reach individual cells simply by diffusing across the body surface. Unlike flatworms, humans would never survive on the low concentrations of oxygen dissolved in water.

High in the mountains, in underground nooks and burrows, in shallow waters and deep in the oceans, you will find oxygen-dependent animals. They use oxygen for aerobic respiration—the only metabolic pathway that generates enough energy for their activities. And they give up carbon dioxide by-products of metabolism to the surroundings. In a process called **respiration**, oxygen moves into the internal environment of such animals and carbon dioxide is released to the external environment. The respiratory system works in conjunction with other organ systems in the body, most notably the circulatory system. Figure 40.2 is a diagram of their interrelationships.

THE NATURE OF RESPIRATORY SYSTEMS

Factors That Affect Gas Exchange

Fick's Law. Respiratory systems are diverse, but they are all alike in their reliance on *diffusion* of gases. Like other substances, oxygen and carbon dioxide diffuse down concentration gradients—or, as we say for gases, down pressure gradients. The more concentrated the molecules of a gas are outside the body, the higher the pressure and the greater the force available to drive individual molecules inside, and vice versa.

Oxygen and carbon dioxide are not the only gases in water or air. At sea level, a given volume of dry air is approximately 78 percent nitrogen, 21 percent oxygen, 0.04 percent carbon dioxide, and 0.96 percent other gases. Each gas obviously exerts only part of the total pressure exerted by the whole mix of gases. Said another way, each exerts a "partial pressure."

At sea level, atmospheric pressure is about 760mm Hg, as measured by a mercury barometer (Figure 40.3). Thus, the partial pressure of oxygen is (760 × 21/100), or about 160mm Hg. The partial pressure of carbon dioxide is about 0.3mm Hg.

Like any gas, oxygen and carbon dioxide tend to diffuse from areas of high to low partial pressure. Respiratory systems take advantage of this tendency, for they work with partial pressure gradients that exist between the internal and external environments. In all cases, gases diffuse across a thin, moist membrane called the **respiratory surface**. The surface must be kept moist at all times, for gases will diffuse into and out of an animal only when they are first dissolved in some fluid.

What determines the amount of oxygen or carbon dioxide diffusing across the respiratory surface in a given time? According to **Fick's law**, the amount depends on the surface area of the membrane and the differences in partial pressure across it. The more extensive the surface area and the larger the pressure gradient, the faster will be the diffusion rate.

Surface-to-Volume Ratio. An animal's surface-to-volume ratio affects diffusion rates. Imagine an animal growing in all directions, like an inflating balloon. As it expands, its surface area does not increase at the same rate as its volume. (As we saw in Figure 4.0, volume increases with the cube of its dimensions, but surface

c

Figure 40.3 Atmospheric pressure as measured by a mercury barometer. At sea level, the level of mercury (Hg) in a glass column is about 760 millimeters (29.91 inches). At this level, the pressure exerted by the column of mercury equals atmospheric pressure outside the column.

760mm Hg

area only increases with the square.) Without further adaptations in the body plan, the animal would die once its diameter exceeded a single millimeter, for the diffusion distance between the respiratory membrane and all of its internal cells would be too great. That is why animals without respiratory organs have flattened or tubelike bodies; most internal cells are kept close to the respiratory surface. Flatworms and roundworms are examples.

Ventilation. There are many more adaptations to the constraints imposed by diffusion rates. When bony fishes move the "lids" over their gills, for example, they are stirring the water around them—and the gases dissolved in it. They are actively **ventilating** the body surface so that the fluid just outside the respiratory membrane does not become depleted of oxygen and loaded with carbon dioxide.

Similarly, the movements of microvilli on the collar cells of sponges help ventilate the body surface and so help improve diffusion rates. So does the action of ciliated cells that line certain cavities and body surfaces of sea stars. And when muscles move your rib cage as you breathe, the movement contributes to pressure gradients across the respiratory membrane of your lungs.

Transport Pigments. As we have seen, diffusion is faster when pressure gradients are steep. Hemoglobin and other pigments associated with the circulatory system are enormously important in this regard. In your own body, each hemoglobin molecule binds loosely with as many as four oxygen molecules in lungs (where oxygen concentrations are high), and it releases them in tissue regions where oxygen concentrations are low. By transporting oxygen away from the respiratory surface, hemoglobin plays a major role in maintaining the required pressure gradient.

Aquatic Environments. A liter of water that is "saturated" with dissolved oxygen still only holds 5 percent as much oxygen as a comparable volume of air. And water is much more dense and viscous than air. Aquatic animals work hard to maintain oxygen pressure gradients across their respiratory surfaces. For example, bony fishes busily ventilating their gills are doing so at great metabolic cost. Whereas a trout might devote 20 percent of its energy output to stirring up water around its gills, a buffalo staring out over the plains might devote a mere 2 percent to breathing.

Aquatic environments influence diffusion rates in other ways. The saltier the water or the higher its temperature, the less oxygen it can hold. The less sunlight there is, the lower the amount of oxygen released into the water by algae and other photosynthetic organisms. And the less the water circulates, the more depleted the oxygen becomes as aquatic animals and decomposers use up what is available. We will return to this topic in Chapter 47.

Land Environments. Compared to water, air has far more oxygen—but it also poses a far greater threat to respiratory systems. Any time the moist respiratory membranes dry out, they stick together and gases no longer can be exchanged across them. Conversely, earthworms and other inhabitants of the underground must contend with variable availability of oxygen in the soil. After a thunderstorm, for example, water may fill the spaces between soil particles. With the flooding, oxygen levels drop and carbon dioxide levels rise. That is why you often see earthworms thrashing about at the soil surface after a heavy rain.

Let's turn now to examples of respiratory systems. In the simplest of these systems, gases are exchanged across the body surface. In others, they are exchanged at specialized respiratory surfaces in gills, tracheas, and lungs.

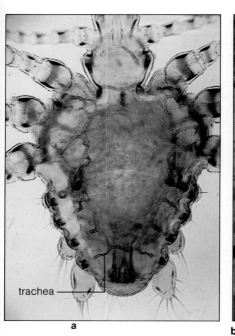

trachea

Figure 40.4 (**a**) Respiratory system of an insect (a louse). (**b**) A closer view of some chitin-reinforced tracheas.

Integumentary Exchange

Many animals do not have massive bodies or high metabolic rates, so their demands for respiration are not great. They rely on **integumentary exchange**, in which oxygen and carbon dioxide simply diffuse across a thin, vascularized layer of moist epidermis at the body surface. Most annelids, some small arthropods, and nudibranchs rely on integumentary exchange. To a large extent, so do frogs and other amphibians. For the water dwellers, the surroundings keep the respiratory surface moist. For land dwellers, mucus and other secretions provide the moisture.

For example, earthworms secrete mucus that helps moisten their integument, which is a single layer of epidermal cells. Oxygen molecules between soil particles dissolve in the mucus and diffuse across the integument. From there, oxygen diffuses into blood capillaries that project, fingerlike, between the epidermal cells. Pressure generated by muscular contractions of the body wall and by the pumping action of tiny "hearts" causes blood to circulate in the narrow, tubelike body. The bulk flow of blood enhances the transport of oxygen to individual cells.

Specialized Respiratory Surfaces of Tracheas, Gills, and Lungs

The integument of many animals is too thick, too hardened, or too sparsely supplied with blood vessels to be a good respiratory surface. Also, animals larger than flatworms cannot depend on their integument alone to provide enough surface area for gas exchange. Without other adaptations in body plan, a larger animal would die. Gases could not diffuse across the integument fast enough to sustain the greater volume of interior cells. This is where tracheas, gills, lungs, and other specialized respiratory organs come in.

Tracheas. Insects and spiders are among the animals with air-conducting tubes called **tracheas**. Most insect tracheas are chitin-reinforced (Figure 40.4). They branch through the body and provide a rather self-contained system of gas conduction and exchange; assistance by a circulatory system is not required. Often a lid (spiracle) spans each opening at the body surface and helps keep the tubes moist by preventing evaporation.

Tracheas branch again and again until they become very fine dead-end tubes, and at their tips they are filled with the liquid that is obligatory for gas exchange. These blind tubes are most abundant in tissues with high oxygen demands. For example, many of them terminate against muscle cells.

Have you ever noticed how foraging bees stop every so often and pump the segments of their abdomen back and forth? The segments extend and retract like a telescope, forcing air into and out of the tracheal system. The stepped-up oxygen intake and carbon dioxide removal help support the high rate of metabolism required for insect flight.

Gills. A typical **gill** has a moist, thin, vascularized layer of epidermis that functions in gas exchange. External gills project from the body of some insects, the larval forms of a few fishes, and a few amphibians. The internal gills of adult fishes are rows of slits or pockets extending from the back of the mouth to the body surface. Water enters the mouth, moves down the pharynx, and flows out across the gills. As Figure 40.5 shows, the water moves *over* fish gills and blood circulates *through* them in opposite directions. Such movement of fluids in opposing directions is called **countercurrent flow**.

An extensive network of blood vessels is associated with the surface of the gills. Water passing over a fish gill first flows over the vessels leading back into the body. Blood inside the vessels contains less oxygen than the surrounding water, so oxygen diffuses inward. Next the water flows over blood vessels from deep body regions. This blood still has less oxygen than the (by now) oxygen-poor water. With the even greater difference in partial pressure, more oxygen diffuses inward. Through this opposing flow mechanism, fishes extract about 80 to 90 percent of the oxygen from water that flows over the gills. That is far more than they would get from a one-way flow mechanism, at far less energy cost.

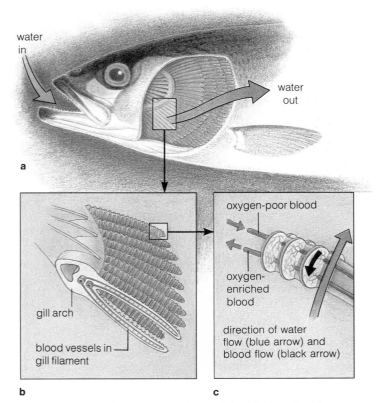

Figure 40.5 Respiratory system of many fishes. (**a**) Location of gills. The bony covering over them has been removed for this sketch. (**b**, **c**) Each gill has extensive capillary beds between two blood vessels. One vessel carries oxygen-poor blood into the gills, the other carries oxygen-enriched blood back into the deeper body tissues. Blood flowing from one vessel to the other runs counter to the direction of water flowing over the gills. The arrangement favors the movement of oxygen (down its partial pressure gradient) into both vessels.

Lungs. A **lung** is an internal respiratory surface in the shape of a cavity or sac. Simple lungs evolved more than 450 million years ago in fishes, and apparently they assisted respiration in oxygen-poor habitats. In some fish lineages, the lungs developed into moist, thin-walled organs called *swim bladders*. (Adjustments of gas volume in a swim bladder help maintain the body's position in the water. Some oxygen is also exchanged with blood and the surrounding tissues.) In other lineages, the lungs became complex respiratory organs, as indicated in Figure 40.6.

The evolution of lungs may be reflected in the respiratory systems of existing vertebrates. African lungfish have gills, but they also use lungs to supplement respiration. In fact, they will drown if they are kept from gulping air at the water's surface. Integumentary exchange still predominates in amphibians, but they, too, supplement respiration with a pair of small lungs. In reptiles, birds, and mammals, paired lungs are the major respiratory surfaces.

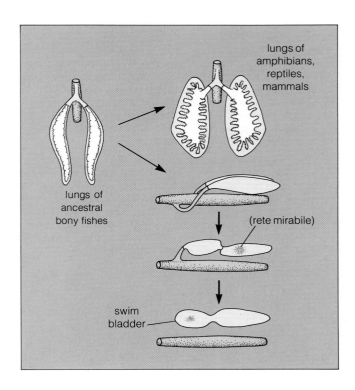

Figure 40.6 Evolution of vertebrate lungs and swim bladders. The esophagus (a tube leading to the stomach) is shaded gold; the respiratory tissues, pink.

Lungs originated as pockets off the anterior part of the gut; they increased the surface area for gas exchange in oxygen-poor habitats. In some lineages, lung sacs became modified into swim bladders: buoyancy devices that help keep the fish from sinking. Adjusting gas volume in the bladders allows fishes to remain at different depths.

Trout and other less specialized fishes have a duct between the swim bladder and esophagus; they replenish air in the bladder by surfacing and gulping air. Most bony fishes have no such duct; gases in the blood must diffuse into the swim bladder. Their swim bladder has a dense mesh of blood vessels (rete mirabile) in which arteries and veins run in opposite directions. Countercurrent flow through these vessels greatly increases gas concentrations in the bladder. Another region of the bladder allows reabsorption of gases by the body tissues.

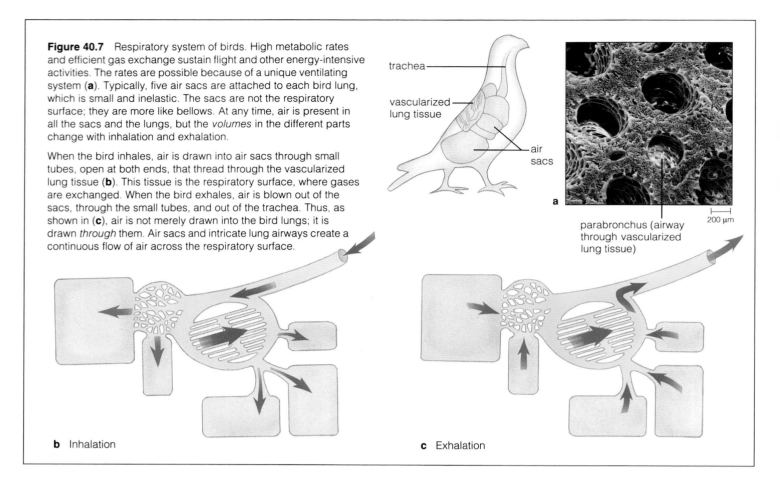

Figure 40.7 Respiratory system of birds. High metabolic rates and efficient gas exchange sustain flight and other energy-intensive activities. The rates are possible because of a unique ventilating system (**a**). Typically, five air sacs are attached to each bird lung, which is small and inelastic. The sacs are not the respiratory surface; they are more like bellows. At any time, air is present in all the sacs and the lungs, but the *volumes* in the different parts change with inhalation and exhalation.

When the bird inhales, air is drawn into air sacs through small tubes, open at both ends, that thread through the vascularized lung tissue (**b**). This tissue is the respiratory surface, where gases are exchanged. When the bird exhales, air is blown out of the sacs, through the small tubes, and out of the trachea. Thus, as shown in (**c**), air is not merely drawn into the bird lungs; it is drawn *through* them. Air sacs and intricate lung airways create a continuous flow of air across the respiratory surface.

trachea

vascularized lung tissue

air sacs

a

parabronchus (airway through vascularized lung tissue)

200 μm

b Inhalation

c Exhalation

In all lungs, *airways* carry gas to and from one side of the respiratory surface, and *blood vessels* carry gas to and from the other side:

1. Air moves by bulk flow into and out of the lungs, and new air is delivered to the respiratory surface.

2. Gases diffuse across the respiratory surface of the lungs.

3. Pulmonary circulation (the bulk flow of blood to and from the lung tissues) enhances the diffusion of dissolved gases into and out of lung capillaries.

4. In other tissues of the body, gases diffuse between blood and interstitial fluid, then between interstitial fluid and individual cells.

Let's focus now on the human respiratory system; its operating principles are the same for most vertebrates. The major exception is the respiratory system of birds, shown in Figure 40.7. Birds possess air sacs that function in ventilating the lungs, where gas exchange occurs.

HUMAN RESPIRATORY SYSTEM

Air-Conducting Portion

The human respiratory system is shown in Figure 40.8. Air enters and leaves through the nose and, to a lesser extent, through the mouth. Hairs and ciliated epithelium lining the two nasal cavities filter out dust and other large particles. Also in the nose, incoming air becomes warmed and additionally picks up moisture from mucus.

The filtered, warmed, and moistened air moves into the **pharynx**, or throat; this is the entrance to both the **larynx** (an airway) and the esophagus (a tube leading to the stomach). When you breathe, a flaplike structure attached to the larynx points up. This is the *epiglottis*, shown in Figure 40.9.

Vocal cords, two thickened folds of the larynx wall, contain muscles that help produce the sound waves necessary for speech. Look again at Figure 40.9. Air forced through the space between the vocal cords (the glottis) gives rise to sound waves. The greater the air pressure on the vocal cords, the louder the sound. The greater the muscle tension on the cords, the higher the pitch of the sound.

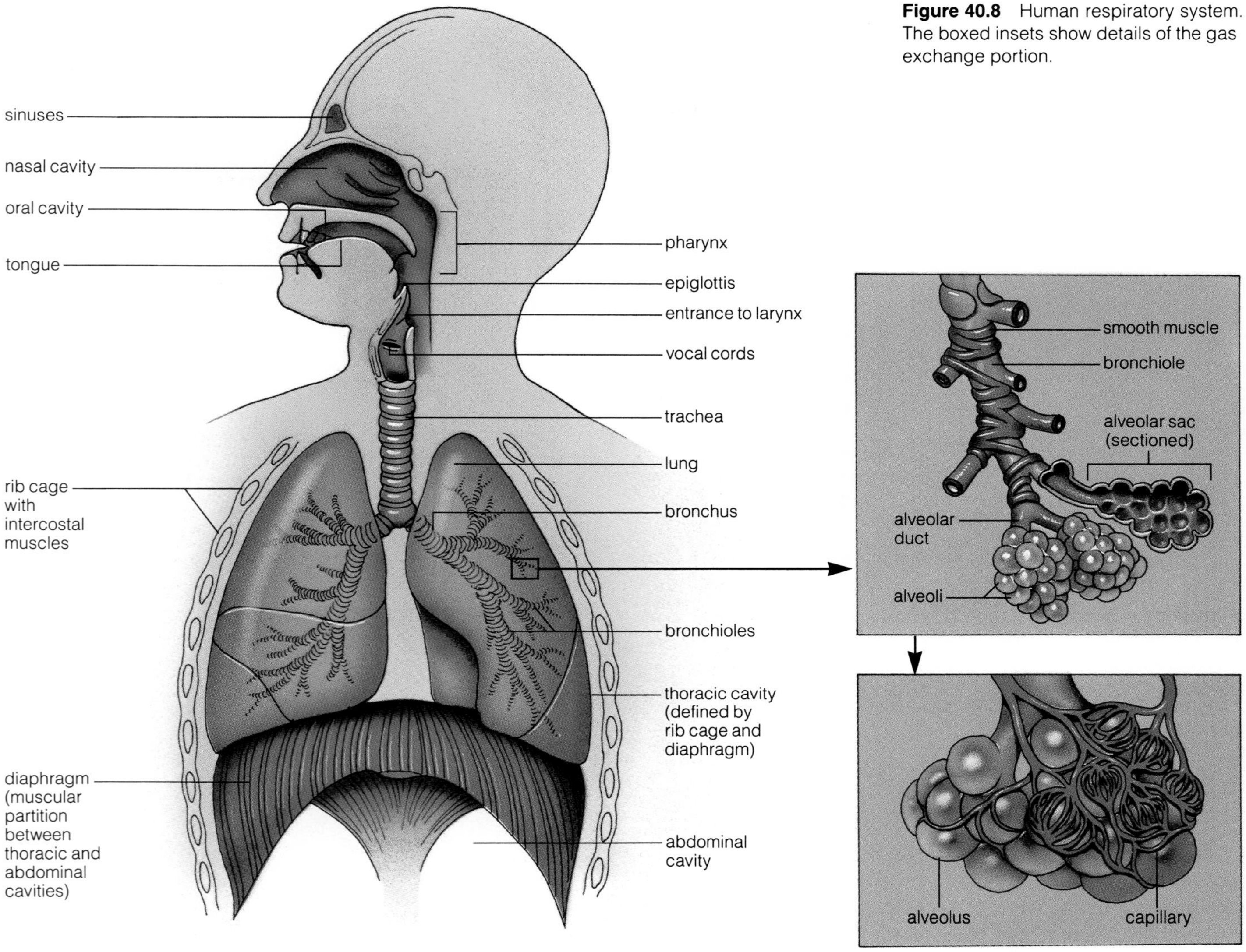

Figure 40.8 Human respiratory system. The boxed insets show details of the gas exchange portion.

sinuses

nasal cavity

oral cavity

tongue

pharynx

epiglottis

entrance to larynx

vocal cords

trachea

lung

rib cage with intercostal muscles

bronchus

bronchioles

thoracic cavity (defined by rib cage and diaphragm)

diaphragm (muscular partition between thoracic and abdominal cavities)

abdominal cavity

smooth muscle

bronchiole

alveolar sac (sectioned)

alveolar duct

alveoli

alveolus

capillary

Figure 40.9 Where the sounds necessary for speech originate. (**a**) Front view of the larynx, showing the location of the vocal cords. (**b**) The two vocal cords as viewed from above when the glottis (the space between them) is closed or opened.

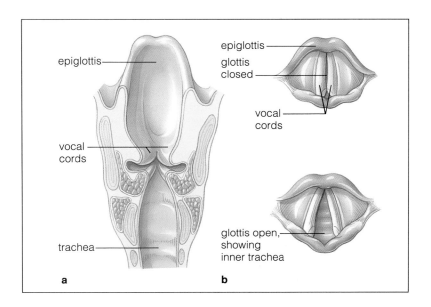

epiglottis

vocal cords

trachea

epiglottis

glottis closed

vocal cords

vocal cords

glottis open, showing inner trachea

a

b

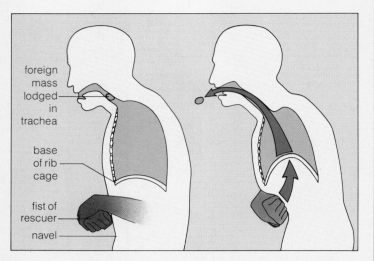

Figure 40.10 The Heimlich maneuver. Each year, several thousand people choke to death when food enters the trachea instead of the esophagus (compare Figure 37.4). Strangulation can occur when the air flow is blocked for as little as four or five minutes. The Heimlich maneuver, an emergency procedure only, often can dislodge the misdirected chunks of food. The idea is to elevate the diaphragm forcibly, causing a sharp decrease in the chest cavity volume and a sudden increase in alveolar pressure. The increased pressure forces air up the trachea and may be enough to dislodge the obstruction.

To perform the Heimlich maneuver, stand behind the victim, make a fist with one hand, then press the fist, thumb-side in, against the victim's abdomen. The fist must be slightly above the navel and well below the rib cage. Next, press the fist into the abdomen with a sudden upward thrust. Repeat the thrust several times if needed. The maneuver can be performed on someone who is standing, sitting, or lying down.

Once the obstacle is dislodged, be sure the person is seen at once by a physician, for an inexperienced rescuer can inadvertently cause internal injuries or crack a rib. It could be argued that the risk is worth taking, given that the alternative is death.

Figure 40.11 Color-enhanced scanning electron micrograph of cilia (gold) in the respiratory tract. Mucus-secreting cells (rust-colored) are interspersed among the ciliated cells. Foreign material sticks to the mucus-coated microvilli at the free surface of these cells, then the cilia sweep the mucus-laden debris back toward the mouth.

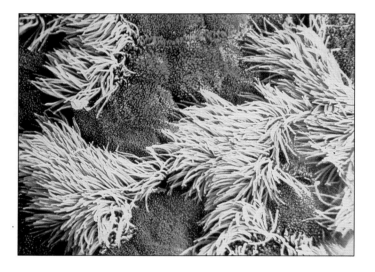

When you swallow, the larynx moves upward and presses the epiglottis down so that it partly covers the opening of the larynx. In this position, the epiglottis helps prevent food from going down the respiratory tract, the consequences of which can be fatal (Figure 40.10).

From the larynx, air moves into the **trachea**—the windpipe—which branches into the two airways leading into the lungs. Each airway is a **bronchus** (plural, bronchi). Its epithelial lining contains cilia and mucus-secreting cells, both with housekeeping roles (Figure 40.11). Bacteria and airborne particles stick in the mucus, then the upward-beating cilia sweep the debris-laden mucus toward the mouth.

Gas Exchange Portion

Humans have two elastic, cone-shaped lungs, separated from each other by the heart. The lungs are located in the rib cage above the *diaphragm*, a muscular partition between the chest cavity and abdominal cavity. They are not attached directly to the wall of the chest cavity. Each is positioned within a *pleural sac*—a thin, doubled-over membrane of epithelium and loose connective tissue.

Imagine pushing a closed fist into a fluid-filled balloon (Figure 40.12). A lung occupies the same kind of position as your fist, and the pleural membrane folds back on itself, as does the balloon. Only an extremely narrow *intrapleural space* separates the two facing surfaces of the membrane, which are coated with a thin film of lubricating fluid. The fluid prevents friction between the pleural membranes while you breathe. When the pleural membrane becomes inflamed and swollen, friction initially occurs, and breathing can be quite painful. As mentioned in Chapter 35, this condition is called *pleurisy*.

Inside the lungs, airways become progressively shorter, narrower, and more numerous. The terminal air-

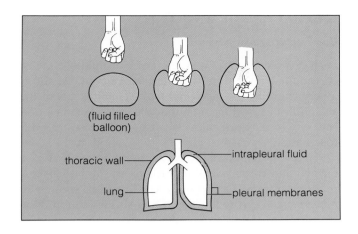

Figure 40.12 Position of the lungs and pleural sac relative to the chest (thoracic) cavity. By analogy, when you push a closed fist into a fluid-filled balloon, the balloon completely surrounds the fist except at your arm. A lung is analogous to the fist; the balloon, to the pleural sac. Here, intrapleural fluid volume is enormously exaggerated for clarity.

ways, the **respiratory bronchioles**, have cup-shaped outpouchings from their walls. Each outpouching is an **alveolus** (plural, alveoli). Most often, alveoli are clustered together, forming a larger pouch called an **alveolar sac** (Figure 40.8). The alveolar sacs are the major sites of gas exchange.

A dense mesh of blood capillaries surrounds the 150 million or so alveoli in each lung. Together, the alveoli provide a tremendous surface area for exchanging gases with the bloodstream. If they were stretched out as a single layer, they would cover the floor of a racquetball court!

AIR PRESSURE CHANGES IN THE LUNGS

Ventilation

Every time you take a breath, you are ventilating the respiratory surfaces of your lungs. When you "breathe," air is inhaled (drawn into the airways), then exhaled (expelled from them). The air movements result from rhythmic increases and decreases in the chest cavity's volume. The changing volumes of the chest cavity reverse the pressure gradients between the lungs and the air outside the body—and gases in the respiratory tract follow those gradients.

Figure 40.13 shows what happens as you start to inhale. The dome-shaped diaphragm contracts and flattens, and muscles lift the ribs upward and outward. Then, as the chest cavity expands, the rib cage moves away slightly from the lung surface. Pressure in the

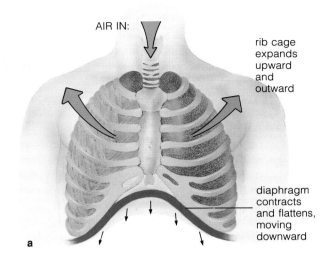

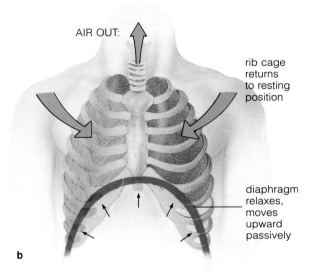

Figure 40.13 Changes in the size of the chest cavity during breathing. Blue line indicates the position of the diaphragm during inhalation (**a**) and exhalation (**b**).

narrow space between each lung and the pleural sac becomes even lower than it was, compared to atmospheric pressure. The pressure difference causes the lung itself to expand more. The expansion allows fresh air to flow down the airways, almost to the respiratory bronchioles.

As you start to exhale, the elastic lung tissue recoils passively. The volume of the chest cavity decreases and compresses the air in alveolar sacs. The alveolar pressure becomes greater than the atmospheric pressure, so air follows the gradient and moves out from the lungs. When the demand for oxygen increases, as during exercise, muscles of the rib cage and abdomen are recruited to hasten airflow. Breathing is now "active."

Lung Volumes

When you are resting, about 500 milliliters of air enter or leave your lungs in a normal breath. Although it takes a great leap of the imagination to compare the flow volume to the amount displaced by oceanic tides, this is nevertheless called the "tidal volume." The maximum volume of air that can move out of your lungs after a single, maximal inhalation is called the "vital capacity." You rarely use more than half the total vital capacity, even when you breathe deeply during strenuous exercise. To do so would exhaust the muscles used in respiration. Even at the end of your deepest exhalation, your lungs still cannot be completely emptied of air; about 1,000 milliliters would remain (Figure 40.14).

How much of the 500 milliliters of inhaled air is actually available for gas exchange? About 150 milliliters of exhaled air remain in the air-conducting tubes between breaths. Thus only (500 − 150), or 350 milliliters of fresh air reach the alveoli with each inhalation. When you breathe, say, 10 times a minute, you are supplying your alveoli with (350 × 10) or 3,500 milliliters of fresh air per minute.

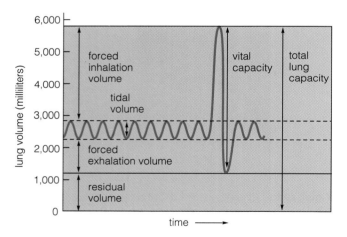

Figure 40.14 Capacities of the human lung. During normal breathing, a tidal volume of air enters and leaves the lungs. Forced inhalation can bring much larger quantities into the lungs, and forced exhalation can release some of the air normally kept in the lungs. A residual volume of gas remains trapped in partially filled alveoli despite the strongest exhalation.

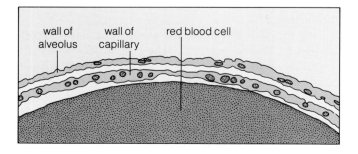

GAS EXCHANGE AND TRANSPORT

Gas Exchange in Alveoli

Each alveolus is only a single layer of epithelial cells, surrounded by a thin basement membrane. At most, a very thin film of interstitial fluid separates gas in the alveoli from blood in the lung capillaries. This is a narrow space, as Figure 40.15 suggests, and gases can diffuse rapidly across it.

Figure 40.16 shows the partial pressure gradients for oxygen and carbon dioxide throughout the human respiratory system. Passive diffusion alone is enough to move oxygen across the respiratory surface and into the bloodstream. And it is enough to move carbon dioxide in the reverse direction.

Driven by its partial pressure gradient, oxygen diffuses from alveolar air spaces, through interstitial fluid, and into the lung capillaries.

Carbon dioxide, driven by its partial pressure gradient, diffuses in the reverse direction.

Gas Transport Between Lungs and Tissues

Blood can carry only so much oxygen and carbon dioxide in dissolved form. The transport of both gases must be enhanced to meet the requirements of humans and other large-bodied animals. Hemoglobin, a pigment in red blood cells, binds and transports oxygen and carbon dioxide. It increases oxygen transport by seventy times. It also increases carbon dioxide transport away from the tissues by seventeen times.

Oxygen Transport.   There is plenty of oxygen and not much carbon dioxide in inhaled air that reaches the alveoli, but the opposite is true of blood entering the lung capillaries. So oxygen diffuses into the blood plasma, then into red blood cells. Once inside those cells, as many as four oxygen molecules can rapidly form a weak, reversible bond with each hemoglobin molecule. A hemoglobin molecule with oxygen bound to it is called **oxyhemoglobin**, or HbO_2. The amount of HbO_2 that forms depends on the partial pressure of oxygen. The higher the pressure, the more oxygen will be picked up, until all hemoglobin-binding sites are saturated.

Figure 40.15   Diagram of a section through an alveolus and an adjacent blood capillary. By comparison to the diameter of the red blood cell, the diffusion distance across the capillary wall, the interstitial fluid, and the alveolar wall is exceedingly small.

HbO₂ holds onto its oxygen rather weakly and will give it up in tissues where the partial pressure of oxygen is lower than in the lungs. It especially does this in tissues where blood is warmer, has an increase in the partial pressure of carbon dioxide, and shows a decrease in pH. All four conditions occur in tissues having greater metabolic activity—hence greater demands for oxygen. That is why more oxygen is released from blood in vigorously contracting muscle tissues, for example.

Carbon Dioxide Transport. The partial pressure of carbon dioxide in metabolically active tissues is greater than it is in blood flowing through the capillaries threading through them. Carbon dioxide diffuses into the capillaries, then it is transported to the lungs. About 7 percent of the carbon dioxide remains dissolved in plasma. About 23 percent binds with hemoglobin, forming **carbamino-hemoglobin** ($HbCO_2$). But most of it—approximately 70 percent—is transported in the form of bicarbonate (HCO_3^-).

The bicarbonate forms when carbon dioxide combines with water in plasma to form carbonic acid. The carbonic acid dissociates (separates) into bicarbonate and hydrogen ions:

$$CO_2 + H_2O \rightleftharpoons H_2CO_3 \rightleftharpoons HCO_3^- + H^+$$

The reactions proceed slowly in plasma, converting only 1 in every 1,000 molecules of carbon dioxide. Thus, the reaction rate is insignificant in plasma. But much of the carbon dioxide diffuses into red blood cells, which contain the enzyme **carbonic anhydrase**. With this enzyme, the reaction rate increases by 250 times. Inside red blood cells, most of the carbon dioxide that is not bound to hemoglobin is converted to carbonic acid and its dissociation products. As a result, the concentration of free carbon dioxide in the blood drops rapidly. Thus, carbonic anhydrase helps maintain the gradient that keeps carbon dioxide diffusing from interstitial fluid into the bloodstream.

What happens to the bicarbonate ions formed during the reaction? They tend to diffuse out of the red blood cells into the blood plasma. What about the hydrogen ions? Hemoglobin acts as a buffer for them and keeps the blood from becoming too acidic. A *buffer*, recall, is a molecule that combines with or releases hydrogen ions in response to changes in cellular pH.

The reactions are reversed in the alveoli, where the partial pressure of carbon dioxide is lower than it is in the surrounding capillaries. The water and carbon dioxide that form as a result of the reactions diffuse into the alveolar sacs. From there the carbon dioxide is exhaled from the body.

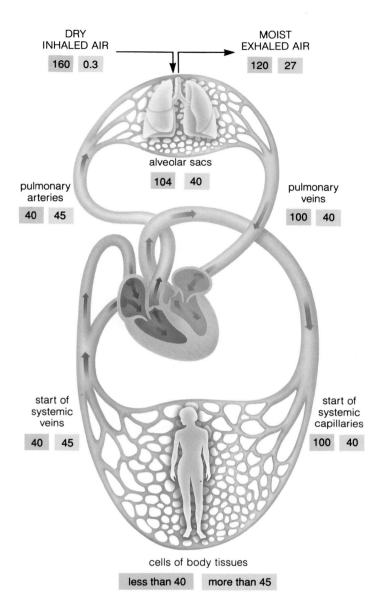

Figure 40.16 Partial pressure gradients for oxygen (blue boxes) and carbon dioxide (pink boxes) through the respiratory tract.

The point to remember about the values shown is that *each gas moves from regions of higher to lower partial pressure*. That is why, for example, you become light-headed when you first visit places at high altitudes. The partial pressure of oxygen decreases with altitude, and your body does not function as well when the pressure gradient between the surrounding air and your lungs is lower than what you normally encounter.

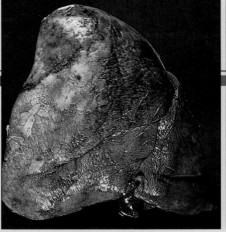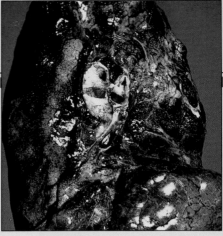

Commentary

When the Lungs Break Down

In large cities, in certain occupations, even near a cigarette smoker, airborne particles and irritant gases are present in abnormal amounts, and they put extra workloads on the lungs. Ciliated epithelium in the bronchioles is especially sensitive to cigarette smoke.

Bronchitis

A disorder called bronchitis can be brought on by smoking and other forms of air pollution that increase secretions of mucus and interfere with ciliary action in the lungs. Mucus and the particles it traps—including bacteria—accumulate in the trachea and bronchi, and this triggers coughing. The coughing persists as long as the irritation does and it aggravates the bronchial walls, which become inflamed. Bacteria or chemical agents start destroying the wall tissue. Cilia are lost from the lining, and mucus-secreting cells multiply as the body works to fight against the accumulating debris. Fibrous scar tissue forms and can obstruct parts of the respiratory tract.

Emphysema

An acute attack of bronchitis can be treated easily if the person is otherwise in good health. With continuing inflammation, however, fibrous scar tissue builds up and the bronchi become clogged with more and more mucus. Enzymes released from the inflammatory cells dissolve the elastic tissues of the alveoli, and the alveolar walls break down. Inelastic fibrous tissue comes to surround the alveoli. The remaining alveoli enlarge and the balance between air flow and blood flow is abnormal. The outcome is emphysema, in which the lungs are so distended and inelastic that gases cannot be exchanged efficiently. Running, walking, even exhaling can be difficult for those with emphysema.

(**a**) *Left:* Normal appearance of human lung tissue. *Right:* Appearance of lung tissue from someone affected by emphysema. (**b**) Cigarette smoke swirling into the human windpipe and down the two bronchial routes to the lungs.

Poor diet, smoking, and chronic colds and other respiratory ailments sometimes make a person susceptible to emphysema later in life. And many who suffer from emphysema do not have a functional gene coding for antitrypsin. This substance inhibits tissue-destroying enzymes produced by the inflammatory cells.

Emphysema can develop slowly, over twenty or thirty years. By the time it is detected, the damage to lung tissue cannot be repaired. On average, 1.3 million people in the United States alone suffer from the disorder.

Effects of Cigarette Smoke

The table below Figure *b* lists some effects of cigarette smoke on the lungs and other organs. Cilia in the bronchioles can be kept from beating for several hours by noxious particles in smoke from one cigarette. The particles also stimulate mucus secretions, which in time can clog the airways. They can kill the infection-fighting phagocytes that normally patrol the respiratory epithelium. "Smoker's cough" is not the only outcome; the coughing can pave the way for bronchitis and emphysema. Marijuana smoke also can cause extensive lung damage.

Controls Over Respiration

Gas exchange is most efficient when the rate of air flow is matched with the rate of blood flow. Both rates can be adjusted locally (in the alveoli of the lungs) and in the tissues of the body as a whole.

Local Controls. Local controls come into play in the lungs themselves when there are imbalances between air flow and blood flow. For example, when your heart is pounding and you don't breathe deeply enough, carbon dioxide levels increase in alveoli. The increase affects smooth muscle in the bronchiole walls. Bronchioles dilate, improving the match between the rates of air and blood flow.

Similarly, a decrease in carbon dioxide levels causes the bronchiole walls to constrict—thereby decreasing the air flow.

b

Risks Associated with Smoking	Benefits of Quitting
Shortened Life Expectancy: Nonsmokers live 8.3 years longer on average than those who smoke two packs daily from the midtwenties on	Cumulative risk reduction; after 10 to 15 years, life expectancy of ex-smokers approaches that of nonsmokers
Chronic Bronchitis, Emphysema: Smokers have 4–25 times more risk of dying from these diseases than do nonsmokers	Greater chance of improving lung function and slowing down rate of deterioration
Lung Cancer: Cigarette smoking the major cause of lung cancer	After 10 to 15 years, risk approaches that of nonsmokers
Cancer of Mouth: 3–10 times greater risk among smokers	After 10 to 15 years, risk is reduced to that of nonsmokers
Cancer of Larynx: 2.9–17.7 times more frequent among smokers	After 10 years, risk is reduced to that of nonsmokers
Cancer of Esophagus: 2–9 times greater risk of dying from this	Risk proportional to amount smoked; quitting should reduce it
Cancer of Pancreas: 2–5 times greater risk of dying from this	Risk proportional to amount smoked; quitting should reduce it
Cancer of Bladder: 7–10 times greater risk for smokers	Risk decreases gradually over 7 years to that of nonsmokers
Coronary Heart Disease: Cigarette smoking a major contributing factor	Risk drops sharply after a year; after 10 years, risk reduced to that of nonsmokers
Effects on Offspring: Women who smoke during pregnancy have more stillbirths, and weight of liveborns averages less (hence, babies are more vulnerable to disease, death)	When smoking stops before fourth month of pregnancy, risk of stillbirth and lower birthweight eliminated
Impaired Immune System Function: Increase in allergic responses, destruction of macrophages in respiratory tract	Avoidable by not smoking

Cigarette smoke contributes to lung cancer. Inside the body, certain compounds in coal tar and cigarette smoke become converted to highly reactive intermediates. These are the real carcinogens; they provoke uncontrolled cell divisions in lung tissues. In its terminal stage, the pain associated with lung cancer is agonizing.

Susceptibility to lung cancer is related to how many cigarettes the individual smokes daily and to how many times and how deeply smoke is inhaled. Cigarette smoking is responsible for at least 80 percent of all lung cancer deaths. It is a disorder that only 10 out of 100 smokers will survive.

Local changes also occur in the diameter of blood vessels that supply different lung regions. If air flow is too great relative to the blood flow, local concentrations of oxygen rise. The increase directly affects smooth muscle in the blood vessel walls, which undergo vasodilation, and thereby increase blood flow to the region. Similarly, if air flow is too small, vasoconstriction leads to a decrease in blood flow to the region, as described on page 673.

By now, it should be clear that gas exchange mechanisms at work in your body are like a fine Swiss watch—intricately coordinated and smooth in their operation. The *Commentary* describes a few respiratory disorders, some serious enough to stop the watch from ticking.

Neural Controls. The nervous system controls oxygen and carbon dioxide levels in arterial blood for the entire body. It does this by adjusting contractions of the diaphragm and muscles in the chest wall, and so adjusts the rate and depth of breathing.

The brain receives input from sensory receptors that can detect rising carbon dioxide levels in the blood. Such increases affect the H^+ concentration in cerebrospinal fluid, and the shift in pH stimulates the receptors. The brain also receives input from sensory receptors in the walls of arteries. Among the receptors are *carotid bodies* where the carotid arteries branch to the brain and the *aortic bodies* in arterial walls near the heart. Among other things, both types of receptors can detect changes in the partial pressure of oxygen dissolved in arterial blood. The brain responds by increasing the rate of ventilation, so more oxygen can be delivered to affected tissues.

Contraction of the diaphragm and muscles that move the rib cage are under the control of neurons in the brain's reticular formation (page 571). One cell cluster in this formation coordinates the signals calling for inhalation; another coordinates the signals for exhalation. The resulting rhythmic contractions are fine-tuned by respiratory centers in other parts of the brain, which can stimulate or inhibit both cell clusters.

RESPIRATION IN UNUSUAL ENVIRONMENTS

Decompression Sickness

In the world's oceans, the water pressure increases greatly with depth. Pressure in itself is not a problem, as long as it is dealt with. For example, to prevent their lungs from collapsing, deep-sea divers rely on tanks of compressed air (that is, air under pressure). Divers also must deal with the additional gaseous nitrogen (N_2) dissolved in their body fluids and tissues, especially adipose tissues.

As a diver ascends, the total pressure of the surrounding water decreases, so N_2 tends to move out of the tissues and into the bloodstream. If the ascent is too rapid, the N_2 comes out of solution faster than the lungs can dispose of it. When this happens, bubbles of N_2 form. Too many bubbles cause pain, especially at the joints. Hence the common name, "the bends," for what is otherwise known as *decompression sickness*. Bubbles that obstruct blood flow to the brain can lead to deafness, impaired vision, and paralysis. At depths of about 150 meters, N_2 poses still another threat, for at high partial pressures it produces feelings of euphoria, even drunkenness, and divers have been known to offer the mouthpiece of their airtank to a fish.

Weddell seals, fin whales, and other marine mammals have built-in respiratory adaptations that allow them to dive deeply with impunity. Members of a more ancient vertebrate lineage do the same, as the *Doing Science* essay describes.

Doing Science

The Leatherback Sea Turtle

Late on a starlit night in May, biologist Molly Lutcavage and several colleagues leave their bungalows on an island in the Caribbean and set out on a turtle patrol. Picking their way down the beach, they keep a steady eye on the pale surfline. Finally a large, dark shape with flippers flailing appears in the surf. A female Atlantic leatherback sea turtle (*Dermochelys coriacea*) is emerging from the sea to excavate a nest in the sand.

Turtles often are portrayed as plodding representatives of the reptilian lineage, which extends 300 million years back in time. Yet sea turtles are graceful swimmers in the world where they spend most of their lives. To add grave injury to insult, all modern sea turtles—be they ridleys, loggerheads, green turtles, or leatherbacks—are now endangered or threatened species. They are on the brink of extinction.

The race is on to learn about the life histories and physiology of sea turtles—information that may help save them from extinction. The leatherbacks are the largest and least understood. Adult females may weigh more than 400 kilograms (880 pounds). An adult male may weigh nearly twice that amount.

Leatherbacks normally leave the water only to breed and lay their eggs. The rest of the time they migrate across vast stretches of the open ocean. And leatherbacks do something no other reptile on earth can do. They can dive as deep as 1,000 meters (3,000 feet) below sea level. Prior to tracking experiments in the mid-1980s, Weddell seals, fin whales, and some other marine mammals were the only natural deep-sea divers known.

By some estimates, a turtle would have to swim for nearly forty minutes to reach such depths—all without taking a breath! How do they dive so deep, for so long, and still have enough oxygen for aerobic metabolism?

Consider that underwater staying power has nothing to do with the lungs. (Even at depths between 80 and 160 meters, the pressure exerted by the surrounding water causes air-filled lungs to collapse.) However, marine mammals have an abundance of myoglobin, the oxygen-binding protein, in their muscle cells. They store plenty of oxygen in their muscles. Compared to other mammals, they also have more blood per unit of body weight, and their blood contains more red blood cells—hence more hemoglobin. Such adaptations allow marine mammals to engage in aerobic respiration without having to come up for air.

Could it be that the leatherback—a bona fide reptile—has mammal-like adaptations for diving?

a An endangered species—female leatherback turtles (*Dermochelys coriacea*), returning to the sea.

Imagine yourself with Lutcavage's team on the sandy beach. You want to draw blood from a leatherback so you can study the oxygen-binding capacity of its hemoglobin. You want to obtain samples of respiratory gases to learn more about how the leatherback actually uses oxygen. You cannot do either when the turtle is swimming in the open ocean. But for the twenty minutes or so when a nesting female is actually depositing eggs on a beach, she enters a trancelike state. During that time, she will not resist being gently handled.

The team gets to work as soon as the female begins her egg-laying behavior. You fit a specially constructed helmet equipped with an expandable plastic sleeve over her head. The fit is snug but not restrictive, so the turtle can breathe normally. With each exhalation, the expired gases flow through a one-way valve into a collection bag. At the same time, you count the number of breaths required to completely fill the bag. (As a baseline, you also tallied the breathing frequency before you placed the helmet over her head.) Then you store the gas sample according to established procedures and set it aside for laboratory analysis.

Working quickly, another member of the team draws blood from a superficial vein in the turtle's neck, then packs the sample in ice to preserve it. Back in the laboratory, the sample will be split into three portions—one for hemoglobin analysis, another for a red blood cell count, and the third for measuring the oxygen level, carbon dioxide level, and pH of the turtle's blood.

Lutcavage and her coworkers gathered samples of blood and respiratory gases from several turtles. They also analyzed skeletal muscle tissue, taken earlier from a drowned turtle. And they pieced together the following picture of how leatherbacks manage their spectacularly deep dives.

From calculations of the turtle's oxygen uptake and total tidal volume during breathing, it became clear that a leatherback cannot inhale and hold enough air in the lungs to sustain aerobic metabolism during a prolonged dive. Other oxygen-supplying mechanisms must be at work.

Analysis also showed that myoglobin levels are high in leatherbacks. In fact, when air in their lungs gives out (or when their lungs collapse) during a dive, large amounts of oxygen would still be available in the muscles. Those muscles therefore could work longer at swimming.

Besides this, leatherbacks have a notable abundance of red blood cells. And the hemoglobin of those cells has a truly high affinity for binding oxygen. (A previous study had revealed the presence of cofactors in leatherback hemoglobin. The cofactors seem to physically alter the positions of potential oxygen-binding sites in ways that maximize oxygen uptake.) As a result, a leatherback's blood may be a remarkable 21 percent oxygen by volume. That is an amount characteristic of a highly active human, not a "plodding turtle." In fact, the oxygen-carrying capacity of leatherbacks is the highest ever recorded for a reptile—and they are close to the capacity of deep-diving mammals. Add to this a myoglobin-based oxygen-storage system, a streamlined body, and massive front flippers, and you have a reptile uniquely adapted for diving.

Studies of leatherback sea turtles are only beginning. Measurements of oxygen consumption by a female nesting on a beach probably do not reveal much about the turtle's normal, day-to-day oxygen use when it swims in the ocean. Researchers are pursuing radio telemetry methods that will let them monitor a turtle's metabolism at sea and during dives. So far, it has proved extremely difficult and expensive to track particular turtles in the vastness of the seas.

Physiological Comparison of a Few Diving Reptiles and Mammals

	Leatherback Turtle	Green Turtle	Crocodile	Killer Whale	Weddell Seal
Maximum diving depth (meters)	Greater than 1,000	Less than 100	Less than 30	260	600
Red blood cell count (percent of total volume)	39	30	28	44	58
Hemoglobin level (grams/ deciliter of blood)	15.6	8.8	8.7	16.0	17–22
Myoglobin level (milligrams/ gram of muscle tissue)	4.9	—*	—*	—*	44.6
Oxygen-carrying capacity (volume percent)	21	7.5–11.9	12.4	23.7	31.6

*Not known.

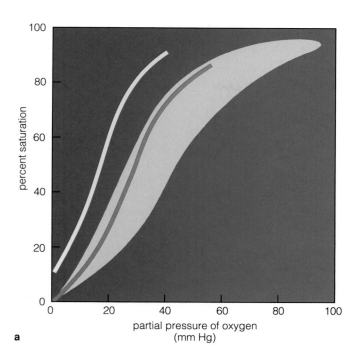

a

b

Figure 40.17 (**a**) Comparison of the binding and releasing capacity of human hemoglobin (red line) and llama hemoglobin (yellow line). The area shaded light blue indicates the range that is typical for most mammals. (**b**) Llamas high up in the Peruvian Andes.

Hypoxia

We conclude this chapter by coming full circle back to the example used to introduce it. The partial pressure of oxygen decreases with increasing altitude. Unlike the occasional climbers of Chomolungma, humans, llamas, and other animals accustomed to living at high altitudes have permanent adaptations to the thinner air. While they were growing up, more air sacs and blood vessels developed in their lungs. The ventricles in their heart became enlarged enough to pump larger volumes of blood. Llamas have an additional advantage. Compared to human hemoglobin, llama hemoglobin has a greater affinity for oxygen. It picks up oxygen more efficiently at the lower pressures characteristic of high altitudes (Figure 40.17).

Visitors who have not had time to adapt to the thinner air at high altitudes can suffer *hypoxia*, or cellular oxygen deficiency. Generally, at 2,400 meters (about 8,000 feet) above sea level, respiratory centers work to compensate for the oxygen deficiency by triggering *hyperventilation* (breathing much faster and more deeply than normal). As we saw at the start of this chapter, altitude sickness sets in at around 3,300 meters (10,000 feet).

Hypoxia also occurs when the partial pressure of oxygen in arterial blood falls because of *carbon monoxide poisoning*. Carbon monoxide, a colorless, odorless gas, is present in automobile exhaust fumes and in smoke from tobacco, coal, or wood burning. It binds to hemoglobin at least 200 times more strongly than oxygen does. Even very small amounts can tie up half of the body's hemoglobin and so impair oxygen delivery to tissues.

SUMMARY

1. Animal cells rely mainly on aerobic respiration, a metabolic pathway that provides enough energy for active life-styles. This pathway requires oxygen and produces carbon dioxide. The process by which the animal body as a whole acquires oxygen and disposes of carbon dioxide is called respiration.

2. Air is a mixture of gases, each exerting a partial pressure. Each gas tends to move from areas of higher to lower partial pressure. Respiratory systems make use of this tendency.

3. Different animals use the integument, gills, tracheas, or lungs as the basis of respiration. In all cases, oxygen and carbon dioxide diffuse across a moist, thin layer of epithelium (the respiratory surface). In vertebrates, airways carry gases to and from one side of the respiratory surface, and blood vessels carry gases to and from the other side.

4. The interconnected airways of the human respiratory system are the nasal cavities, pharynx, larynx, trachea, bronchi, and bronchioles. Alveoli, located at the end of the airways, are the primary gas exchange portion of the system.

5. During inhalation, the chest cavity expands, the pressure in the lungs falls below atmospheric pressure, and air flows into the lungs. During normal exhalation, these processes are reversed.

6. Driven by its partial pressure gradient, oxygen brought into the lungs diffuses from alveolar air spaces into the pulmonary capillaries. Then it diffuses into red blood cells and binds weakly with hemoglobin. When oxygen-enriched blood reaches body tissues, hemoglobin gives up the oxygen, which diffuses out of the capillaries, across interstitial fluid, and into cells.

7. Hemoglobin combines with or releases oxygen in response to shifts in oxygen levels, carbon dioxide levels, pH, and temperature.

8. Driven by its partial pressure gradient, carbon dioxide diffuses from cells, across interstitial fluid, and into the bloodstream. Most reacts with water to form bicarbonate, but the reactions are reversed in the lungs, where carbon dioxide diffuses from the lungs into the air spaces of the alveoli.

Review Questions

1. Label the component parts of the human respiratory system: *705*

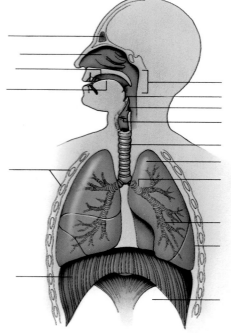

2. What is the main requirement for gas exchange in animals? What types of systems are used for gas exchange in (a) water-dwelling animals and (b) land-dwelling animals? *700–701*

3. What is oxyhemoglobin? Where does it form? *708*

4. What drives oxygen from alveolar air spaces, through interstitial fluid, and across capillary epithelium? What drives carbon dioxide in the reverse direction? *708*

5. How does hemoglobin help maintain the oxygen partial pressure gradient during gas transport in the body? What reactions enhance the transport of carbon dioxide through the body? *708–709*

6. Gas exchange is most efficient when the rates of air flow and blood flow are balanced. Give an example of a local control that comes into play in the lungs when the two rates are imbalanced. Do the same for a neural control. *710–712*

Self-Quiz (Answers in Appendix IV)

1. _____ is used and _____ is produced in aerobic metabolism.

2. Operation of respiratory systems depends on the tendency of a _____ to diffuse down its _____ .

3. Respiratory systems require thin, moist layers of _____ across which gases can easily _____ .

4. Respiratory systems differ in their _____ for increasing gas exchange efficiency and the means for matching _____ to blood flow.

5. Which of the following is *not* related to the function of a respiratory system?
   a. air is a mixture of gases
   b. each gas in air exerts a partial pressure
   c. each gas in air tends to move from areas of higher to lower partial pressure
   d. all of the above are directly applicable

6. The basis of respiration in different animals might be _____ .
   a. the body surface
   b. gills
   c. tracheas
   d. lungs
   e. all of the above
   f. b, c, and d above

7. During inhalation, _____ .
   a. the pressure in the thoracic cavity is greater than the pressure within the lungs
   b. the pressure in the thoracic cavity is less than the pressure within the lungs
   c. the diaphragm moves upward and becomes more curved
   d. the chest cavity volume decreases

8. Oxygen diffusing into pulmonary capillaries also diffuses into _____ and binds with _____ .
   a. white blood cells; carbon dioxide
   b. red blood cells; carbon dioxide
   c. white blood cells; hemoglobin
   d. red blood cells; hemoglobin

9. Due to its partial pressure gradient, carbon dioxide diffuses from cells, into interstitial fluid, and into the _____ ; in the lungs, carbon dioxide diffuses into the _____ .
   a. alveoli; bronchioles
   b. bloodstream; bronchioles
   c. alveoli; bloodstream
   d. bloodstream; alveoli

10. Match these respiratory components with their descriptions.
   _____ bronchi
   _____ alveoli
   _____ trachea
   _____ larynx
   _____ pharynx

   a. microscopically small air sacs where gases are exchanged
   b. contains vocal cords
   c. throat
   d. air-conducting tube; human windpipe
   e. connects trachea to lungs

Selected Key Terms

alveolar sac 707
alveolus 707
aortic body 712
blood vessel 704
bronchus 706
buffer 709
carbamohemoglobin 709
carbonic anhydrase 709
carotid body 712
countercurrent flow 702
diaphragm 706
diffusion 700
epiglottis 704
Fick's law 700
gill 702

integumentary exchange 702
larynx 704
lung 703
oxyhemoglobin 708
pharynx 704
pleural sac 706
respiration 700
respiratory bronchiole 707
respiratory surface 700
swim bladder 703
tidal volume 708
trachea 706
ventilating 701
vocal cord 704

Readings

American Cancer Society. 1980. *Dangers of Smoking; Benefits of Quitting and Relative Risks of Reduced Exposure.* Revised edition. New York: American Cancer Society.

Vander, A., J. Sherman, and D. Luciano. 1990. *Human Physiology: The Mechanisms of Body Function.* Fifth edition. New York: McGraw-Hill. Clear introduction to the respiratory system.

West, J. 1989. *Respiratory Physiology: The Essentials.* Fourth edition. Baltimore: Williams & Wilkins. Excellent, brief introduction to respiratory functions. Paperback.

Tale of the Desert Rat

Judging from the fossil record, about 375 million years ago some animals left the ancient, shallow seas behind and began invading the land. They had a demanding evolutionary legacy—their cells were geared to operating in a salty fluid of relatively stable temperature. Why did their invasion succeed? Largely because those animals carried water and solutes inside them—an "internal environment," so to speak—that could bathe and otherwise service their cells.

Living on land presented new challenges. Winds and radiant energy from the sun could dehydrate the body. Water was not always plentiful and most of it was fresh, not salty. Temperatures changed more dramatically from day to night and from one season to the next. Life on land required adaptations that could protect the volume, composition, and temperature of the internal environment.

Things are no different today. No matter which land-dwelling animal you observe, you discover fascinating aspects of body plan, body functions, and behavior that counter threats to the stability of the internal environment.

Think about a kangaroo rat, living in an isolated desert of New Mexico (Figure 41.1). Winter rains are brief, then for months on end the sun bakes the sand and the only free water is the water sloshing in the canteens of researchers or tourists. Yet with nary a sip of free water, the kangaroo rat maintains its internal environment with exquisite precision until the next season's rains.

Kangaroo rats wait out the daytime heat in burrows, then forage in the cool of night for dry seeds or the occasional succulent. They are not sluggish about this. They hop rapidly over sizeable distances as they search for seeds and flee from snakes and other predators. All that hopping depends on a good supply of ATP energy *and* water for metabolic reactions.

Seeds provide the energy—and they replenish water in the internal environment. Seeds are rich in carbohydrates, which are chockful of energy. When living

Figure 41.1 A kangaroo rat, master of water conservation in the deserts.

Table 41.1	Normal Balance Between Water Gain and Water Loss in Humans and in Kangaroo Rats			
Organism	Water Gain (milliliters)		Water Loss (milliliters)	
Adult human (measured on daily basis)	Ingested in solids:	850	Urine:	1,500
	Ingested as liquids:	1,400	Feces:	200
	Metabolically derived:	350	Evaporation:	900
		2,600		2,600
Kangaroo rat (measured over 4 weeks)	Ingested in solids:	6.0	Urine:	13.5
	Ingested as liquids:	0	Feces:	2.6
	Metabolically derived:	54.0	Evaporation:	43.9
		60.0		60.0

cells break down carbohydrates, the reactions yield water. This "metabolic water" represents about 12 percent of the water that your own body gains each day. It represents a whopping *90 percent* of the total intake for kangaroo rats.

Beyond this, when kangaroo rats return to their burrow after a night of foraging, they empty their cheek pouches of seeds. There, in the cool burrow, the seeds soak up water vapor that is exhaled with each rat breath—and when the seeds are eaten, that water is recycled to the internal environment. In such ways, kangaroo rats use seeds to replenish their water stores.

Besides replenishing and recycling water, kangaroo rats are natural experts at conserving it. As the rat breathes, some water vapor condenses on the cool nasal epithelium, like beads of water condensing on a glass of iced tea, then diffuses back into the internal environment. Kangaroo rats have no sweat glands, so they don't lose water by perspiration. Like all other mammals, they do have kidneys. Water and solutes from the blood flow continuously through the kidneys, where adjustments are made with respect to how much water and which solutes are reabsorbed or disposed of (as urine). Of all the water your own body loses on an average day, nearly two-thirds of it exits by way of the kidneys. The kangaroo rat's kidneys are twice as efficient at reducing urinary water loss, for reasons that will become apparent in this chapter.

In such ways, the kangaroo rat gets and gives up water and solutes. Like other land animals, it survives only as long as the daily gains *balance* the daily losses, as Table 41.1 indicates. How the balancing acts are accomplished will be our initial focus in this chapter. Then we will consider how mammals survive in hot, cold, and sometimes capricious climates on land.

KEY CONCEPTS

1. To maintain a hospitable internal environment, all animals must make controlled adjustments to the external environment as well as obligatory exchanges with it.

2. Animals continually take in water, nutrients, and ions, and they produce metabolic wastes. In vertebrates, the kidneys help balance the intake and output of water and dissolved substances.

3. Urine forms in kidney nephrons. Its volume and composition result from the processes of filtration, reabsorption, and secretion.

4. Body temperatures of animals depend on the balance between heat produced through metabolism, heat absorbed from the environment, and heat lost to the environment.

5. Body temperature is maintained within a favorable range through controls over metabolic activity and adaptations in structure, physiology, and behavior.

MAINTAINING THE EXTRACELLULAR FLUID

When we speak of the internal environment, we are talking about both blood plasma and interstitial fluid, the solute-rich solutions in the body's tissues. Together, they are the *extracellular fluid* that services all living cells of the body (page 542). Maintaining the volume and composition of this fluid is absolutely essential for cell survival. Humans and many other animals have a well-developed urinary system that serves this function. Figure 41.2 shows how the urinary system is linked with other organ systems in the human body.

Water Gains and Losses

Ordinarily, and on a daily basis, humans and other mammals take in just as much water as they lose. Table 41.1 gives two examples of this balancing act. The body *gains* water through two processes:

1. Absorption of water from liquids and solid foods in the gut.

2. Metabolism—specifically, the breakdown of carbohydrates, fats, and other organic molecules in reactions that yield water as a by-product.

Among land-dwelling mammals, thirst behavior influences the gain of water. When water levels decrease, the brain compels the individual to seek out water holes, streams, cold drinks in the refrigerator, and so on. The mechanisms involved are described later in the chapter.

The mammalian body *loses* water mostly by the following processes:

1. Excretion by way of the urinary system.

2. Evaporation from the respiratory surface.

3. Evaporation through the skin.

4. Sweating.

5. Elimination by way of the gut.

The process of greatest importance in controlling water loss is **urinary excretion**. By this process, organs called kidneys help eliminate excess water and excess or harmful solutes from the internal environment. The evaporative processes listed are called "insensible water losses" because the individual is not consciously aware that they are taking place. Temperature-control centers in the brain govern sweating. Normally, the large intestine reabsorbs nearly all water in the gut, so very little water leaves the body in feces.

Solute Gains and Losses

The preceding chapter described how oxygen enters the internal environment by crossing respiratory surfaces. Aside from oxygen, many diverse solutes enter the internal environment by three processes:

1. Absorption from the gut. The absorbed substances include *nutrients* such as glucose, which are used as energy sources and in biosynthesis reactions. They also include drugs, food additives, and *mineral ions*, such as sodium and potassium ions.

2. Secretion of hormones and other substances.

3. Metabolism, which produces carbon dioxide and other *waste products* of degradative reactions.

Carbon dioxide, the most abundant waste product of metabolism, is eliminated from the body at respiratory surfaces. Aside from carbon dioxide, these are the other major metabolic wastes that must be eliminated from the body:

1. *Ammonia*, formed in "deamination" reactions whereby amino groups are stripped from amino acids. If allowed to accumulate in the body, ammonia can be highly toxic.

2. *Urea*, produced in the liver in reactions that link two ammonia molecules to carbon dioxide. Urea is the main nitrogen-containing waste product of protein breakdown and is relatively harmless.

3. *Uric acid*, formed in reactions that degrade nucleic acids. If allowed to accumulate, uric acid can crystallize and sometimes collect in the joints.

4. *Phosphoric acid* and *sulfuric acid*, also produced during protein breakdown.

Usually, protein breakdown produces small amounts of other metabolic wastes besides the ones just listed. Some are highly toxic and may be responsible for many of the symptoms associated with kidney failure.

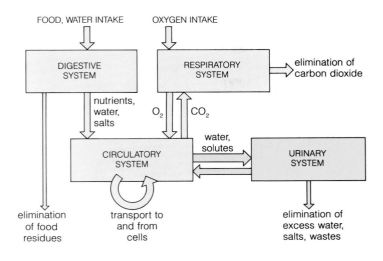

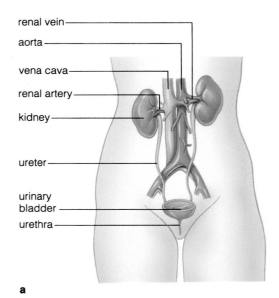

Figure 41.2 Links between the urinary system and other organ systems that maintain operating conditions in the internal environment.

a

Urinary System of Mammals

In all mammals, a pair of organs called **kidneys** continuously filter water, mineral ions, organic wastes, and other substances from the blood. Only a tiny portion of the water and solutes entering the kidneys leaves as a fluid called **urine**. In fact, all but about 1 percent is returned to the blood. But the composition of the fluid that *is* returned has been adjusted in vital ways. *Through their action, kidneys regulate the volume and solute concentrations of extracellular fluid.*

Each kidney has an outer cortex wrapped around a central region, the medulla. A tough coat of connective tissue encloses both. That coat is the renal capsule (from the Latin *renes*, meaning kidneys).

Internally, each kidney is divided into several lobes (Figure 41.3). Each lobe contains blood vessels and numerous slender tubes called **nephrons**. Water and solutes filtering out of the blood enter the nephrons. Most of the filtrate is reabsorbed from nephrons, but some continues on through tubelike collecting ducts and into the kidney's central cavity (renal pelvis). This fluid is the urine.

Urine flows from each kidney into a *ureter*, then into a storage organ, the *urinary bladder*. It leaves through a long tube, the *urethra*, which leads to the outside. The two kidneys, two ureters, urinary bladder, and urethra constitute the **urinary system** of mammals (Figure 41.3).

Urination, or urine flow from the body, is a reflex response. As a urinary bladder fills, tension increases in its strong, smooth-muscled walls. The increased tension causes muscles that prevent the flow of urine into the urethra to relax. At the same time, the bladder walls contract and force fluid through the urethra. The reflex response is involuntary but can be consciously blocked.

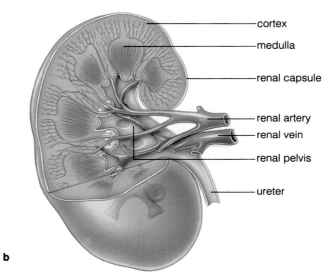

b

Figure 41.3 (**a**) Components of the human urinary system. (**b**) Closer look at the kidney.

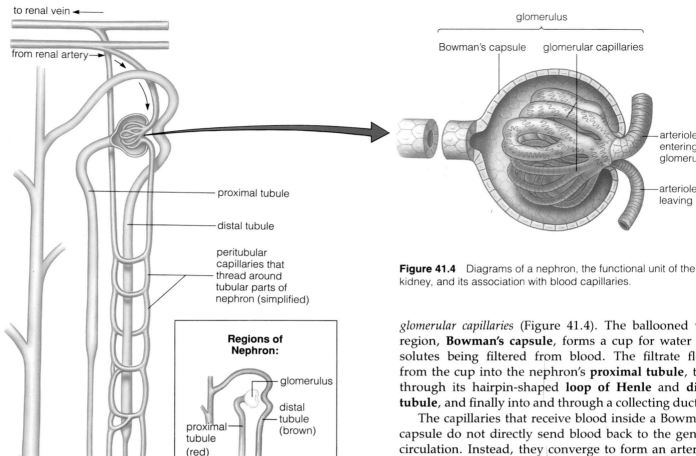

to renal vein ←

from renal artery →

proximal tubule

distal tubule

peritubular capillaries that thread around tubular parts of nephron (simplified)

collecting duct

Regions of Nephron:

glomerulus

proximal tubule (red)

distal tubule (brown)

loop of Henle (yellow)

glomerulus

Bowman's capsule   glomerular capillaries

arteriole entering glomerulus

arteriole leaving

Figure 41.4 Diagrams of a nephron, the functional unit of the kidney, and its association with blood capillaries.

You may have heard about *kidney stones*, these being deposits of uric acid, calcium salts, and other substances that settled out of urine and collected in the renal pelvis. At times a stone becomes lodged in the ureter or, on rare occasions, in the urethra. It can interfere with urine flow and cause pain. Kidney stones usually pass naturally from the body during urination. If they do not, they must be removed by medical or surgical procedures.

Nephron Structure

Each fist-sized kidney of humans has more than a million nephrons. The nephron wall consists only of a layer of epithelial cells, but the cells and junctions between them are not all the same. Some wall regions are highly permeable to water and solutes. Still other regions bar the passage of solutes *except* at active transport systems built into the plasma membrane (page 85).

Filtration starts at the **glomerulus**, where the nephron wall balloons around a cluster of blood capillaries called

glomerular capillaries (Figure 41.4). The ballooned wall region, **Bowman's capsule**, forms a cup for water and solutes being filtered from blood. The filtrate flows from the cup into the nephron's **proximal tubule**, then through its hairpin-shaped **loop of Henle** and **distal tubule**, and finally into and through a collecting duct.

The capillaries that receive blood inside a Bowman's capsule do not directly send blood back to the general circulation. Instead, they converge to form an arteriole that branches into *another* set of capillaries. This set, the *peritubular capillaries*, threads around the rest of the nephron and recaptures water and essential solutes (Figure 41.4). Eventually these capillaries merge to form veins, which carry blood out of the kidney.

URINE FORMATION

Urine-Forming Processes

By definition, urine is a fluid by which the body rids itself of water and solutes in excess of the amounts necessary to maintain the extracellular fluid. Figure 41.5 outlines the three processes required for urine formation. These are filtration, tubular reabsorption, and secretion.

Filtration starts and ends at the glomerulus. In this process, blood pressure (generated by heart contractions) forces water and solutes out of the glomerular capillaries and into the cupped region inside Bowman's capsule. The blood is said to be filtered here because blood cells, proteins, and other large solutes are left behind as water and smaller solutes (such as glucose, sodium, and urea) are forced out of the capillaries. The filtrate itself will flow on, into the proximal tubule.

Reabsorption takes place at tubular parts of the nephron. In this process, water and solutes move across the tubular wall and *out* of the nephron (by diffusion or active transport), then into adjacent capillaries. Most of the filtrate's water and usable solutes are reclaimed here and are returned to the general circulation. Table 41.2 gives some examples of daily reabsorption values.

Secretion also occurs across the tubular walls, *but in the opposite direction*. Excess hydrogen ions, potassium ions, and a few other substances move *out* of the capillaries and into cells making up the nephron walls. Then those cells secrete the substances into fluid within the tubule. Secretion, a highly regulated process, also rids the body of uric acid, some breakdown products of hemoglobin and other proteins, and some other metabolic wastes. Additionally, it can rid the body of many drugs, including penicillin, and other foreign substances.

A concentrated or dilute urine forms through the processes of filtration, reabsorption, and secretion.

The urine contains waste products as well as water and solutes in excess of the amounts necessary to maintain the extracellular fluid.

Factors Influencing Filtration

Each day, more blood flows through the kidneys than through the tissues of any other organ except the lungs. Each minute, about 1-1/2 quarts of blood course through them! That is nearly one-fourth of the cardiac output. How can kidneys handle blood flowing through on such a massive scale? There are two mechanisms.

First, the arterioles delivering blood to a glomerulus have a wider diameter—and less resistance to flow—than most arterioles. The hydrostatic pressure caused by heart contractions therefore does not drop as much when blood flows through them. As a result, pressure in the glomerular capillaries is higher than in other capillaries. *Second*, glomerular capillaries are highly permeable. They do not allow blood cells or protein molecules to escape, but compared to other capillaries, they are 10 to 100 times more permeable to water and small solutes. Because of the higher hydrostatic pressure and greater capillary permeability, the kidneys can filter an average of 45 gallons (180 liters) per day.

At any given time, the *rate* at which kidneys filter a given volume of blood depends on the blood flow to them and on how fast their tubules are reabsorbing water. Reabsorption, as you will see shortly, is partly under hormonal control. Blood flow to the kidneys is influenced by neural controls over blood flow through the body as a whole. Recall that when you exercise, more blood than usual must be diverted to your heart and skeletal muscles to sustain the increased activity.

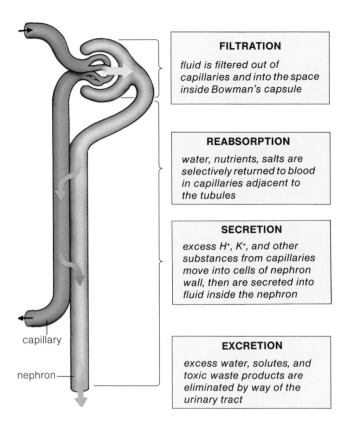

FILTRATION

fluid is filtered out of capillaries and into the space inside Bowman's capsule

REABSORPTION

water, nutrients, salts are selectively returned to blood in capillaries adjacent to the tubules

SECRETION

excess H⁺, K⁺, and other substances from capillaries move into cells of nephron wall, then are secreted into fluid inside the nephron

EXCRETION

excess water, solutes, and toxic waste products are eliminated by way of the urinary tract

capillary

nephron

Figure 41.5 Processes involved in urine formation and excretion.

Table 41.2 Average Daily Reabsorption Values for a Few Substances

	Filtered	Excreted	Proportion Reabsorbed
Water	180 liters	1.8 liters	99%
Glucose	180 grams	None, normally	100%
Sodium ions	630 grams	3.2 grams	99.5%
Urea	54 grams	30 grams	44%

Blood is diverted away from the kidneys when neural signals call for vasoconstriction of the arterioles leading into them. Because the flow volume to the kidneys is reduced, the filtration rate decreases.

Local chemical signals also influence filtration rates. When arterial blood pressure decreases, locally produced chemicals stimulate vasodilation of the arterioles leading into the glomeruli, so more blood flows in. When blood pressure rises, the arterioles are stimulated to constrict, and less blood flows in. This helps keep blood flow to the kidneys relatively constant, despite changes in blood pressure.

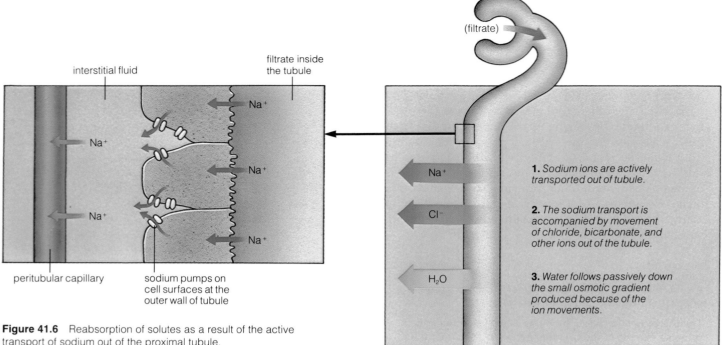

Figure 41.6 Reabsorption of solutes as a result of the active transport of sodium out of the proximal tubule.

Labels in figure:
interstitial fluid
filtrate inside the tubule
Na⁺
peritubular capillary
sodium pumps on cell surfaces at the outer wall of tubule
(filtrate)
Na⁺
Cl⁻
H₂O

1. Sodium ions are actively transported out of tubule.

2. The sodium transport is accompanied by movement of chloride, bicarbonate, and other ions out of the tubule.

3. Water follows passively down the small osmotic gradient produced because of the ion movements.

Reabsorption of Water and Sodium

Only tiny fractions of the water and sodium that enter the kidney are excreted in the urine. Yet even those minuscule amounts must be carefully regulated so that the internal environment remains constant.

Normally, for example, you cannot upset the balance for long by drinking too much water at lunch. The kidneys simply do not reclaim the excess; it will be excreted. Similarly, you cannot upset the balance by not taking in enough water, at least over the short term. The kidneys simply will conserve water already in the body; only a small volume of urine will be excreted.

Similarly, eating a large bag of salty potato chips for lunch will put excess sodium into your bloodstream, but the excess will be excreted. Conversely, if your doctor were to place you on a low-sodium diet, your body will work to conserve sodium. A larger fraction of the sodium in the filtrate passing through nephrons will be reabsorbed, and less will be excreted in the urine. The same thing happens when too much sodium is lost through diarrhea or excessive sweating. How are regulatory events such as this accomplished? Let's take a look.

Proximal Tubule. Of all the water and sodium filtered in the kidneys, more than half is promptly reabsorbed at the proximal tubule—the part of the nephron closest to the glomerulus. As Figure 41.6 shows, epithelial cells making up the tubule wall have transport proteins at their outer surface. Nearly all cells have pro-

teins of this sort, which function as sodium "pumps." In this case, the proteins actively transport sodium ions from the filtrate into the interstitial fluid. An outward movement of chloride, bicarbonate, and other ions accompanies the sodium transport.

The solute concentration inside the tubule drops only slightly because of this outward movement. At the same time, however, the movement affects the osmotic gradient between the filtrate and the interstitial fluid. The wall of the proximal tubule happens to be quite permeable to water. So now water follows the small osmotic gradient—it moves passively out of the tubule.

In this fashion, the volume of fluid remaining within the tubule decreases greatly, but the concentration of solutes—sodium especially—changes very little.

Urine Concentration and Dilution. The situation changes after fluid moves on through the proximal tubule and enters the loop of Henle. This hairpin-shaped structure plunges into the kidney medulla. In the interstitial fluid surrounding the hairpin, the solute concentration increases progressively with depth.

The descending limb of the loop is permeable to water. Water moves out, by osmosis, and the solute concentration in the fluid remaining inside increases until it matches that in the interstitial fluid. In the ascending limb, water cannot cross the tubule wall, but sodium is actively transported out. As sodium (and chloride) move out of the filtrate, the solute concentration rises outside the tubule and falls inside. This increase in sol-

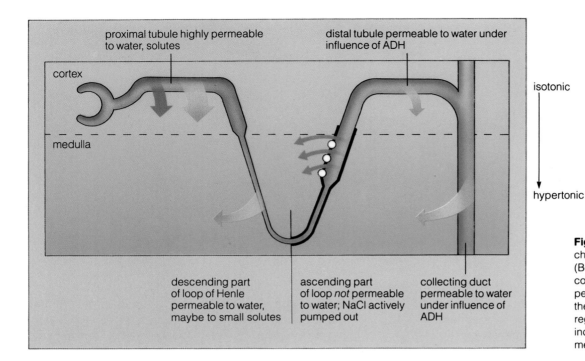

proximal tubule highly permeable to water, solutes

distal tubule permeable to water under influence of ADH

cortex

medulla

isotonic

hypertonic

descending part of loop of Henle permeable to water, maybe to small solutes

ascending part of loop *not* permeable to water; NaCl actively pumped out

collecting duct permeable to water under influence of ADH

Figure 41.7 Permeability characteristics of the nephron. (Both the distal tubule and the collecting duct have very limited permeability to solutes; most of the solute movements in these regions are related directly or indirectly to active transport mechanisms.)

ute concentration outside the tubule favors reabsorption of water from the descending limb.

Through this interaction between the ascending and descending limbs of the loop of Henle, a very high solute concentration develops in the deeper portions of the medulla. At the same time, the interaction lowers the solute concentration in the fluid traveling up the ascending limb of the loop. So the tubular fluid finally delivered to the distal tubule in the kidney cortex is quite dilute, with a low sodium concentration. As you will now see, the stage is set for the excretion of either highly dilute or highly concentrated urine—or anywhere in between.

Hormone-Induced Adjustments. Because so much water and sodium are reabsorbed in the proximal tubule and loop of Henle, the volume of dilute urine reaching the start of the distal tubule has been greatly reduced. Yet if even that reduced volume were excreted without adjustments, the body would rapidly become depleted of both water and sodium. Controlled adjustments are made at cells located in the walls of distal tubules and collecting ducts. Two hormones serve as the agents of control. **ADH** (antidiuretic hormone) influences water reabsorption, and **aldosterone** influences sodium reabsorption.

Let's first consider the role of ADH in adjusting the rate of water reabsorption. The hypothalamus controls the release of ADH from the posterior pituitary gland. It triggers ADH secretion when the solute concentration of extracellular fluid rises above a set point (com-

pare page 542). This can happen when water intake is restricted or when the extracellular fluid volume is reduced, as by severe bleeding (hemorrhage). ADH acts on distal tubules and collecting ducts, making their walls more permeable to water (Figure 41.7). Thus, in the kidney cortex, water is reabsorbed from the dilute fluid inside the tubules. The volume of fluid inside is reduced somewhat, and now it passes down through the collecting ducts, which plunge down into the medulla. Remember that solute concentrations in the surrounding interstitial fluid are high in this region. Even more water is reabsorbed here—so only a small volume of very concentrated urine is excreted.

Conversely, when water intake is excessive, the solute concentration in extracellular fluid falls. ADH secretion is inhibited. Without ADH, the walls of the distal tubules and collecting ducts become impermeable to water. A large volume of dilute urine can be excreted, and in this way the body rids itself of the excess water.

ADH enhances water reabsorption at distal tubules and collecting ducts when the body must conserve water. When excess water must be excreted, ADH secretion is inhibited.

Let's now consider the role of aldosterone in adjusting the rate of sodium reabsorption. When the body loses more sodium than it takes in, the volume of extracellular fluid falls. Sensory receptors in the walls of blood vessels and the heart detect the decrease, and

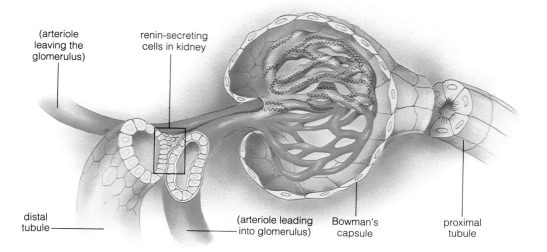

Figure 41.8 Location of renin-secreting cells that play a role in sodium reabsorption.

(arteriole leaving the glomerulus)

renin-secreting cells in kidney

distal tubule

(arteriole leading into glomerulus)

Bowman's capsule

proximal tubule

renin-secreting cells in the kidneys are called into action. Those cells are part of the "juxtaglomerular apparatus." The name refers to a region of contact betwen the arterioles of the glomerulus and the distal tubule of the nephron (Figure 41.8).

Renin acts on molecules of an inactive protein that circulates in the bloodstream. In effect, enzyme action lops off part of the molecule. Then the fragment is converted to a hormone, angiotensin II. The hormone stimulates cells of the adrenal cortex, the outer portion of a gland perched on top of each kidney (page 588). In response to stimulation, target cells secrete aldosterone. This hormone causes cells of the distal tubules and collecting ducts to reabsorb sodium faster. The outcome? Less sodium is excreted in the urine.

Conversely, when the body contains too much sodium, aldosterone secretion is inhibited. Less sodium is reabsorbed, and more is excreted.

When the body cannot rid itself of excess sodium, it inevitably retains excess water, and this leads to a rise in blood pressure. Abnormally high blood pressure (hypertension) can adversely affect the kidneys as well as the vascular system and brain (page 671). One of the ways to control blood pressure is to restrict the intake of sodium chloride—table salt.

Aldosterone enhances sodium reabsorption at distal tubules and collecting ducts when the body must conserve sodium.

Thirst Behavior

You are not entirely at the mercy of events in the kidney when your body is in need of water. The same stimuli that lead to ADH secretion and the stepped-up reabsorption of water in the kidneys can stimulate thirst behavior. Suppose you eat an entire box of salty popcorn at the movies. Soon the salt is moving into your bloodstream. The solute concentration in your extra-cellular fluid rises, the hypothalamus detects the increase, and it calls for ADH secretion. Besides this, your **thirst center** is stimulated. Signals from this cluser of nerve cells in the hypothalamus can inhibit saliva production. Your brain interprets the resulting sensation of dryness in your mouth as "thirst" and leads you to seek out drinking fluids.

In fact, a "cottony mouth" is one of the early signs that your body is becoming dehydrated. Dehydration commonly results after severe cases of hemorrhage, burns, or diarrhea. It also results after profuse sweating and after the body has been deprived of water for a prolonged time. At such times thirst behavior becomes exceptionally intense.

Acid–Base Balance

So far, we have focused on the kidney's primary function—that is, how it controls water and sodium reabsorption, and so influences the total volume and distribution of body fluids. Another kidney function also has profound impact on the health of the individual. *Together with the respiratory system and other organ systems, the kidneys help keep the extracellular environment from becoming too acidic or too basic (alkaline).*

The overall acid–base balance is maintained by controls over ion concentrations, especially hydrogen ions (H^+). The controls are exerted through (1) buffer systems, (2) respiration, and (3) excretion by way of the kidneys.

Normally, the extracellular pH for the human body is between 7.35 and 7.45. Maintaining that value means neutralizing or eliminating a variety of acidic and basic

Commentary

Kidney Failure and Dialysis

An estimated 13 million people in the United States alone suffer from kidney disorders, as when diabetes or immune responses damage the nephron and interfere with urine formation.

When the kidneys malfunction, control of the volume and composition of the extracellular fluid is disturbed, and toxic by-products of protein breakdown can accumulate in the bloodstream. Nausea, fatigue, loss of memory and, in advanced cases, death may follow. A *kidney dialysis machine* can restore the proper solute balances. Like the kidney itself, the machine helps maintain extracellular fluid by selectively removing and adding solutes to the bloodstream.

"Dialysis" refers to the exchange of substances across a membrane between solutions of differing compositions. In *hemodialysis*, the machine is connected to an artery or a vein, then blood is pumped through tubes made of a mate- rial similar to sausage casing or cellophane. The tubes are submerged in a warm-water bath. The precise mix of salts, glucose, and other substances in the bath set up the correct gradients with the blood. In *peritoneal dialysis*, fluid of the proper composition is put into the abdominal cavity, left in place for a period of time, and then drained out. Here, the lining of the cavity (the peritoneum) serves as the dialysis membrane.

Hemodialysis generally takes about four hours; blood must circulate repeatedly before solute concentrations in the body are improved. The procedure must be performed three times a week. It is used as a temporary measure in patients with reversible kidney disorders. In chronic cases, the procedure must be used for the rest of the patient's life or until a functional kidney is transplanted. With treatment and controlled diets, many individuals are able to resume fairly normal activity.

---

substances entering the blood from the gut and from normal metabolism. Recall that acids lower the pH and bases raise it. In individuals on an ordinary diet, normal cell activities produce an excess of acids. The acids dissociate into H^+ and other fragments, and this lowers the pH. The effect is minimized when excess H^+ ions react with different kinds of buffers. The bicarbonate–carbon dioxide buffer system is an example. The reactions can be summarized this way:

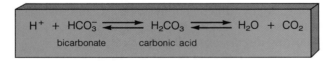

$$H^+ + HCO_3^- \rightleftharpoons H_2CO_3 \rightleftharpoons H_2O + CO_2$$

bicarbonate   carbonic acid

In this case, the H^+ is neutralized and the carbon dioxide that forms is excreted by the lungs.

Keep in mind that this buffer system and others do not *eliminate* acid; they only neutralize H^+ temporarily. *Only the urinary system can eliminate excess amounts of H^+ and restore the body's buffers.*

Notice, in the preceding equation, that the reaction arrows also run in reverse. The reverse reactions unfold in cells of the nephron wall. *First*, the HCO_3^- that forms in those cells is moved into interstitial fluid, then into the capillaries around the nephrons. The capillaries deliver the HCO_3^- to the general circulation, where it helps buffer excess acid. *Second*, the H^+ that forms in the

cells is secreted into the fluid inside the nephron. There, it can combine with bicarbonate ions to form CO_2, which is returned to the blood and excreted by the lungs. It also can combine with phosphate ions or ammonia (NH_3), which are excreted in urine. In such ways, hydrogen ions can be permanently removed from extracellular fluid.

Through such mechanisms, the kidneys help maintain the health of the individual. When such mechanisms fail, serious problems arise (see the *Commentary* above).

On Fish, Frogs, and Kangaroo Rats

Now that you have an idea of how your own body maintains water and solute levels, you can gain insight into what goes on in some other vertebrates, including that kangaroo rat described at the start of the chapter.

The tissues of herring, snapper, and other bony fishes that live in the seas have about three times less solutes than seawater does. Such fishes continuously lose water (by osmosis) to their hypertonic environment, and continual drinking brings in replacements. (Experimentally prevent a fish from drinking and it will die from dehydration within a few days.) Marine fishes excrete ingested solutes against concentration gradients. Fish kidneys do not have loops of Henle, so it is impossible to excrete

Figure 41.9 Salt-water balance by the salmon. Salmon are among the fishes that are able to live in both saltwater and freshwater. They hatch in streams and later move downstream to the seas, where they feed and mature. Then they return to their streams to spawn (page 136).

For most salmon, salt tolerance arises through changing concentrations of hormones, which seem to be triggered in some way by increasing daylength in spring. Prolactin, a pituitary hormone, plays a role in sodium retention in freshwater. When a freshwater fish has its pituitary gland removed, it will die from sodium loss—but that fish will live if prolactin is administered to it. By contrast, cortisol is a steroid hormone secreted by the adrenal cortex (page 588), and it is crucial in developing salt tolerance in salmon. Cortisol secretions are associated with increases in sodium excretion, in the sodium-potassium pumping activity by cells in the gills, and in absorption of ions and water in the gut. In young salmon, cortisol secretion increases prior to the seaward movement—and so does salt tolerance.

urine having a higher solute concentration than that of the body fluids. Most excess solutes are pumped out through membranes of fish gills, the cells of which actively transport sodium ions out of the blood.

In freshwater, a hypotonic medium, bony fishes and amphibians tend to gain water and lose solutes. They do not drink water. Rather, water moves by osmosis into the body, through thin gill membranes, or, in frogs and other adult amphibians, through the skin. Excess water leaves by way of well-developed kidneys, which excrete a large volume of dilute urine. Some solutes also are excreted, but the losses are balanced by solutes gained from food and by the active transport of sodium ions across the gills, into the body. Figure 41.9 describes the wonderful balancing act in a type of fish that makes its home in both freshwater *and* seawater.

And about that kangaroo rat! Its remarkable ability to restrict water loss results from its long loops of Henle. The solute concentration in the interstitial fluid around those loops becomes very high. The osmotic gradient between the fluid and the urine is so steep, most of the water reaching the equally long collecting ducts is reabsorbed. Thus kangaroo rats give up only a tiny volume of concentrated urine—three to five times more concentrated than that of humans.

MAINTAINING BODY TEMPERATURE

Maintaining the volume and composition of the internal environment is serious business. Maintaining its temperature is equally serious. Consider that temperatures in the world just outside the body often change quickly and that exercise can send metabolic rates soaring. Such changes trigger slight increases or decreases in the normal **core temperature**. "Core" refers to the body's inter-

nal temperature, as opposed to temperatures of the tissues near its surface.

Temperatures Suitable for Life

In a manner of speaking, the animal body runs on enzymes—and enzyme activity is affected by temperature. The enzymes of most animals commonly remain functional within the 0°–40°C range (Table 41.3). Above 41°C or so, they do not function as well because denaturation occurs. (Denaturation disrupts the chemical interactions holding a molecule in its required three-dimensional shape.) Also, the rate of enzyme activity generally decreases by at least half when the temperature drops by ten degrees. Clearly, then, metabolism can be upset if body temperatures exceed or fall below the proper range.

How do animals keep their body temperature fairly constant? They do so by balancing heat gains and heat losses. Many different physiological and behavioral responses can restore the normal temperature, as the following discussions will make clear.

Heat Gains and Heat Losses

Enzyme-mediated reactions proceed simultaneously in the millions or billions of cells of a large-bodied animal. Heat is an inevitable by-product of all that metabolic activity. (Even as you sit quietly, reading this book, you are producing roughly 1 kilocalorie per hour per kilogram of body weight.) If that heat were to accumulate internally, the body temperature would steadily rise. You probably know that a warm body tends to lose heat to a cooler environment. Under such circumstances, its temperature will hold steady if the rate of heat loss strikes a balance with the rate of metabolic heat produc-

tion. Of course, the balance can tip, as when the body produces more heat than it loses. In general, the body's heat content depends on the balance between heat gains and losses, as summarized here:

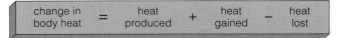

| change in body heat | = | heat produced | + | heat gained | − | heat lost |

Here, the gains and losses refer to exchanges that take place at the outer surface of the body, which includes part of the respiratory tract.

Three processes—radiation, conduction, and convection—can move heat away from or to the body. A fourth process—evaporation—can only move heat away from the body.

Radiation is a process by which heat is gained when the body is exposed to intense radiant energy (as from the sun or a heat lamp) or to any surface warmer than its own surface temperature. Indoors, well over a third of your total heat loss typically occurs by way of radiation.

Conduction is the direct transfer of heat between objects that are in direct contact. Because heat moves down thermal gradients, we lose heat by conduction when we sit on cold ground, and we gain heat when we sit on warm sand at the beach.

Convection is the transfer of heat by way of moving fluid, such as air or water currents. The process involves conduction (heat moves down the thermal gradient between the body surface and the air or water next to it). It also involves mass transfer, with currents carrying heat away from or toward the body. When your skin temperature is higher than the air temperature, you lose heat by convection. Even when there is no breeze, your body loses heat by creating its own convective current. Air becomes less dense as it is heated and rises away from the body. Indoors, you typically lose as much heat by convection as you do by radiation.

Figure 41.10 Evaporative water loss and sweating. Horses, humans, and some other mammals have sweat glands that move water and specific solutes through pores to the skin surface. In fact, an average-sized human can produce 1 to 2 liters of sweat per hour—for hours on end. For every liter of sweat that evaporates, the body loses 600 kilocalories of heat energy. During extreme exercise, this mechanism balances the high rates of heat production in skeletal muscle. In itself, sweat dripping from the skin does *not* dissipate body heat by evaporation. When you drip with sweat while exercising on a hot, humid day, the rate of evaporation does not keep pace with the rate of sweat secretion—the high water content of the surrounding air slows evaporation.

Table 41.3	Temperatures Favorable for Metabolism, Compared with Environmental Temperatures
Temperatures generally favorable for metabolism:	0°C to 40°C (32°F to 104°F)
Air temperatures above land surfaces:	−70°C to +85°C (−94°F to +185°F)
Surface temperatures of open oceans:	−2°C to +30°C (+28.4°F to +86°F)

Evaporation is the conversion of a liquid to a gaseous state. The conversion requires energy, which is supplied from the heat content of the liquid (page 28). Evaporation of water from the body surface has a cooling effect because the water molecules being released (as water vapor) carry away some of the energy with them. For land animals, some evaporative water loss occurs at the moist respiratory surfaces and across the skin. Animals that sweat, pant, or lick their fur also lose water by this process (Figure 41.10). The rate of evaporation depends on the humidity and on the rate of air movement. If air next to the body is already saturated with water (that is, when the local relative humidity is 100 percent), water will not evaporate. If air next to the body is hot and dry, evaporation may be the only means of countering heat production and heat gains (from radiation and convection).

Heat can be brought into or carried away from the body by the processes of radiation, conduction, and convection. Evaporation carries heat away from it.

The factors that dictate how much heat is exchanged by these processes can be adjusted by behavioral and physiological means.

Classification of Animals Based on Temperature

Ectotherms. We humans have high metabolic rates that sustain our active way of life. But most animals have low metabolic rates, and on top of that, they are poorly insulated. This means they rapidly absorb and gain heat, especially when they have small bodies. They maintain their core temperature mostly by heat gains from the environment, not from metabolism. Such animals are **ectotherms**, which means "heat from outside." Lizards and other reptiles are examples.

Ectotherms are not entirely at the mercy of their environment, however, for they can make behavioral adjustments to changing external temperatures. We call this **behavioral temperature regulation**.

Lizards move about, putting themselves in places where they minimize heat or cold stress. To warm up, they move out of shade and keep orienting their body to expose the maximum surface area to the sun's infrared radiation. They gain heat by conduction when they bask on rocks that absorbed heat from the sun earlier in the day. In such ways, the lizard body can warm up as fast as 1°C per minute.

Lizards lose heat just as rapidly when the sun goes down and temperatures drop. Then, metabolic activity decreases and they become almost immobilized. Before that happens, they usually crawl into crevices or under rocks, where heat loss is not as great and where they are not as vulnerable to predators.

Endotherms and Heterotherms. Like most birds and mammals, we are **endotherms**, which means "heat from within." In endotherms, body temperature is controlled mainly by (1) metabolic activity and (2) controls over heat conservation and dissipation. We also make behavioral adjustments that supplement the physiological controls.

Most endotherms have an active life-style, made possible by high metabolic rates. It is a costly adaptation. A foraging mouse uses up to thirty times more energy than a foraging lizard of the same weight. Yet such energy outlays have advantages, for they are the main reason why endotherms can be active under a wide range of temperatures. Cold nights or cold seasons don't stop them from foraging, for example, or escaping from predators, or digging a burrow.

Endotherms conserve or dissipate heat associated with high metabolic rates by employing a variety of adaptations. Think about how fur, feathers, and layers of fat help reduce heat loss. Think about how clothing can reduce heat loss or heat gain. Or think about the ways that bodies are shaped and insulated. Some mammals in cold regions have more massive bodies than closely related species in warmer regions. Compared to the streamlined, thin-legged jackrabbits of the American Southwest, for example, you might think that the arctic hare is rather bulky. However, its more massive, rounder body has a greater volume of cells for generating heat and less surface area for losing it.

Like ectotherms, endotherms also adjust behaviorally to heat stress. During the day, the core temperatures of kangaroo rats and some other desert mammals of north temperate regions are much lower than temperatures of the air and the ground surface. Outside temperatures often are lower during the night as well as during winter. However, the soil well below the surface never heats up much, and it is here that most desert rodents and other mammals find refuge from the daytime heat. Typically those animals forage by night and spend the hottest part of the day in burrows or in the shade of bushes or rock outcroppings.

Some birds and mammals fall between the ectothermic and endothermic categories. Part of the time, these so-called **heterotherms** allow their body temperature to fluctuate as ectotherms do, and at other times they control heat exchanges as endotherms do. Hummingbirds have very high metabolic rates for their size, and they devote much of the day to locating and sipping nectar as an energy source for metabolism. Because hummingbirds do not forage at night, they could rapidly run out of energy unless their metabolic rates decreased considerably. At night, they may enter a sleeplike state and become almost as cool as their surroundings.

Thermal Strategies Compared. In general, ectotherms are at an advantage in the warm, humid tropics. They do not have to expend much energy to maintain body temperature, and more energy can be devoted to other tasks, including reproduction. Indeed, reptiles far exceed mammals in numbers and species diversity in the tropics. However, endotherms have the advantage and are more abundant in moderate to cold settings. High metabolic rates allow some endotherms to occupy even the polar regions, where you would never find a lizard.

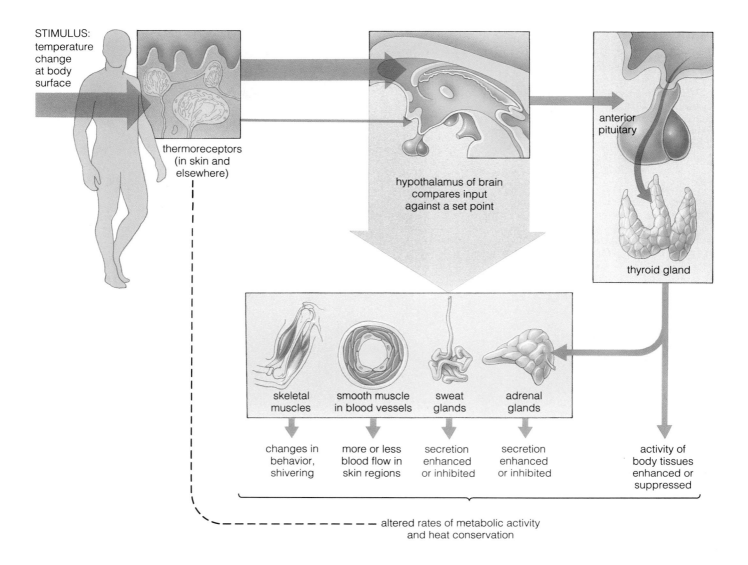

STIMULUS: temperature change at body surface

thermoreceptors (in skin and elsewhere)

hypothalamus of brain compares input against a set point

anterior pituitary

thyroid gland

skeletal muscles

smooth muscle in blood vessels

sweat glands

adrenal glands

changes in behavior, shivering

more or less blood flow in skin regions

secretion enhanced or inhibited

secretion enhanced or inhibited

activity of body tissues enhanced or suppressed

altered rates of metabolic activity and heat conservation

Figure 41.11 Homeostatic controls over the internal temperature of the human body.

Temperature Regulation in Birds and Mammals

Responses to Cold Stress. Table 41.4 lists the major responses to decreases in outside temperatures. They include peripheral vasoconstriction, the pilomotor response, shivering, and nonshivering heat production.

Among mammals, **peripheral vasoconstriction** is a normal response to a drop in outside temperature. As Figure 41.11 suggests, the hypothalamus governs this response. Thermoreceptors at the body surface detect the decrease in temperature and fire off signals to the hypothalamus. In turn, the hypothalamus sends out commands to smooth muscles in the walls of blood vessels in the skin. When the muscles contract, vasoconstriction occurs—and the bloodstream's convective delivery of heat to the body's surface is reduced. How effective is the response? To give an example, when your fingers or toes become cold, all but 1 percent of the blood that would otherwise flow to their skin is curtailed.

Table 41.4	Responses to Cold Stress
Core Temperature	Responses Made
36°–34°C (about 95°F)	Shivering response, increase in respiration. Increase in metabolic heat output. Constriction of peripheral blood vessels; blood is routed to deeper regions. Dizziness and nausea set in.
33°–32°C (about 91°F)	Shivering response stops. Metabolic heat output drops.
31°–30°C (about 86°F)	Capacity for voluntary motion is lost. Eye and tendon reflexes inhibited. Consciousness is lost. Cardiac muscle action becomes irregular.
26°–24°C (about 77°F)	Ventricular fibrillation sets in (page 672). Death follows.

Figure 41.12 A tragic episode of hypothermia.

In 1912, the ocean liner *Titanic* set out from Europe on her maiden voyage across the cold Atlantic waters to America. In that same year, a huge chunk of the leading edge of a Greenland glacier broke off and began floating out to sea. Late at night on April 14, off the coast of Newfoundland, the iceberg and the *Titanic* made their ill-fated rendezvous. The *Titanic* was the largest ship afloat and was believed to be unsinkable. Survival drills had been neglected, and there were not enough lifeboats to hold even half the 2,200 passengers. The *Titanic* sank in about 2-1/2 hours.

Within two hours, rescue ships were on the scene—yet 1,517 bodies were recovered from a calm sea. All were wearing life jackets. None had drowned. Probably every one of those individuals had died from hypothermia— from a drop in body temperature below tolerance levels.

Table 41.5	Summary of Mammalian Responses to Cold Stress and to Heat Stress	
Environmental Stimulus	Main Responses	Outcome
Drop in temperature	Vasoconstriction of blood vessels in skin; changes in behavior (e.g., curling up the body to reduce surface area exposed to the environment)	Heat is conserved
	Increased muscle activity; shivering; nonshivering heat production	Heat production increases
Rise in temperature	Vasodilation of blood vessels in skin; sweating; changes in behavior; panting	Heat is dissipated from body
	Decreased muscle activity	Heat production decreases

In another response to a drop in outside temperature, smooth muscle controlling the erection of hair or feathers is stimulated to contract. This is a **pilomotor response**. The plumage or pelt fluffs up, and this creates a layer of still air that reduces convective and radiative heat losses from the body. Heat loss can be further restricted by behavioral responses that reduce the amount of body surface exposed for heat exchange—as when cats curl up into a ball or when you hold both arms tightly against your body.

When other responses are not enough to counter cold stress, the hypothalamus calls for an increase in skeletal muscle activity that leads to **shivering**. The word refers to rhythmic tremors in which the muscles contract about ten to twenty times per second. Within a short time, heat production throughout the body increases several times over. Shivering comes at a high energy cost and is not effective for very long.

Heat production also can be increased without shivering. Prolonged or severe cold exposure can lead to a hormonal response that elevates the rate of metabolism (Figure 41.11). This **nonshivering heat production** is prominent in brown adipose tissue, a specialized tissue of hibernating animals and in rodents and other animals that become acclimatized to cold. Human infants have this tissue; adults have very little unless they are cold-adapted. Korean diving women, for example, who spend six hours a day in cold water, have well-developed brown adipose tissue.

When defenses against cold are not adequate, the result is *hypothermia*, a condition in which the core temperature falls below normal. In humans, a drop of only a few degrees affects brain function and leads to

Figure 41.13 A jackrabbit (*Lepus californicus*) cooling off on a hot summer day in the mountains of Arizona. Notice the dilated blood vessels in its large ears. Both the large surface area of the ears and the extensive vascularization are useful for dissipating heat (by way of convection and radiation).

Figure 41.14 (*Right*) Response to heat stress by mammals without sweat glands. Some animals resort to behavioral mechanisms, such as licking their fur or panting (breathing hard through the mouth). Panting enhances respiratory heat loss, the amount depending on the increases in ventilation and in watery secretions from salivary glands. The endurance of sled dogs during a hard race depends partly on their capacity to dissipate metabolic heat through evaporation from the tongue and respiratory tract.

confusion; further cooling can lead to coma and death (Figure 41.12). Some victims of extreme hypothermia, children particularly, have survived prolonged immersion in cold water of even lower temperatures. In fact, many animals can recover from profound hypothermia. However, cells that become frozen may be destroyed unless thawing is precisely controlled (this sometimes can be done in hospitals). Tissue destruction through localized freezing is called *frostbite*.

Responses to Heat Stress. Table 41.5 summarizes the main responses to heat stress as well as cold stress. Here again, the hypothalamus has roles in responses to increases in core temperature. In **peripheral vasodilation**, hypothalamic signals cause blood vessels in skin regions to dilate. More blood flows from deeper body regions to skin regions, where the excess heat it carries is dissipated (Figure 41.13).

Evaporative heat loss is another response that can be influenced by the hypothalamus, which can activate

sweat glands. Your skin has 2-1/2 million or more sweat glands, and considerable heat is dissipated when the water they give up to the skin surface evaporates. With extreme sweating, as might occur in a marathon race, the body loses an important salt—sodium chloride—as well as copious amounts of water. Such losses may change the character of the internal environment to the extent that the runner may collapse and faint.

What about mammals that sweat very little or not at all? Some of them, including dogs, pant. "Panting" refers to shallow, rapid breathing that increases evaporative water loss from the respiratory tract (Figure 41.14). Cooling occurs when the water evaporates from the nasal cavity, mouth, and tongue.

Sometimes peripheral blood flow and evaporative heat loss are not enough to counter heat stress, and *hyperthermia* results. This is a condition in which the core temperature increases above normal. For humans and other endotherms, an increase of only a few degrees above normal can be dangerous.

Fever. During a *fever*, the hypothalamus actually resets the body's "thermostat" that dictates what the core temperature should be. The normal response mechanisms are brought into play, but they are carried out to maintain a higher temperature! At the onset of fever, heat loss decreases, and heat production increases. At the time the person feels chilled. When the fever "breaks," peripheral vasodilation and sweating increase as the body attempts to reduce the core temperature to normal; then, the person feels warm.

Fever may be an important defense mechanism against infections, and perhaps against cancer. Following infection, macrophages and other cells secrete signaling molecules, including interleukin-1 and interferons. The secretions somehow influence the hypothalamus. We know that aspirin and other drugs that interfere with the synthesis of prostaglandins can block their influence. (Actually, prostaglandins injected directly into the hypothalamus can induce fever.) The controlled increase in body temperature during a fever seems to enhance the body's immune response. Therefore, the widespread practice of administering aspirin and other drugs may interfere with beneficial effects of fever. But, without question, these drugs are necessary when the temperature in severe fevers approaches dangerous levels.

SUMMARY

Control of Extracellular Fluid

1. By balancing water and solute gains with water and solute losses, the body maintains its internal environment.

2. Water is gained by absorption from the gastrointestinal tract and by metabolism. A thirst mechanism controls water gain. Water is lost by evaporation from the lungs and skin, elimination from the gastrointestinal tract, and excretion of urine. Controls over water loss deal mainly with varying the composition and volume of urine.

3. In mammals, urine formation occurs in a pair of kidneys. Each human kidney contains about 1 million tubelike blood-filtering units called nephrons. A nephron has a cup-shaped beginning (Bowman's capsule), proximal tubule, loop of Henle, and distal tubule. It leads into a collecting duct.

4. Kidney function depends on intimate links between the nephron and the bloodstream. Blood flows from an arteriole into a set of capillaries inside Bowman's capsule, then into a second set of capillaries that thread around the tubular parts of the nephron, then back to the bloodstream by way of another arteriole.

5. Each day, massive volumes of fluid are filtered at the glomeruli. Most of the water and solutes filtered are reabsorbed, and the excess is excreted as urine. Urine composition and volume depends on three processes:

 a. Filtration of blood at the glomerulus of a nephron, with blood pressure providing the force for filtration.

 b. Reabsorption. Water and solutes move out of tubular parts of the nephron and back into adjacent blood capillaries.

 c. Secretion. Excess ions and a few foreign substances move out of those capillaries and into the nephron, so that they are disposed of in urine.

6. Reabsorption of many solutes may occur passively, following concentration gradients. But in other instances, active transport (with its expenditure of energy) is required. Sodium reabsorption is an important example of active transport. Water reabsorption is always passive, occurring along its osmotic gradient.

7. The hormone ADH is secreted when the body must retain water; it acts on the nephron walls and makes them permeable to water. When the body must rid itself of excess water, ADH secretion is inhibited. The hormone aldosterone controls sodium reabsorption.

Control of Body Temperature

1. Body temperature of animals is determined by the balance between metabolically produced heat and the heat absorbed from and lost to the environment.

2. Animals exchange heat with the environment by these processes:

 a. Radiation: the emission of energy in the form of infrared and other wavelengths which, following absorption at the surface of the animal body or some other object, is converted to heat energy.

 b. Conduction: the direct transfer of heat energy between two objects in direct contact with each other.

 c. Convection: heat transfer by air or water currents; involves conduction and mass transfer of heat-bearing currents away from or toward the animal body.

 d. Evaporation: the conversion of a liquid to the gaseous state; requires energy, supplied from the heat content of the liquid.

3. The body temperature of different animals depends on the rate of metabolic activity and anatomical, behavioral, and physiological adaptations.

 a. Ectotherms are animals whose body temperature is determined more by heat exchange with the environment than by metabolic heat.

 b. The body temperature of endotherms is determined largely by metabolic activity and by precise controls over heat produced and heat lost.

 c. Heterotherms allow their body temperature to fluctuate at some times, and at other times they control heat balance.

Review Questions

1. All animals have mechanisms for maintaining body fluid concentration and composition. In your own body, which organs cooperate in these tasks? *721*

2. Describe what happens during (a) filtration, (b) reabsorption, and (c) secretion in the kidney's nephron/capillary unit. What do these three processes influence? *722–723*

3. Which hormone is involved in the control of water reabsorption? Which hormone plays a major role in sodium reabsorption? *725–726*

4. Which type of ion is especially important in maintaining the body's acid–base balance? *726*

Self-Quiz *(Answers in Appendix IV)*

1. In vertebrates, maintaining the volume and composition of extracellular fluid depends on three kidney functions: _____, _____, and _____.

2. Urine formation occurs in _____.
   a. glomeruli   c. nephrons
   b. loops of Henle   d. ureters

3. The body gains water by _____.
   a. gastrointestinal absorption   c. both a and b
   b. metabolism   d. neither a nor b

4. The body loses water by _____.
   a. evaporation from lungs   c. excretion of urine
   and skin   d. all of the above
   b. elimination from
   gastrointestinal tract

5. Each human kidney contains about _____ nephrons.
   a. 1,000   c. 100,000
   b. 10,000   d. 1,000,000

6. The processes responsible for urine composition and volume occur in this order: _____.
   a. filtration, reabsorption, secretion
   b. secretion, reabsorption, filtration
   c. reabsorption, secretion, filtration
   d. secretion, filtration, reabsorption

7. Which of the following descriptions does *not* match the process named?
   a. *filtration:* blood enters Bowman's capsule of glomerulus
   b. *reabsorption:* water and solutes selectively returned to blood capillaries
   c. *secretion:* excess ions and some other substances move from blood capillaries into nephron
   d. all of the above match
   e. none of the above match

8. Concentration gradients that are maintained between nephrons and surrounding interstitial fluid are responsible for _____.
   a. blood filtration
   b. secretion
   c. reabsorption
   d. hypertension

9. The hormone ADH controls _____.
   a. nephron production
   b. sodium reabsorption
   c. secretion
   d. water retention

10. Match the salt–water balance concepts.
    _____ aldosterone   a. blood filter of a nephron
    _____ nephron   b. controls sodium reabsorption
    _____ thirst mechanism   c. occurs at blood capillaries
    _____ reabsorption   around the nephrons
    _____ glomerulus   d. site of urine formation
                        e. controls water gain

Selected Key Terms

ADH *725*
aldosterone *725*
Bowman's capsule *722*
conduction *729*
convection *729*
core temperature *728*
distal tubule *722*
ectotherm *730*
endotherm *730*
evaporation *729*
excretion (urinary) *720*
extracellular fluid *720*
filtration *722*
glomerular capillaries *722*
glomerulus *722*
heterotherm *730*
hyperthermia *733*
hypothermia *732*
kidney *721*
loop of Henle *722*

nephron *721*
nonshivering heat
  production *732*
peripheral
  vasoconstriction *731*
peripheral vasodilation *733*
peritubular capillary *722*
pilomotor response *732*
proximal tubule *722*
radiation *729*
reabsorption *722*
secretion *723*
shivering response *732*
thirst center *726*
ureter *721*
urethra *721*
urinary bladder *721*
urinary system *721*
urine *721*

Readings

Schmidt-Nielsen, K. 1990. *Animal Physiology.* Fourth edition. New York: Cambridge. Chapters 8 and 9 provide an excellent introduction to water–solute balances in animals.

Smith, H. 1961. *From Fish to Philosopher.* New York: Doubleday. Available in paperback.

Valtin, H. 1983. *Renal Function: Mechanisms Preserving Fluid and Solute Balance in Health.* Second edition. Boston: Little, Brown. Paperback.

Vander, A., J. Sherman, and D. Luciano. 1990. "The Kidneys and Regulation of Water and Inorganic Ions" in *Human Physiology.* Fifth edition. New York: McGraw-Hill.

From Frog to Frog and Other Mysteries

With a full-throated croak that only a female of its kind could find seductive, a male frog proclaims the onset of warm spring rains, of ponds, of sex in the night. By August the summer sun will have parched the earth, and his pond dominion will be gone. But tonight is the hour of the frog! Through the dark, a female moves toward the vocal male. They meet, they dally in behavioral patterns characteristic of their species. He clamps his forelegs about her swollen abdomen and gives it a prolonged squeeze. Out into the surrounding water streams a ribbon of hundreds of eggs. As the eggs are being released, the male expels a milky cloud of swimming sperm. Each sperm penetrates an egg, and soon afterward, their nuclei fuse. With this fusion, fertilization is completed. A single fertilized egg, the zygote, has formed.

For the leopard frog *Rana pipiens*, a drama now begins to unfold that has been reenacted each spring, with only minor variations, for many millions of years. Within a few hours after fertilization, the single-celled zygote begins dividing into two cells, then four, then many more to produce the early embryo. In less than twenty hours, it has become a ball of tiny cells, no larger than the zygote.

And now the cells begin to migrate, change shape, and interact. Some cells at the embryo's surface sink inward, forming a dimple. Their cellular descendants will give rise to internal tissue layers. Other cells at the surface lengthen or flatten out; together they form a groove. Their cellular descendants eventually will give rise to the nervous system. Through interactions between surface cells and interior cells, eyes start to develop. Within the embryo, a heart is forming and will soon start to beat rhythmically.

A tail takes shape; a mouth forms. These developments, appearing one after another, are signs of a process going on in *all* the cells that were so recently developed from a single zygote. The cells are becoming different from one another in both appearance and function!

Within twelve days after fertilization, the embryo has become a larval form—a tadpole—that swims and feeds on its own (Figure 42.1). After several

a

Figure 42.1 Development of the leopard frog, *Rana pipiens*. (**a**) A male clasping a female in a behavior called amplexus. When the female releases her eggs into the water, the male releases his sperm over the eggs. (**b**) Frog embryos. (**c**) A larval form called a tadpole. (**d**) Transitional form between the tadpole and the young adult frog (**e**).

months, legs start to grow. The tail shortens, then disappears. The small mouth, once suitable for feeding on algae, develops jaws and now snaps shut on insects and worms. Eventually an adult frog leaves the water for life on land. With luck it will avoid predators, disease, and other threats in the months ahead. In time it may even find a pond filled by the new season's rains, and the cycle will begin again.

How does the single-celled zygote of a frog or any other complex animal become transformed into all the specialized cells and structures of the adult? With this question we turn to one of life's greatest mysteries—to the development of new individuals in the image of their parents. We are just starting to understand the underlying mechanisms.

b

c

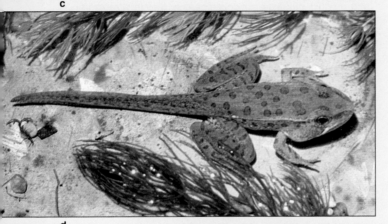

d

e

KEY CONCEPTS

1. Most animals reproduce sexually. Elaborate reproductive structures and forms of courtship and parental behavior have accompanied the separation into male and female sexes. The biological cost of those structures and behaviors is offset by the advantages afforded by diversity among the offspring. At least some of the diverse phenotypes should prove to be adaptive in changing or new environments.

2. Development commonly proceeds through six stages: gamete formation, fertilization, cleavage, gastrulation, organ formation, and growth and tissue specialization.

3. The fate of embryonic cells is determined partly at cleavage (when daughter cells inherit qualitatively different regions of cytoplasm) and partly by cell interactions in the developing embryo. Cell differentiation and morphogenesis depend on those events.

THE BEGINNING: REPRODUCTIVE MODES

In earlier chapters we looked at the cellular basis of **sexual reproduction**, in which offspring are produced by way of meiosis, gamete formation, and fertilization. We also looked at **asexual reproduction**, in which offspring are produced by means other than gamete formation. Let's turn now to some structural, behavioral, and ecological aspects of these reproductive modes.

Think about a new sponge produced asexually, by budding from the parent body (page 409). Or think about a flatworm engaged in fission, in this case dividing lengthwise into two parts that give rise to two new flatworms. In both cases, the offspring are genetically identical to the parents. Having the same genes as parents is useful when parents are well adapted to the surroundings—and when the surroundings remain stable.

But most animals live under changing, unpredictable conditions. They rely mainly on sexual reproduction, with sperm from a male fertilizing eggs from a female.

a

b

c

d   fetus in sac

e

Complete separation into male and female sexes is biologically costly. Getting sperm and eggs together depends on large energy investments in specialized reproductive structures and forms of behavior. Even so, the cost is offset by the variation in traits among the resulting offspring, at least some of which are likely to survive and reproduce in a changing environment.

Consider the question of *reproductive timing*. How do mature sperm become available exactly when eggs mature in a separate individual? Timing depends on energy outlays for sensory structures and rather involved hormonal controls in both parents. Both parents must produce mature gametes in response to the same cues, such as changes in daylength that mark the onset of the best season for reproduction. Moose, for instance, become sexually active in late summer or early fall—and their offspring are born the following spring, when the weather improves and food will be plentiful for many months.

Consider also the challenge of finding and recognizing a potential mate of the same species. Different

Figure 42.2 Examples of where the animal embryo develops, how it is nourished, and how (if at all) it is protected. Snails (**a**) are *oviparous* (egg-producing) but are not doting parents; their fertilized eggs are unprotected. Egg-laying mammals, including the duck-billed platypus shown in Figure 26.1, are oviparous, yet they also secrete milk to nourish the juveniles. Birds, too, are oviparous. Their fertilized eggs, which have large yolk reserves, develop and hatch outside the mother's body.

(**b**) The fertilized egg of humans and most other mammals is retained inside the mother's body and nourished by her tissues until the time of birth. Such animals are *viviparous* (*viva-*, alive; *-parous*, to produce). The kangaroo (**c**) and opossum (**d**), both marsupials, are viviparous. But their young emerge in unfinished form and undergo further fetal development in a pouch on the ventral surface of the mother's body, where they are nourished from mammary glands.

Some fishes, lizards, and many snakes are *ovoviviparous*. Their fertilized eggs develop within the mother's body. Such eggs are not nourished by the mother's tissues; they are sustained by yolk reserves. (**e**) A copperhead is one of the ovoviviparous snakes. Her liveborn are still contained in the relics of egg sacs.

liveborn snake in egg sac

species meet the challenge by investing energy in the production of chemical signals, structural signals such as feathers of certain colors and patterns, and sensory receptors able to pick up the signals that are being sent. Besides this, the males often expend considerable energy executing courtship routines (see Figure 50.3).

Fertilization also comes at a substantial cost. Many invertebrates and bony fishes simply release eggs and motile sperm into the water, and the chance of successful fertilization would not be good if adults produced only one sperm or one egg each season. Such species invest a great deal of energy in producing gametes by the hundreds of thousands. Nearly all land-dwelling animals depend on **internal fertilization**, the union of sperm and egg *inside* the body of the female parent. They invest energy on elaborate reproductive organs, such as a penis (by which a male deposits its sperm within the female) and a uterus (a chamber in the female where the embryo grows and develops).

Finally, energy is set aside for *nourishing some number of offspring*. Nearly all animal eggs contain **yolk**, a substance made up of proteins and lipids that can nourish the embryo. Eggs of some species have more yolk than others. Sea urchin eggs are small, they are released in large numbers, and the biochemical investment in yolk for each one is limited. There is a premium on rapid development, given the presence of predators that end up consuming most of the eggs. For example, sea stars engage in feeding frenzies when sea urchin eggs become available. No need of abundant yolk here; sea urchin offspring reach a self-feeding, free-moving larval stage in less than a day.

Bird eggs are also released from the mother, but they have large yolk reserves that nourish the embryo through a longer period of development, inside an eggshell that forms after fertilization. Human eggs have almost no yolk. After fertilization, the egg attaches to the uterus inside the mother's body, where the embryo is nourished and supported by physical exchanges with her tissues during an extended pregnancy (Figure 42.2).

As these examples suggest, animals show great diversity in reproduction and development. However, some patterns are widespread in the animal kingdom, and they will serve as a framework for our reading.

Separation into male and female sexes is an energetically costly mode of reproduction. Specialized structures and forms of behavior are required for getting sperm together with eggs and for lending nutritional support to offspring.

This reproductive mode is advantageous in unpredictable environments, for at least some traits of the diverse offspring may prove adaptive under new or changing conditions.

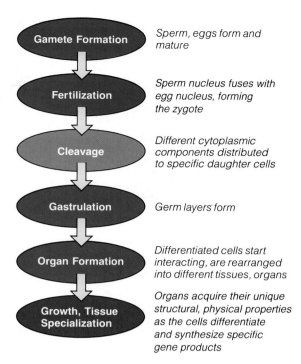

Figure 42.3 Overview of stages of animal development.

STAGES OF DEVELOPMENT

Figure 42.3 lists the stages of animal development. In the first stage, **gamete formation**, sperm or eggs form and mature within the parents. **Fertilization**, the second stage, starts when a sperm penetrates an egg. It ends when the sperm nucleus fuses with the egg nucleus and gives rise to the zygote, the first cell of the new individual. Next comes **cleavage**, when mitotic cell divisions typically convert the zygote to a ball of cells, the *blastula*. Cells increase in number but the embryo does not increase in size during this stage. The daughter cells collectively occupy the same volume as did the zygote.

As cleavage draws to a close, the pace of cell division slackens and gives way to **gastrulation**, a stage of major cell rearrangements. The organizational framework for the whole body is laid out as cells become arranged into two or three primary tissues, or *germ layers*. The human body arises from three such layers:

endoderm	*inner layer; gives rise to inner lining of gut and organs derived from it*
mesoderm	*intermediate layer; gives rise to muscle, the organs of circulation, reproduction, and excretion, most of the internal skeleton, and connective tissue layers of the gut and integument*
ectoderm	*surface layer; gives rise to tissues of nervous system and outer layer of integument*

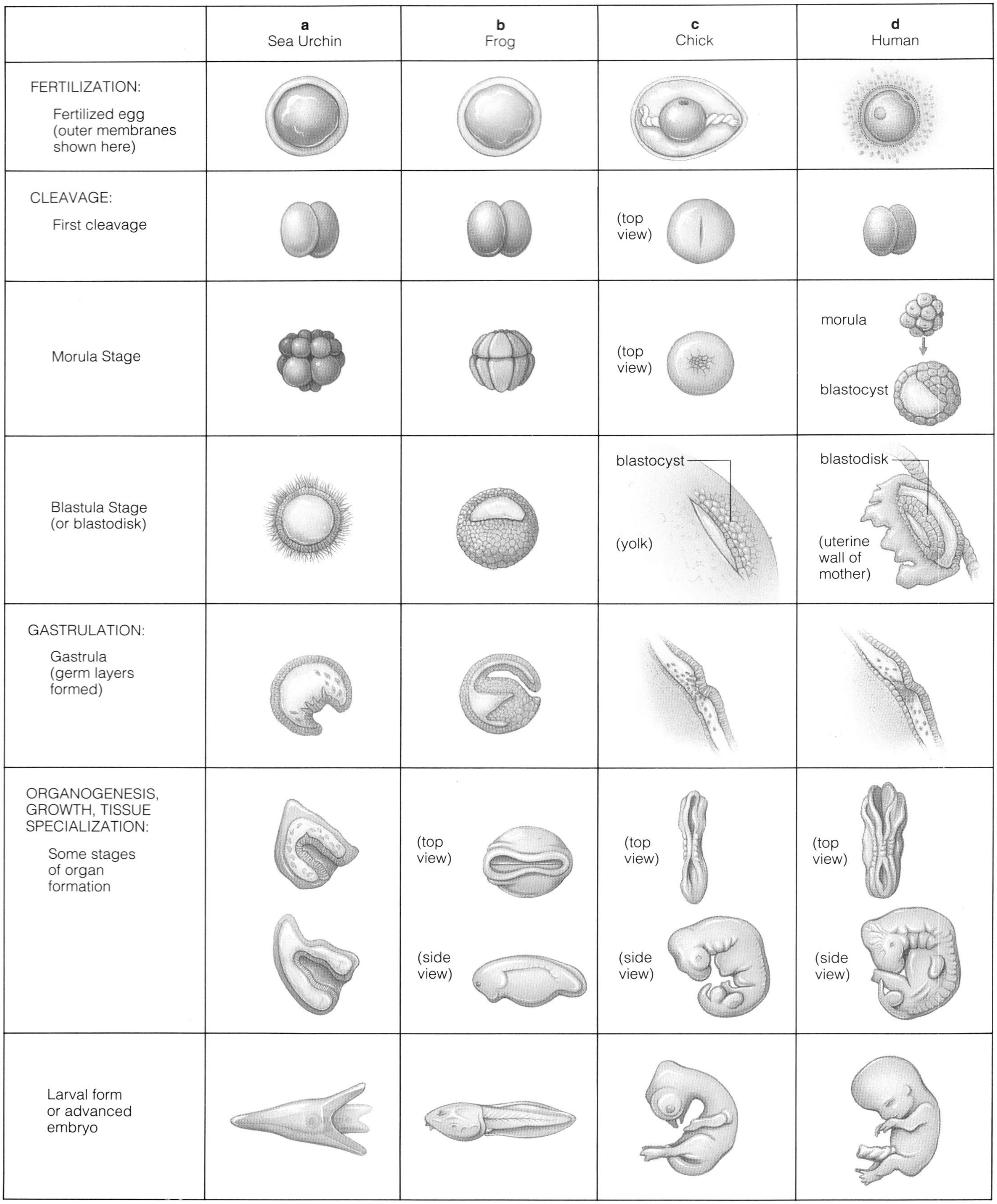

	a Sea Urchin	**b** Frog	**c** Chick	**d** Human
FERTILIZATION: Fertilized egg (outer membranes shown here)				
CLEAVAGE: First cleavage			(top view)	
Morula Stage			(top view)	morula blastocyst
Blastula Stage (or blastodisk)			blastocyst (yolk)	blastodisk (uterine wall of mother)
GASTRULATION: Gastrula (germ layers formed)				
ORGANOGENESIS, GROWTH, TISSUE SPECIALIZATION: Some stages of organ formation		(top view) (side view)	(top view) (side view)	(top view) (side view)
Larval form or advanced embryo				

Figure 42.4 (Left) Comparison of embryonic development in four different animals. The drawings are not to the same scale; however, they show the developmental patterns that are common to all four types. For clarity, the membranes surrounding the embryo are not shown from cleavage onward. Blastula and gastrula stages are shown in cross-section. They are described in more detail in this chapter and the next.

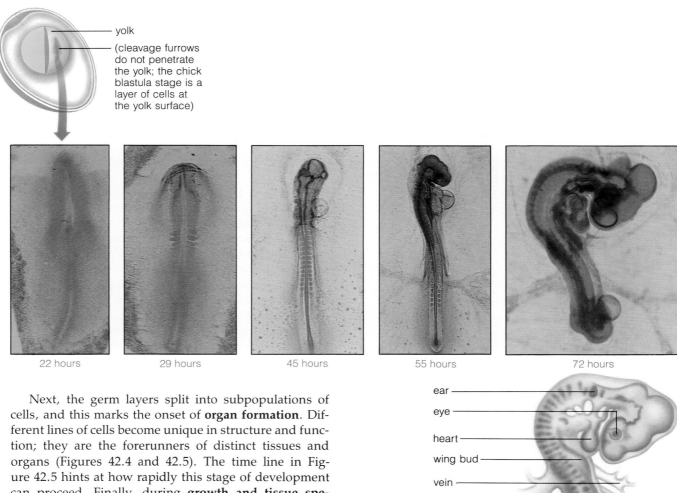

yolk

(cleavage furrows do not penetrate the yolk; the chick blastula stage is a layer of cells at the yolk surface)

22 hours

29 hours

45 hours

55 hours

72 hours

Next, the germ layers split into subpopulations of cells, and this marks the onset of **organ formation**. Different lines of cells become unique in structure and function; they are the forerunners of distinct tissues and organs (Figures 42.4 and 42.5). The time line in Figure 42.5 hints at how rapidly this stage of development can proceed. Finally, during **growth and tissue specialization**, organs acquire specialized properties. This stage continues into adulthood.

Adult animals of different species obviously do not all look alike. Neither do the embryonic forms, even though they commonly pass through all the stages just outlined. As Figure 42.4 suggests, by the end of each stage, the embryo has become more complex than it was before. This is important to think about, for structures that develop during one stage serve as the foundation for the stage following it.

Each stage of embryonic development builds on structures that were formed during the stage preceding it. Development cannot proceed properly unless each stage is successfully completed before the next begins.

ear

eye

heart

wing bud

vein

artery

leg bud

tail bud

Figure 42.5 Onset of organ formation in a chick embryo during the first 72 hours of development. The heart begins to beat at some time between 30 and 36 hours.

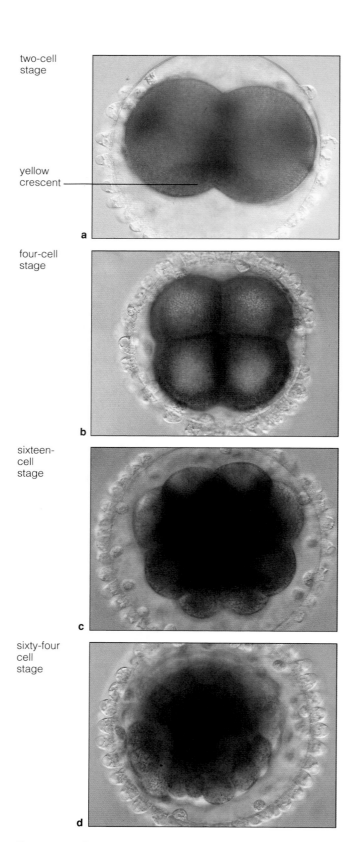

two-cell
stage

yellow
crescent

a

four-cell
stage

b

sixteen-
cell
stage

c

sixty-four
cell
stage

d

Figure 42.6 Cytoplasmic localization in the embryo of a tunicate (*Styela*). Cleavage divides regions of cytoplasm into particular cells. The fertilized eggs of these animals have a pigmented region (the yellow crescent). This region becomes localized into a small group of cells that will give rise to the muscles of the tunicate larva.

PATTERNS OF DEVELOPMENT

Key Mechanisms: An Overview

In large part, developmental complexity is mapped out in the cytoplasm of an immature egg, or *oocyte*. Later on, as more cells form in the developing embryo, individual cells and groups of cells start to interact. Their interactions are a basis for **cell differentiation**, with cells in different locations in the embryo becoming specialized in prescribed ways at specific times. As the embryo continues its development, differentiated cells become organized into tissues and organs, a process called **morphogenesis**. Before we turn to specific examples, let's define two of the most important mechanisms underlying these developmental events. They are called cytoplasmic localization and embryonic induction.

Cytoplasmic Localization. The future shape and arrangement of body parts in an embryo depend largely on what goes on in a maturing oocyte, even before fertilization. An oocyte is larger and more complex than a sperm. As it matures, organelles, proteins, RNA, and other components accumulate and become distributed in prescribed locations in the cytoplasm. These "cytoplasmic determinants" are maternal instructions that will have roles in the way body parts become arranged in orderly fashion in the developing embryo.

During cleavage, daughter cells end up in specific locations in the early embryo. Simply by virtue of their location, those cells inherit different cytoplasmic determinants (Figure 42.6). This **cytoplasmic localization** helps seal the developmental fate of the descendants of those cells.

Embryonic Induction. As the embryo develops, one or more groups of cells interact physically or chemically in ways that affect the development of at least one of those groups. This mechanism is called **embryonic induction**. For example, certain groups of cells produce a hormone, growth factor, or some other substance that diffuses to other cell groups and triggers the synthesis of a particular protein or some other change in their activities. Such interactions give groups of cells their developmental marching orders, so to speak.

Development involves cell differentiation and morphogenesis. Both processes depend on the segregation of cytoplasmic determinants during cleavage and interactions among embryonic cells.

With these key mechanisms in mind, let's now look more closely at how early developmental events unfold.

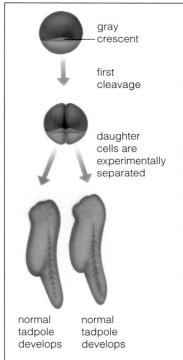

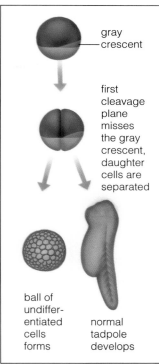

a Experiment 1

b Experiment 2

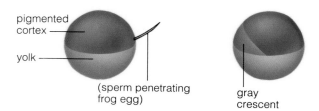

Figure 42.7 Two experiments illustrating how qualitative differences in the cytoplasm of a fertilized egg help determine the fate of cells in a developing embryo. Frog eggs contain granules of dark pigment in their cortex (the plasma membrane and the cytoplasm just below it). The granules are concentrated near one pole of the egg; yolk is concentrated near the other pole. At fertilization, a portion of the granule-containing cortex shifts away from the yolk, and this exposes lighter colored cytoplasm in a crescent-shaped gray area:

pigmented cortex

yolk

(sperm penetrating frog egg)

gray crescent

Normally, the first cleavage divides the gray crescent between two daughter cells. (**a**) Even if the two daughter cells are separated from each other experimentally, each may still give rise to a complete tadpole. (**b**) But if a fertilized egg is manipulated so that the first cleavage plane misses the gray crescent entirely, only one of the two daughter cells gets the gray crescent and it alone will develop into a normal tadpole. The daughter cell deprived of substances in the gray crescent will only give rise to a ball of undifferentiated cells.

Developmental Information in the Egg

A sperm cell produced during spermatogenesis is little more than paternal DNA, packaged with a few components that are necessary for moving the DNA to an egg (page 762). The oocyte is another story. It is the scene of dynamic activity, including increases in volume. Specialized proteins (including enzymes and yolk proteins), mRNA transcripts, and other molecules accumulate in its cytoplasm and will direct the development of the early embryo. For example, many of the mRNA molecules being transcribed from the maternal DNA are translated at once into enzymes, histones, and other proteins that will be used for chromosome replications in the early embryo. Ribosomal subunits and other cytoplasmic components necessary for protein synthesis are stockpiled. In addition, many mRNA transcripts accumulate in different regions of the egg cytoplasm. They will be allocated to different daughter cells during cleavage, then activated to direct the synthesis of specific sets of proteins.

Also present in the maturing oocyte are microtubules and other cytoskeletal elements, oriented in specific directions. These elements will influence the first cell divisions in the embryo. Whenever a cell divides in two, it does so at a prescribed angle relative to the adjacent cells, based partly on the orientation of microtubules of the mitotic spindle (page 141). The amount and distribution of yolk within the egg cytoplasm also will influence cleavage. Even the nucleus has a characteristic position in a frog egg and imparts "polarity" to it. The *animal pole* is simply the one closest to the nucleus. Opposite is the *vegetal pole*, where substances such as yolk accumulate. All animal eggs show some degree of polarity—that is, two identifiable poles—which will influence the structural patterns that emerge as the embryo develops.

Fertilization

When a sperm penetrates an egg, it triggers structural reorganization in the egg cytoplasm. You can observe indirect signs of this reorganization in frog eggs, which contain granules of dark pigment in their cortex. (The "cortex" is the plasma membrane and the cytoplasm just beneath it.) The animal pole has more pigment granules and is darker than the vegetal pole. Sperm penetration causes microtubules to move the granules and then the cortex itself. The outcome is a **gray crescent**, an area of intermediate pigmentation near the equator, on the side of the egg *opposite* the sperm penetration site (Figures 42.7 and 42.8). This area establishes the body axis of the frog embryo, and gastrulation will begin there. This is another example of how the organization of the egg's cytoskeleton represents critical information for the embryo.

animal pole

(gray crescent)

vegetal pole

(gray crescent)

Figure 42.8 Micrographs and diagrams of the early embryonic development of a frog. For these micrographs, the jellylike layer surrounding the egg has been removed, with the exception of (**i**). (**a**) Within about an hour after fertilization, the gray crescent establishes the body axis for the embryo, and gastrulation will begin here. (**b**–**f**) Cleavage leads to a blastula, a ball of cells in which a cavity (blastocoel) has appeared.

(**g**, **h**) Cells move about and become rearranged during gastrulation. Tissue layers form; a primitive gut cavity (archenteron) develops. (**i**, **j**) Neural developments now take place, and the fluid-filled body cavity in which vital organs will be suspended appears. (**k**) Differentiation proceeds, moving the embryo on its way to becoming a functional larval form.

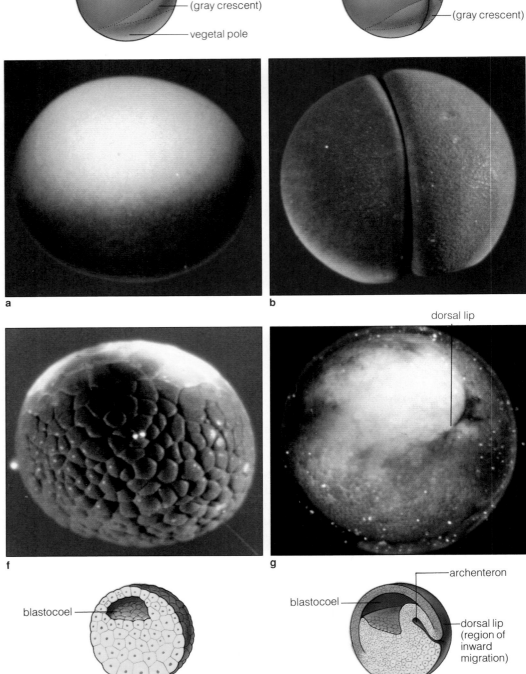

a

b

dorsal lip

f

g

blastocoel

archenteron

blastocoel

dorsal lip (region of inward migration)

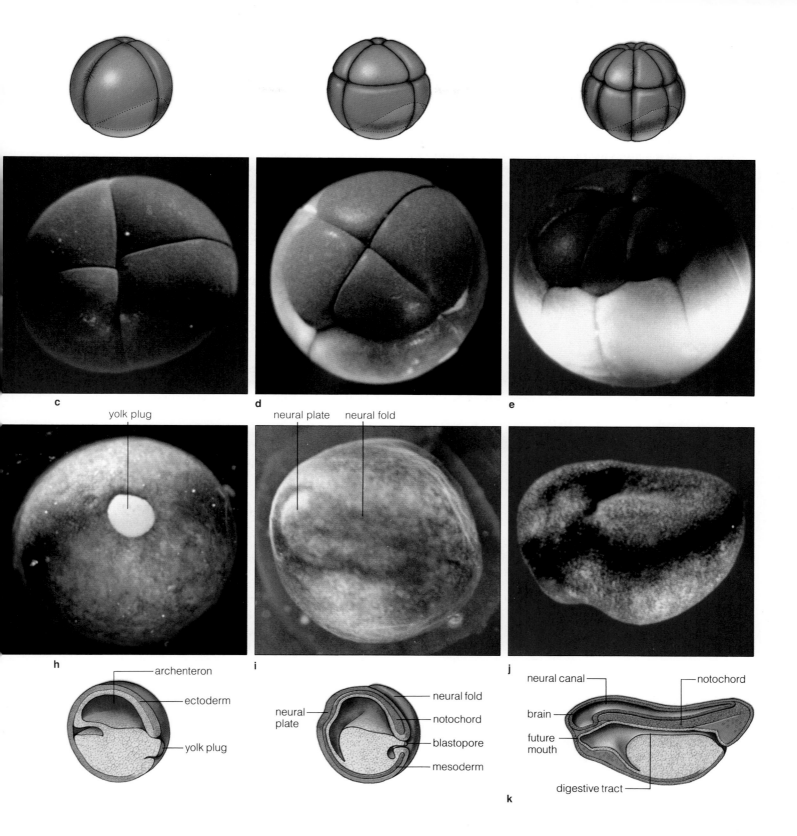

c

d

e

yolk plug

neural plate   neural fold

h

archenteron

ectoderm

yolk plug

i

neural
plate

neural fold

notochord

blastopore

mesoderm

j

neural canal   notochord

brain

future
mouth

digestive tract

k

Cleavage

Fertilization is followed by mitotic cell divisions that convert the zygote into the early multicelled embryo. A cleavage furrow forms during each mitotic cell division (page 145), and it defines the plane where the cytoplasm will be pinched in two. Usually the embryo does not increase in size while the initial cleavages are proceed-

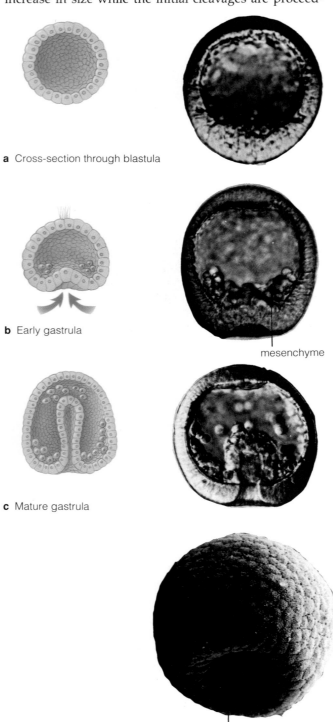

a Cross-section through blastula

b Early gastrula

mesenchyme

c Mature gastrula

d blastopore

ing. All the daughter cells collectively occupy the same volume as did the zygote, although they differ in size, shape, and activity.

Successive cleavages commonly produce a **blastula**, a ball of cells with a fluid-filled cavity (blastocoel) inside. The cleavage patterns themselves vary among different animal groups, however, often as a result of dramatic differences in the density and distribution of yolk. Cleavage of a sea urchin egg, which has little yolk, produces a hollow, single-layered sphere of cells (Figure 42.9). The concentrated yolk of an amphibian egg impedes cleavage near the vegetal pole, and a blastocoel forms near the animal pole. The abundant yolk of reptile, bird, and most fish eggs restricts cleavage to a tiny, caplike region at the animal pole. In these eggs, cleavage produces two flattened layers with a thin blastocoel between them, perched on the yolk surface (Figures 42.4c and 42.5c).

The abundant yolk of insect eggs influences cleavage in a truly distinctive way. Consider a *Drosophila* egg (Figure 42.10). At first the nucleus divides repeatedly and cleavage furrows do not form. Within hours, the fertilized egg is a bag of nuclei crowded together in the yolky cytoplasm. Then most of the nuclei migrate toward the egg surface, where a layer of cells forms. This layer (the blastoderm) is the embryo.

The blastula stage of mammals, called a blastocyst, differs from that of other animals in a key respect. By this stage, *two* distinct cell regions can be discerned. Some cells, which form a hollow sphere, are not part of the embryo. Their cellular descendants will form a portion of the placenta. (That structure, composed of embryonic and maternal tissues, will be vital for nurturing the developing embryo.) Other cells are clustered together and attached to the inner surface of the blastocyst wall. As described in the next chapter, the embryo proper develops from this "inner cell mass."

As these examples make clear, the particular cleavage planes that form in an animal zygote help dictate the size and spatial positions of the resulting cells. They also dictate which cytoplasmic determinants each cell will inherit. Together, these factors will affect how each cell will interact with others during the gastrula stage—and so on through subsequent stages of development.

Figure 42.9 (**a**) Blastula of a sea urchin (*Lytechinus*). The inward migration of surface cells during gastrulation is apparent in the cross-sections in (**b**) and (**c**). Some of the cells, designated mesenchyme, will ultimately develop into a third germ layer, the mesoderm. (**d**) This scanning electron micrograph shows a surface view of individual cells and the inward cell migrations of an early gastrula.

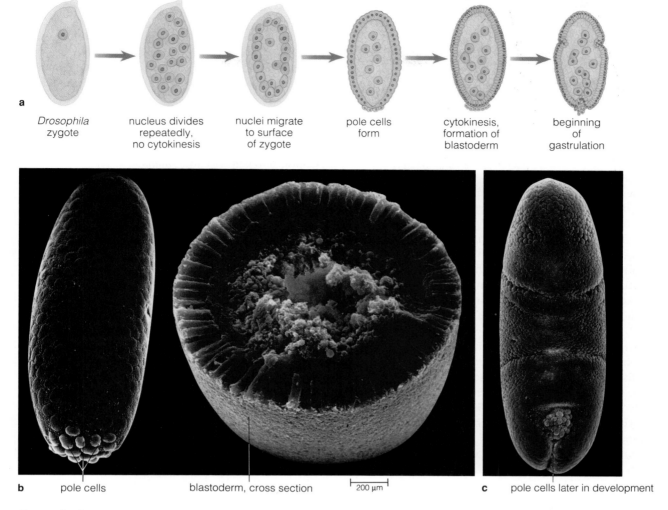

a   *Drosophila* zygote   nucleus divides repeatedly, no cytokinesis   nuclei migrate to surface of zygote   pole cells form   cytokinesis, formation of blastoderm   beginning of gastrulation

b   pole cells   blastoderm, cross section   200 μm   **c**   pole cells later in development

Gastrulation

As cleavage draws to a close, the pace of cell division slackens. The cells begin to move about and change their positions relative to one another. Gastrulation, the stage of cell rearrangements, has begun.

There is little (if any) increase in size during gastrulation, but the cell rearrangements dramatically change the embryo's appearance. In sea urchins, for example, surface cells migrate inward, and some form the lining of an internal cavity (archenteron). That cavity will eventually become the gut (Figure 42.9). In other species, cell rearrangements establish a long axis for further development after gastrulation. For example, a tube will form along this axis in vertebrate embryos. As you will see shortly, the neural tube is the forerunner of the brain and spinal cord.

This last example brings us to the significance of gastrulation. Think about the structural organization of animals. With few exceptions, animals have an internal region of cells, tissues, and organs that function in digestion and absorption of nutrients. They have a surface

Figure 42.10 (**a**) Delayed cleavage in a *Drosophila* zygote. The nucleus divides repeatedly, but cleavage is postponed until the nuclei migrate toward the zygote's surface. Then, a layer of single cells forms above the yolky cytoplasm. That layer, the blastoderm, is the early embryo (**b**).

(**c**) Cytoplasmic localization is known to affect a key aspect of *Drosophila* development. While the blastoderm is forming, cytoplasmic components called polar granules migrate to and become isolated in pole cells that form at one end of the zygote. In certain mutant embryos, the polar granules do not migrate properly—and reproductive cells never do form. The adult fly is sterile.

tissue region that protects internal parts and is equipped with sensory receptors for detecting changes in the outside world. Most animals have an intermediate region of tissues organized into many internal organs, such as those concerned with movement, support, and blood circulation. The three regions develop from the three germ layers—the endoderm, ectoderm, and mesoderm. This three-layered organization is typical of most animals, and it arises through gastrulation.

CELL DIFFERENTIATION

Through cell differentiation, a single fertilized egg gives rise to diverse types of specialized cells. All cells produce a number of the same kinds of proteins. But each differentiated cell type also produces some proteins that are *not* found in other cell types. Those proteins are the basis of distinctive cell structures and functions. Differentiated cells have the same number and kind of genes (they are all descended from the same zygote). Through gene controls, however, restrictions are placed on *which* genes will be expressed in a given cell (page 236).

For example, while you were still an embryo and your eyes were developing, some cells started synthesizing quantities of crystallin, a family of proteins that would be used in the construction of transparent fibers in the lens. No other cell type in your developing body could activate the genes necessary to do this. The transparent fibers caused the lens cells to elongate and flatten and gave them their unique optical properties. The crystallin-producing cells are only one of many differentiated types.

By conservative estimates, adult mammals end up with populations of at least 150 different cell types, each with its distinctive structure, products, and functions. (It also has populations of *stem cells*, which can self-replicate, differentiate, or both. Descendants of these stem cells continually replace worn-out or dead cells in skin, blood, and the gut mucosa.)

With only rare exceptions, the fully differentiated state of a given cell type is reached without any loss of genetic information. How do we know this? Consider John Gurdon's experiments with the South African clawed frog, *Xenopus laevis*. Gurdon and his coworkers removed or inactivated the nucleus of an unfertilized frog egg. They also isolated intestinal cells from *Xenopus* tadpoles and carefully ruptured the plasma membrane, leaving the nucleus and most of the cytoplasm intact. Then they inserted the nucleus of the ruptured tadpole cell into the enucleated egg. In some cases, the transplanted nucleus—which was from a highly differentiated intestinal cell—directed the developmental program leading to a whole frog! Clearly the intestinal cell nucleus still contained the same genes as the original zygote nucleus.

As another example, consider what happens when a human embryo spontaneously splits during the first cleavage into two separate cells. The result is not two half-embryos but *identical twins*, or two complete, normal individuals having the same genetic makeup. (By contrast, nonidentical twins occur when two different eggs are fertilized at the same time by two different sperm.) Spontaneous splitting at cleavage is actually the normal pattern of development for armadillos. For those animals, the embryo splits at the four-cell stage to produce quadruplets, every time.

MORPHOGENESIS

For each animal, tissues and organs become organized with great precision into patterns characteristic of the species. What kinds of events give rise to that organization? Considerable research in embryology has shown that the main ones are these:

1. Cell division
2. Cell migrations
3. Changes in cell size and shape
4. Localized growth
5. Controlled cell death

Let's consider a few examples of the morphogenetic changes that such cellular events can bring about.

Cell Migrations

During morphogenesis, cells and entire tissues migrate from one site to another. In **active cell migration**, cells move about by means of pseudopods. Pseudopods are temporary projections from the main cell body, some rather bulbous, others fingerlike. The cells migrate over prescribed pathways, reaching a prescribed destination and establishing contact with cells already there. These movements must be extremely accurate. Through such migrations, for instance, the forerunners of nerve cells make billions of precise connections that enable the human nervous system to function.

How do the cells "know" where to move? In part, they move in response to chemical gradients, a behavior called *chemotaxis*. The gradients are probably created when specific substances are released from target tissues. This type of movement was shown in Figure 22.18, which described the development of the slime mold *Dictyostelium discoideum*.

Cells also move in response to *adhesive cues*, provided by recognition proteins at the surface of other cells and by molecules of the extracellular matrix. In vertebrate embryos, pigment cells are following such cues when they move along blood vessels but not along the axons of neurons. Similarly, Schwann cells are following adhesive cues when they migrate along the axons of neurons but not along blood vessels. It seems likely that the synthesis, secretion, deposition, and removal of specific extracellular substances help coordinate the active migration of cells in a developing embryo.

How do the migrating cells know when to stop moving? Again, adhesive cues appear to be involved. Migrating cells move to locations where adhesive interactions are strongest. Once they become arranged in a manner that maximizes their adhesion, further migration is impeded.

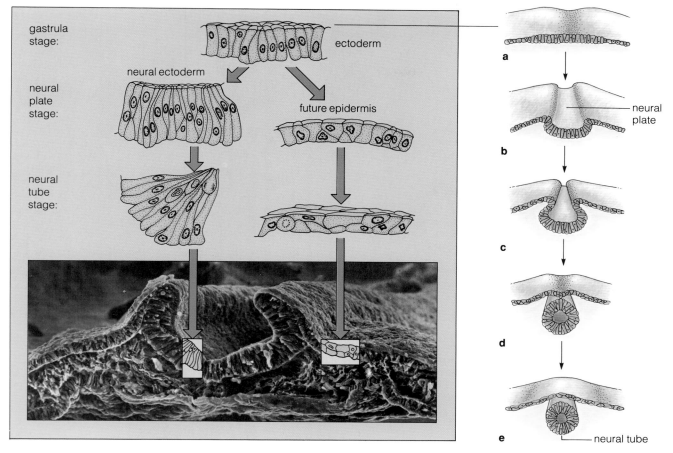

Figure 42.11 Example of morphogenesis—the changes in cell shape that underlie the formation of a neural tube (the forerunner of the brain and spinal cord). As gastrulation draws to a close, the ectoderm is a uniform sheet of cells (**a**). Some ectodermal cells elongate, forming a neural plate, then they constrict at one end to become wedge-shaped. The changes in cell shape cause the ectodermal sheet to fold over the neural plate to form the neural tube (**b–e**). Other ectodermal cells flatten while these changes are occurring; they will become part of the epidermis. The scanning electron micrograph and diagrams show the neural tube and epidermis forming in a chick embryo.

Changes in Cell Size and Shape

Another morphogenetic movement is the inward or outward folding of sheets of cells. Such folding is brought about by coordinated changes in cell shapes. Think about what happens after the three germ layers form in the embryos of amphibians, reptiles, birds, and mammals. As Figure 42.11 shows, ectodermal cells at the midline of the embryo elongate and form a **neural plate**, the first indication that a region of ectoderm is on its way to developing into nervous tissue. The change in each cell's shape is supported by the elongation of microtubules in its cytoplasm. Next, cells near the middle become wedge-shaped. The shape arises when a ring of microfilaments in the cytoplasm constricts one end of each elongated cell. Collectively, the changes in cell shape cause the neural plate to fold over and meet at the embryo's midline to form the **neural tube**. This tube is destined to become

the brain and spinal cord (page 566). Meanwhile, other ectodermal cells have flattened; they will become the epidermis above the tube.

Localized Growth and Cell Death

Morphogenesis depends on **localized growth**, which contributes to changes in the sizes, shapes, and proportions of body parts. How localized growth occurs in some tissues more than others is not fully understood, but regulatory genes are almost certainly involved. Recall that there are few differences between chimpanzee and human DNA. Yet adult chimpanzees and humans differ quite a bit in the proportions of their body parts, including the skull, even though their fetuses develop at about the same rate, in parallel ways (page 296). Possibly during human evolution, there were mutations in regulatory genes that affected proportional changes in the skull.

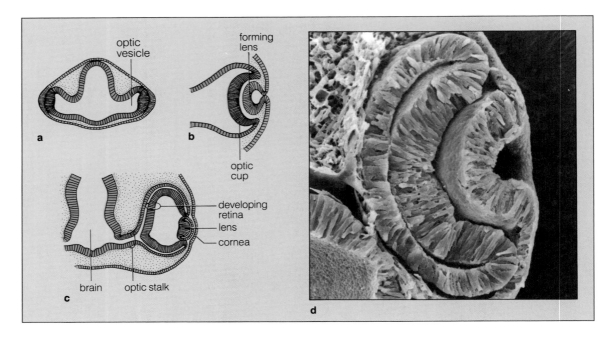

Figure 42.12 Eye formation. The retina develops as an outgrowth of the brain; the lens, as an ingrowth of the ectoderm. (**a**) An optic vesicle grows out of the side of the brain. When it contacts the head ectoderm, it induces the elongation and inward folding of the ectodermal cells to form a lens vesicle (**b**). Meanwhile, the optic vesicle is induced to sink inward, forming the optic cup. (**c**) The cup's inner layer will form the retina. (**d**) Scanning electron micrograph of an optic cup and lens in a chick embryo.

Morphogenesis also depends on **controlled cell death**. The term refers to the elimination of tissues and cells that are used for only short periods in the embryo or the adult. Controlled cell death is genetically programmed, as the following examples suggest.

Perhaps you have noticed that kittens and puppies are born with their eyes sealed shut. The eyelids form as an unbroken layer of skin. Just after birth, the cells stretching in a thin line across the middle of each eyelid die on cue. As the dead cells degenerate, a slit forms in the skin, and then the upper and lower lids part company.

In a human embryo, hands and feet start out as paddle-shaped structures. Skin cells between the lobes of four "paddles" die on cue, leaving separate toes and fingers (Figure 9.1). This mechanism is genetically programmed. Between the time a death signal is sent and the actual time of death, protein synthesis declines dramatically in the doomed cells.

Finally, the embryos of ducks also have paddlelike appendages, but cell death normally does not occur in them; that is why ducks have webbed feet instead of separated toes. In some mice and some humans, a gene mutation blocks cell death in the paddles, and the digits remain webbed.

Pattern Formation

For morphogenesis to proceed smoothly in a developing embryo, cells must sense their position relative to one another and use the information to produce ordered, spatial arrangements of differentiated tissues. The term **pattern formation** refers to the mechanisms responsible for the specialization of tissues and their positioning in space. As we have seen, the most important of those mechanisms are cytoplasmic localization and embryonic induction.

Vertebrate Eye Formation. Hans Spemann's research into the development of the vertebrate eye provides us with a classic example of embryonic induction. The retina of an eye originates from the forebrain but its lens, which focuses light onto the retina, originates from the epidermis (Figure 42.12). Spemann experimented with a salamander embryo in which optic cups had already begun to grow out of the forebrain. He surgically removed one of the optic cups and inserted it under the ectoderm of the belly region. Belly epidermal cells that came in contact with the transplanted optic cup were induced to form a lens—which fit perfectly into the transplanted optic cup!

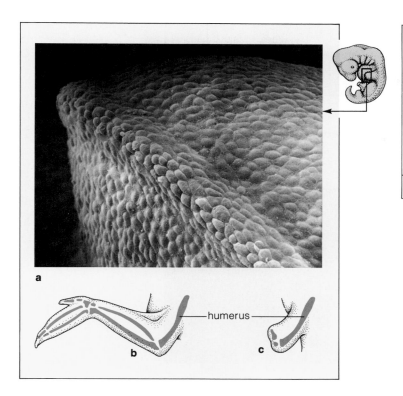

Figure 42.13 (**a**) Scanning electron micrograph of the normal ectodermal ridge of a wing bud in a chick embryo. (**b**) Normal pattern of bone formation. (**c**) Pattern of bone formation when ectodermal cells at the ridge are surgically removed before the wing bud is fully grown. All new development ceases. Only the bones that have been determined by the time of the surgery will differentiate; notice the deficiencies in the wing skeleton.

Chick Wing Formation. The importance of embryonic induction is further illustrated by the precise positioning of bones that occurs when a chick wing develops. Through cell divisions in its mesoderm and the ectodermal covering, the wing bud grows out from the body. When certain groups of ectodermal cells are removed from the tip of a half-grown wing bud, terminal wing bones never develop (Figure 42.13). Apparently, as new cells are added to the outwardly growing tip, they "assess" which bones were formed previously, then they become the next bones in line. If the ectodermal ridge is surgically removed, there will be no new mesodermal cells to form the remaining bones.

Pattern Formation in *Drosophila*. As is the case for other arthropods, the *Drosophila* body is segmented. Segmentation is one consequence of cytoplasmic localization. Before cleavage, cytoplasmic components direct the migration of different nuclei in the *Drosophila* zygote toward different areas of the zygote's surface (Figure 42.10).

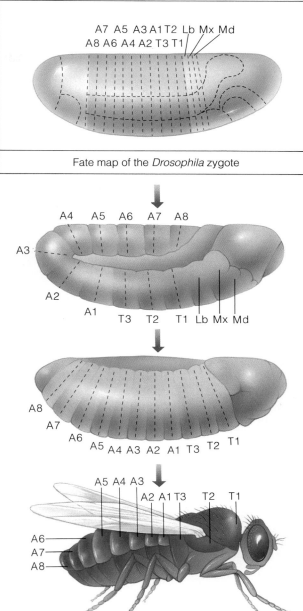

Figure 42.14 Fate map of the *Drosophila* zygote. Dashed lines indicate the regions that will develop into specialized segments, many of which have highly specialized appendages. The developmental fate of the different segments begins with cytoplasmic localization in the zygote. The drawings are not to scale.

In certain experiments, small regions of the cytoplasm were destroyed in the zygote and segments failed to develop properly. Such experiments have helped developmental biologists draw a "fate map" on the surface of the *Drosophila* zygote. The map is used to determine the origin of each kind of differentiated cell in the different body segments (Figure 42.14).

Figure 42.15 Imaginal disks of a *Drosophila* larva. Each disk will give rise to a specific part of the adult fly.

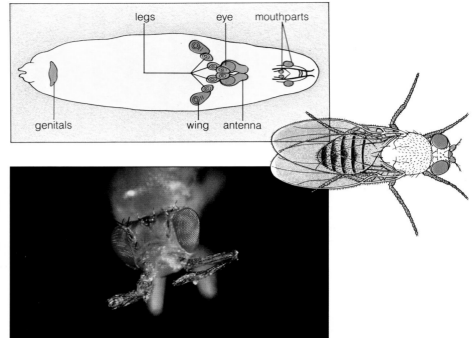

Figure 42.16 Experimental evidence of embryonic induction in *Drosophila*. A cluster of cells (imaginal disk) that gives rise to antennae was surgically removed and exposed to abnormal tissue environments. Then it was reinserted into a different location in a larva undergoing metamorphosis. Legs appeared on the head where antennae would normally be.

Figure 42.17 Controls over the development of a silkworm moth, *Platysamia cecropia*. A hatched larva eats until it completes five near-doublings in size through rapid epidermal cell divisions and enlargements. The cells also secrete chitin to the epidermal surface, and this forms a cuticle that sets an upper limit on increases in mass. When the limit is reached, cell division idles and the larva molts (sheds its cuticle). Cell divisions resume, more chitin is secreted, and a larger cuticle is produced. The larva grows and molts repeatedly. Then chitin is deposited over the whole insect, legs and all; this is the pupal stage.

The insect spins a cocoon around itself for three days, then the body undergoes massive cell destruction and tissue reorganization. The contents of degraded cells form a nutrient-rich soup that sustains the growth of imaginal disks. The disks had been growing slowly, but now they rapidly give rise to what will become adult tissues and organs. The pupa lasts eight winter months. In spring, cell death and tissue changes transform the pupa into the adult.

Transformation of a larva into a moth is an example of metamorphosis. The adult "plan" is already laid out in immature tissues. Neural and endocrine controls dictate when the plan will be fulfilled. Cells in the larval brain secrete a hormone that acts on paired prothoracic glands. The glands produce and secrete precursors of *b-ecdysone*, a hormone that activates gene expression and stimulates molting. Hormones also act on two glands (the corpora allata) behind the brain. The glands secrete *juvenile hormone* (JH) which, at high levels, prolongs the juvenile state.

The *absence* of JH triggers cell differentiation into the adult. High levels of both JH and ecdysone promote larval growth and development. The amount of JH declines steadily and disappears by the fifth larval stage. The pupa forms when ecdysone levels are high and JH levels are low. Then, developmental restraints are lifted from cells of the imaginal disks.

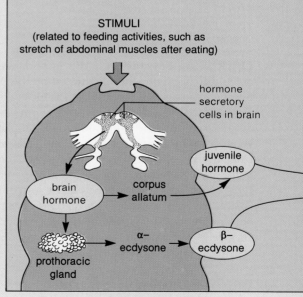

STIMULI
(related to feeding activities, such as stretch of abdominal muscles after eating)

hormone secretory cells in brain

juvenile hormone

brain hormone

corpus allatum

α-ecdysone

β-ecdysone

prothoracic gland

anterior end of larva

Other *Drosophila* experiments provide evidence of embryonic induction. Some cells in *Drosophila* larvae are arranged in clusters, each of which will give rise to specific body parts of the adult (Figure 42.15). The clusters are *imaginal disks*. Normally, the fate of the cells in each disk is sealed early in development. However, when an "antenna" disk is surgically removed and exposed to abnormal tissue environments before reinsertion into a larva undergoing metamorphosis, it may later differentiate into a leg (Figure 42.16). The experiment provides evidence that inducer signals influence which genes will be "read" when cells of the disks start the program of differentiation.

The same thing happens as a result of homeotic mutations. Apparently, a **homeotic mutation** affects a regulatory gene that activates sets of genes concerned with development. These single-gene mutations lead either to alterations in the inducer substance itself or to abnormal responses to it. One homeotic mutation activates the wrong set of genes in the cells of an "antenna" disk in *Drosophila*, causing those cells to differentiate into a leg.

Inducer Signals. What acts as an inducer during pattern formation? How is it transmitted from the inducer tissue to the responding tissue? Several experiments indicate that chemical signals diffuse from one tissue to the other. When the two tissues destined to become the epithelium and connective tissue of a pancreas are surgically separated from each other in a mouse embryo, the future epithelial tissue does not differentiate properly. However, suppose the two tissues are grown in a culture medium, separated only by a filter that is permeable to large molecules (but not to cells or cell structures). Then, the responding tissue differentiates properly. Although physical contact with the inducing tissue had been prohibited, signals still reached it.

This completes our discussion of the types of mechanisms underlying specific steps along the road leading from the embryo to the adult form. We conclude with the example in Figure 42.17, which shows how separate mechanisms can be integrated and controlled to bring about the transformation. This figure describes how controls govern the dramatic metamorphosis of a silkworm larva into a moth.

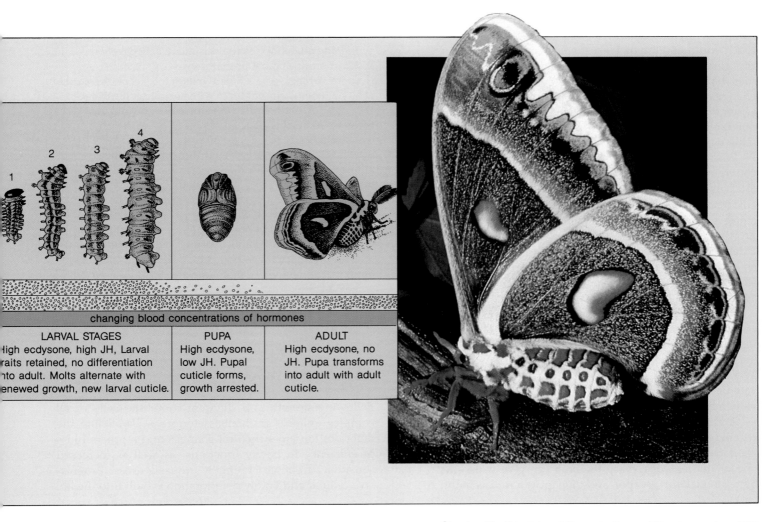

changing blood concentrations of hormones

LARVAL STAGES	PUPA	ADULT
High ecdysone, high JH, Larval traits retained, no differentiation into adult. Molts alternate with renewed growth, new larval cuticle.	High ecdysone, low JH. Pupal cuticle forms, growth arrested.	High ecdysone, no JH. Pupa transforms into adult with adult cuticle.

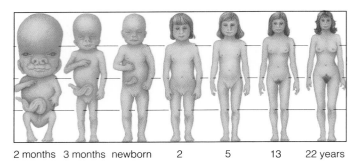

2 months 3 months newborn 2 5 13 22 years

Figure 42.18 Diagram of changes in the proportions of the human body during prenatal and postnatal growth.

POST-EMBRYONIC DEVELOPMENT

Once all the organs necessary for feeding and other vital activities have formed and are functioning, the new individual is ready to lead a more or less independent existence. Embryonic development is over, and now the young animal follows a prescribed course of further growth and development that leads to the **adult**, the sexually mature form of the species.

In nematodes and a few other animals, the transition from embryo to adult is straightforward. The young simply are miniatures of what is to come. All it takes to reach adulthood are increases in size and maturation of the gonads. The post-embryonic development of reptiles, birds, and mammals involves increases in size and changes in body proportions, as Figure 42.18 suggests. Other organs besides the gonads may not be fully developed. Newly hatched birds, for example, have only partially developed primary feathers just under the skin.

For insects and some other kinds of animals, the course of development is "indirect," for a larval stage intervenes between the embryo and the adult. First the embryo grows into a larva (a sexually immature, free-living and free-feeding animal), then the larva itself grows and changes into the sexually mature form. Extended larval stages are typical of animals that release small and relatively yolkless eggs into water. Short larval stages are common among animals that lay large, yolky eggs in water.

In some species, the transformation from larva to adult is gradual, with the immature form simply growing in size. In other species, the transformation involves massive tissue reorganization and drastic remodeling into the adult form, as in frogs (Figure 42.1). The reactivated growth and transformation of a larva into the sexually mature adult is called **metamorphosis**.

A different kind of reactivated growth occurs during **regeneration**: the replacement of body parts that have been lost by accident. For example, if a predator grasps one of the legs of a crab, the crab can give it up (the better to make an escape), and then grow a replacement.

Commentary

Death in the Open

By Lewis Thomas (Printed by permission from the author and the New England Journal of Medicine, January 11, 1973, 288:92–93)

Everything in the world dies, but we only know about it as a kind of abstraction. If you stand in a meadow, at the edge of a hillside, and look around carefully, almost everything you can catch sight of is in the process of dying, and most things will be dead long before you are. If it were not for the constant renewal and replacement going on before your eyes, the whole place would turn to stone and sand under your feet.

There are some creatures that do not seem to die at all; they simply vanish totally into their own progeny. Single cells do this. The cell becomes two, then four, and so on, and after a while the last trace is gone. It cannot be seen as death; barring mutation, the descendants are simply the first cell, living all over again. The cycles of the slime mold have episodes that seem as conclusive as death, but the withered slug, with its stalk and fruiting body, is plainly the transient tissue of a developing organism; the free-swimming amoebocytes use this mode collectively in order to produce more of themselves.

There are said to be a billion billion insects on the earth at any moment, most of them with very short life expectancies by our standards. Someone has estimated that there are 25 million assorted insects hanging in the air over every temperate square mile, in a column extending upward for thousands of feet, drifting through the layers of atmosphere like plankton. They are dying steadily, some by being eaten, some just dropping in their tracks, tons of them around the earth, disintegrating as they die, invisibly.

Who ever sees dead birds, in anything like the huge numbers stipulated by the certainty of the death of all birds? A dead bird is an incongruity, more startling than an unexpected live bird, sure evidence to the human mind that something has gone wrong. Birds do their dying off somewhere, behind things, under things, never on the wing.

AGING AND DEATH

Following growth and differentiation, the cells of all complex animals gradually deteriorate. Paralleling the deterioration are structural changes and a gradual loss of efficiency in bodily functions, as well as increased sensitivity to environmentally induced stress. Progressive cellular and bodily deterioration is built into the life

Animals seem to have an instinct for performing death alone, hidden. Even the largest, most conspicuous ones find ways to conceal themselves in time. If an elephant missteps and dies in an open place, the herd will not leave him there; the others will pick him up and carry the body from place to place, finally putting it down in some inexplicably suitable location. When elephants encounter the skeleton of an elephant in the open, they methodically take up each of the bones and distribute them, in a ponderous ceremony, over neighboring acres.

It is a natural marvel. All of the life on earth dies, all of the time, in the same volume as the new life that dazzles us each morning, each spring. All we see of this is the odd stump, the fly struggling on the porch floor of the summer house in October, the fragment on the highway. I have lived all my life with an embarrassment of squirrels in my backyard, they are all over the place, all year long, and I have never seen, anywhere, a dead squirrel.

I suppose that is just as well. If the earth were otherwise, and all the dying were done in the open, with the dead there to be looked at, we would never have it out of our minds. We can forget about it much of the time, or think of it as an accident to be avoided, somehow. But it does make the process of dying seem more exceptional than it really is, and harder to engage in at the times when we must ourselves engage.

In our way, we conform as best we can to the rest of nature. The obituary pages tell us of the news that we are dying away, while birth announcements in finer print, off at the side of the page, inform us of our replacements, but we get no grasp from this of the enormity of the scale. There are 4 billion of us on the earth in this year, 1973, and all 4 billion must be dead, on a schedule, within this lifetime. The vast mortality, involving something over 50 million each year, takes place in relative secrecy. We can only really know of the deaths in our households, among our friends. These, detached in our minds from all the rest, we take to be unnatural events, anomalies, outrages. We speak of our own dead in low voices; struck down, we say, as though visible death can occur only for cause, by disease or violence, avoidably. We send off for flowers, grieve, make ceremonies, scatter bones, unaware of the rest of the 4 billion on the same schedule. All of that immense mass of flesh and bone and consciousness will disappear by absorption into the earth, without recognition by the transient survivors.

Less than half a century from now, our replacements will have more than doubled in numbers. It is hard to see how we can continue to keep the secret, with such multitudes doing the dying. We will have to give up the notion that death is a catastrophe, or detestable, or avoidable, or even strange. We will need to learn more about the cycling of life in the rest of the system, and about our connection in the process. Everything that comes alive seems to be in trade for everything that dies, cell for cell. There might be some comfort in the recognition of synchrony, in the information that we all go down together, in the best of company.

cycle of all organisms in which differentiated cells show extensive specialization. The process is called **aging**.

Aging in humans leads to structural changes such as loss of hair and teeth, increased skin wrinkling and fat deposition, and decreased muscle mass. Less obvious are gradual physiological changes. For example, metabolic rates decline in kidney cells, so the body cannot respond as effectively as it once did to changes in extracellular fluid volume and composition. Another change involves collagen, the fibrous protein that is present in the extracellular spaces of nearly all tissues. Collagen may represent as much as 40 percent of your body's proteins. With increasing age, new collagen fibers that are being produced are structurally altered—and such structural changes are bound to have widespread physical effects.

No one knows what causes aging, although researchers have given us some interesting things to think about. More than two decades ago, Paul Moorhead and Leonard Hayflick cultured normal embryonic cells from humans. They discovered that all of the cell lines proceeded to divide about fifty times, then the entire population died off. Hayflick took some of the cultured cells and froze them for a period of years. Afterward, he allowed them to thaw and placed them in a culture medium. The cells proceeded to complete the cycle of fifty doublings—whereupon they all died on schedule.

Such experiments suggest that normal cell types have a **limited division potential**, whereby mitosis is genetically programmed to decline at a particular stage of the life cycle. But does the change in mitosis *cause* aging or is it a *result* of the aging process? Consider that neurons, which do not divide at all after an early developmental stage, still deteriorate gradually during the life of an animal. Their predictable deterioration might indicate that similar changes may be occurring in dividing cells throughout the body.

Some researchers believe that cells gradually lose the capacity for DNA self-repair, perhaps as a result of an accumulation of environmental insults. Over time, DNA mutations certainly could thwart the production of enzymes and other proteins required for proper cell functioning. Consider that cells depend on smooth exchanges of materials between the cytoplasm and the extracellular fluid. And collagen, recall, is present in extracellular spaces throughout the body. If deteriorating regions of DNA code for defective collagen molecules, it is conceivable that the movement of oxygen, nutrients, hormones, and so forth to and from cells could be hampered, with repercussions extending through the entire body.

Finally, consider what might happen as a result of deterioration of genes coding for membrane proteins that serve as self-markers (page 685). If these identification markers change, does the immune system then perceive the body's own cells as "foreign" and mount an attack on them? According to one theory, such autoimmune responses might intensify over time, thereby producing the increased vulnerability to disease and stress associated with aging.

SUMMARY

1. Asexual reproduction in animals occurs most often by fission or budding. Sexual reproduction requires the fusion of nuclei of two gametes.

2. Animal development commonly proceeds through six stages:

a. Gametogenesis, during which the egg and sperm mature within the reproductive organs of the parents.

b. Fertilization, which begins when a sperm penetrates an egg and is completed when the sperm and egg nuclei fuse.

c. Cleavage, when the fertilized egg (zygote) undergoes mitotic cell divisions that form the early multicelled embryo (the blastula). Most genes are inactive during cleavage, and the destiny of cell lineages is established in part by the sector of cytoplasm inherited at this time.

d. Gastrulation, when the organizational framework of the whole animal is laid out. Endoderm, ectoderm, and usually mesoderm form; all the tissues of the adult body will develop from these germ layers.

e. Organogenesis, the onset of organ formation. The different organs start developing by a tightly orchestrated program of cell differentiation and morphogenesis.

f. Growth and tissue specialization, when organs enlarge overall and acquire their specialized chemical and physical properties. The maturation of tissues and organs continues into post-embryonic stages.

3. Cell differentiation and morphogenesis have their foundations in (a) the distribution of localized cytoplasmic components during cleavage and (b) cell interactions that begin during organogenesis.

4. Cell differentiation is a process whereby initial populations of genetically equivalent cells give rise to subpopulations of phenotypically different types of cells.

5. Starting at organogenesis, selective restrictions are placed on gene expression in different cell types. As a result, some cells remain in an undifferentiated state, and others proceed through differentiation and develop into specialized cell types.

6. Morphogenesis is a process whereby different cell types become organized into all the specialized tissues and organs of the body. It involves the growth, shaping, and arrangement of body parts according to prescribed patterns, and it is brought about by cell divisions, cell migrations, changes in cell shapes, localized growth, and controlled cell death.

7. Embryonic cells sense their position relative to one another and respond to this information by producing ordered arrangements of differentiated tissues. This kind of developmental activity is called pattern formation.

8. Following embryonic development, some animals grow directly into the adult form. Other animals proceed through indirect development, in which a larval stage is interposed between the embryo and the adult. Reactivated growth and development of the larva into the sexually mature adult form is called metamorphosis.

9. All complex animals gradually show changes in structure and a decline in efficiency (aging). Although aging is part of the life cycle of all animals having extensively specialized cell types, its cause is unknown.

Review Questions

1. Define and describe the main features of the following developmental stages: fertilization, cleavage, gastrulation, and organogenesis. *739, 741*

2. Cell differentiation and morphogenesis are two processes that are critical for development. Can you define them? Can you state briefly which two mechanisms serve as the basic foundation for cell differentiation and morphogenesis? *742*

3. What specific types of mechanisms bring about morphogenetic changes? *748–750*

4. During cleavage, the embryonic cells exhibit few regional differences, but in later stages, the rudimentary tissues and organs are discrete and the different cell groups do not merge. Suggest mechanisms that may be responsible for their discrete character. *748*

5. Experimentally, it is possible to divide an amphibian egg so that the gray crescent is wholly within one of the two cells formed. If the two cells are separated from each other, only the cell with the gray crescent will form an embryo with a long axis, notochord, nerve cord, and back musculature. The other cells form a shapeless mass of immature gut and blood cells. What do you think is the explanation of these outcomes? *742, 743*

Self-Quiz *(Answers in Appendix IV)*

1. Sexual reproduction by animals requires separation into male and female sexes. This reproductive strategy _____.
   - a. saves time
   - b. saves effort
   - c. saves energy
   - d. is costly

2. Asexual reproduction has advantages in _____ environments; sexual reproduction has the greater advantage in _____ environments.
   - a. predictable; unpredictable
   - b. land; aquatic
   - c. urban; rural
   - d. both a and b are correct

3. Development cannot proceed properly unless each stage is successfully completed before the next begins, starting with _____.
   - a. gamete formation
   - b. fertilization
   - c. cleavage
   - d. gastrulation
   - e. organ formation
   - f. growth, tissue specialization

4. During cleavage, daughter cells end up in different regions of the embryo and so inherit different components of the fertilized egg. This mechanism is called _____.
   - a. cytoplasmic localization
   - b. embryonic induction
   - c. cell differentiation
   - d. morphogenesis

5. Gene controls place restrictions on which genes will be expressed in different cells of a developing embryo. The immediate outcome is _____.
   - a. cytoplasmic localization
   - b. embryonic induction
   - c. cell differentiation
   - d. morphogenesis

6. During _____, differentiated cells become organized into tissues and organs.

7. In _____, one or more groups of cells interact physically or chemically in ways that affect the development of at least one of those groups.
   - a. cytoplasmic localization
   - b. embryonic induction
   - c. cell differentiation
   - d. morphogenesis

8. The mitotic cell divisions of cleavage produce an embryonic stage known generally as a _____.
   - a. centromere
   - b. gastrula
   - c. blastula
   - d. larva

9. Morphogenesis involves _____.
   - a. cell divisions and migrations
   - b. changes in cell size and shape
   - c. localized growth
   - d. controlled cell death
   - e. all of the above are correct

10. Match the development stage with its description.
    - _____ cleavage
    - _____ gametogenesis
    - _____ organ formation
    - _____ growth, tissue specialization
    - _____ gastrulation
    - _____ fertilization

    - a. egg and sperm mature in parents
    - b. sperm, egg nuclei fuse
    - c. formation of germ layers
    - d. most genes are inactive but fate of cell lineages sealed partly by cytoplasmic localization
    - e. organs, tissues acquire specialized properties
    - f. starts when germ layers split into subpopulations of cells

Selected Key Terms

active cell migration *748*
adult *754*
aging *755*
animal pole *743*
asexual reproduction *737*
blastula *739*
cell differentiation *742*
cleavage *739*
controlled cell death *750*
cytoplasmic localization *742*
ectoderm *739*
embryonic induction *742*
endoderm *739*
fertilization *739*
gamete formation *739*
gastrulation *739*
germ layer *739*
growth and tissue specialization *741*
limited division potential *756*
localized growth *749*
mesoderm *739*
metamorphosis *754*
morphogenesis *742*
neural plate *749*
neural tube *749*
oocyte *742*
organ formation *741*
pattern formation *750*
regeneration *754*
reproductive timing *738*
sexual reproduction *737*
vegetal pole *743*
yolk *739*

Readings

Balinsky, B. 1981. *An Introduction to Embryology.* Fifth edition. Philadelphia: Saunders.

Browder, L., C. Erickson, and W. Jeffrey. 1991. *Developmental Biology.* Third edition. Philadelphia: Saunders.

Carlson, B. 1988. *Patten's Foundations of Embryology.* Fifth edition. New York: McGraw-Hill.

Gilbert, S. 1991. *Developmental Biology.* Third edition. Sunderland, Massachusetts: Sinauer. Excellent introduction to animal development; stunning illustrations.

Raff, R., and T. Kaufman. 1983. *Embryos, Genes, and Evolution.* New York: Macmillan.

Saunders, J. 1982. *Developmental Biology: Patterns, Problems, Principles.* New York: Macmillan. Excellent introduction to embryology.

The Journey Begins

At first, nothing spectacular happens to the fertilized egg. It doesn't burgeon abruptly in size, like an inflating balloon. It doesn't even grow. Then comes the first clue that a spectacular journey is about to begin. That single cell starts *carving itself up*. By self-contained machinery, the cell cleaves itself in one direction, then another, and another until it has transformed itself into a hollow ball of about sixty tiny cells. The miniature cells all look alike, even the small bunch of cells huddled together against part of the ball's inner surface. But now a space is opening up inside the huddle. Fluid moves in; the space widens. The fluid is lifting the cell mass away from the inner surface, like tissue fluid does to the skin above a blister.

By this time the cells of the "blister" are spread out as an oval-shaped disk, like a pancake that didn't quite make it into a circle when the batter was poured. Make that two stacked pancakes—the cells of the disk are arranged as two layers. The disk itself is tiny. At less than a millimeter across, it could stretch out across the head of a straightpin with room to spare. But that speck is an embryo, and it is already going places!

Now a third layer is forming between the original two. Within those three layers, cells are following commands to change shape, to get moving. Many adjacent cells of one sheetlike layer are all constricting on one side of the cell body but not the other. Their coordinated constriction makes part of the sheet curve up and fold over! Other cells are elongating, making their component layer thicker. And still other cells are migrating to new destinations! The effect is stunning— a mere pancake of an embryo has become a pale, translucent crescent with surface bumps, tucks, and hollows.

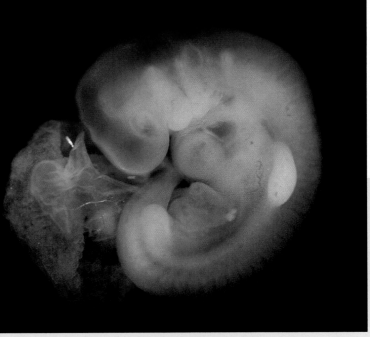

Figure 43.1 *Below left,* the billowing entrance to an ovarian duct—the tubelike road to the uterus. There, on that road, a sperm traveling from the opposite direction encountered an egg, and a remarkable developmental journey began. *Above,* the embryo just four weeks after the moment of fertilization.

Day after day, the structural molding goes on. By the third week after fertilization, the tubelike forerunner of the nervous system appears. Sheets of cells roll up, forming tubes of a future circulatory system. By the fourth week, a patch of cells starts beating rhythmically in the thickening wall of one tube. They are descendants of the embryonic heart. From now on, through birth, adulthood, until the time of death, those cells will continue beating as the heart's pacemaker. On the embryo's sides, budding regions of tissue mark the beginning of upper and lower limbs.

At this stage it is clear that the embryo is a vertebrate. Already it has a blunt head end, an obvious bilateral symmetry, a series of tissue segments along its back, and an embryonic tail. But is it a human? A minnow? A duckling? At this stage only an expert could answer.

Five weeks into the journey, the limb buds have paddles—the start of hands and feet. Round dots appear under the transparent skin and gradually darken and grow larger; the eyes are forming. Around and below them, cells migrate in concert, interacting in ways that sculpt out a nose, cheeks, and a mouth. By now, the embryo is recognizably a human in the making.

This was your beginning. Later embryonic and fetal events filled in the details, rounded out contours, added flesh and fat and hair and nails to your peanut-sized body. For all of those events to unfold, gametes first had to form in your parents and meet in a moist tunnel leading away from one of your mother's ovaries (Figure 43.1). That beginning, and all the subsequent events that unfolded inside your mother's body, is the story of this chapter.

KEY CONCEPTS

1. Humans have a well-developed reproductive system, consisting of a pair of primary reproductive organs (testes in males, ovaries in females), accessory glands, and ducts. Testes produce sperm; ovaries produce eggs. Both types of gonads also produce sex hormones, the secretion of which is under the control of the hypothalamus and pituitary gland.

2. Human males produce sperm continuously from puberty onward. The hormones testosterone, LH, and FSH control male reproductive functions.

3. The reproductive capacity of human females is cyclic and intermittent, with eggs being released and the uterine lining being prepared for pregnancy on a monthly basis. The hormones estrogen, progesterone, FSH, and LH control this cyclic activity.

4. As for other vertebrates, human development proceeds through six stages: gamete formation, fertilization, cleavage, gastrulation, organ formation, and growth and tissue specialization.

In the preceding chapter, we looked at some general principles of animal reproduction and development. Here, we will focus on humans as a way of presenting an integrated picture of the structure and function of reproductive organs, gamete formation, and the stages of development from fertilization to birth. As part of this picture, we will also consider some of the mechanisms controlling reproduction.

HUMAN REPRODUCTIVE SYSTEM

For both men and women, the reproductive system consists of a pair of primary reproductive organs (gonads), accessory glands, and ducts. Male gonads are **testes** (singular, testis), and female gonads are **ovaries**. Testes produce sperm; ovaries produce eggs. Both also secrete sex hormones, which influence reproductive functions and the development of secondary sexual traits. Such

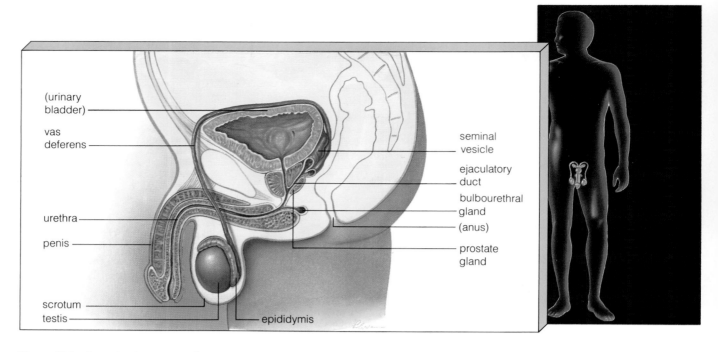

Figure 43.2 Reproductive system of the human male.

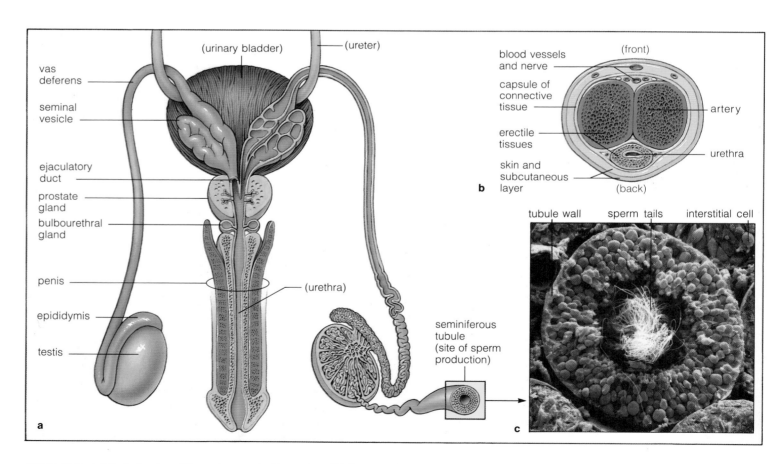

Figure 43.3 (**a**) Posterior view of the male reproductive system. (**b**) Cross-section of the penis. (**c**) Scanning electron micrograph showing the cells inside a seminiferous tubule.

Table 43.1	Organs and Accessory Glands of the Male Reproductive Tract
Organs:	
Testis (2)	Production of sperm, sex hormones
Epididymis (2)	Sperm maturation site and storage
Vas deferens (2)	Rapid transport of sperm
Ejaculatory duct (2)	Conduction of sperm
Penis	Organ of sexual intercourse
Accessory Glands:	
Seminal vesicle (2)	Secretions large part of semen
Prostate gland	Secretions part of semen
Bulbourethral gland (2)	Production of lubricating mucus

traits are distinctly associated with maleness and femaleness, although they do not play a direct role in reproduction. Examples are the amount and distribution of body fat, hair, and skeletal muscle.

Gonads look the same in all early human embryos. But after seven weeks of development, activation of genes on the sex chromosomes and hormone secretions trigger their development into testes *or* ovaries. The gonads and accessory organs are already formed at birth, but they do not reach full size and become functional until twelve to sixteen years later.

Male Reproductive Organs

Where Sperm Form. Figure 43.2 shows the male reproductive system and Table 43.1 lists its components. The testes start forming as buds from the wall of the embryo's abdominal cavity. Before birth they descend into the scrotum, an outpouching of skin below the pelvic region. Sperm develop properly when the scrotum's interior is kept a few degrees cooler than the body's normal temperature. Through controlled contractions of muscles in the scrotum, the temperature stays at 95°F or so. When it is cold outside, contractions draw the pouch closer to the (warmer) body. When it is warm outside, the muscles relax and lower the pouch.

Each testis is partitioned into as many as 300 wedge-shaped lobes. Each lobe contains two to three highly coiled tubes, the **seminiferous tubules**, and this is where sperm develop (Figure 43.3). Although a testis is only about 5 centimeters long, 125 meters of tubes are packed into it! Connective tissue between the tubes contains *Leydig cells* with endocrine functions. They secrete primarily the sex hormone testosterone.

Just inside the walls of these tubes we find undifferentiated diploid cells called spermatogonia. Ongoing cell divisions force the cells away from the walls, toward

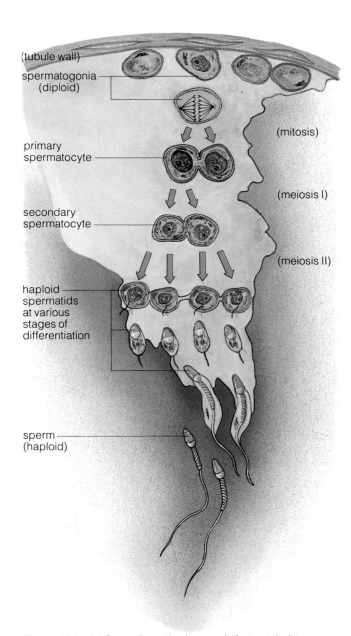

Figure 43.4 (a) Sperm formation in a seminiferous tubule. Undifferentiated diploid cells (spermatogonia) are closest to the tubule walls. They are forced away from it by ongoing mitotic cell divisions and are transformed into primary spermatocytes. Following meiosis I, they become secondary spermatocytes. Each chromosome in these haploid cells still consists of two sister chromatids. Sister chromatids separate from each other during meiosis II. The resulting spermatids gradually develop into mature sperm. The entire process takes about nine to ten weeks.

the tube's interior. During their forced departure, the cells are gradually transformed into primary spermatocytes. As Figure 43.4 shows, they undergo meiosis I, the result being secondary spermatocytes. Although the cells are now haploid, keep in mind that each chromosome they contain is still in the duplicated state; it consists of two sister chromatids (page 153). The sister chromatids of each chromosome are separated from each other during meiosis II. The resulting cells are haploid spermatids,

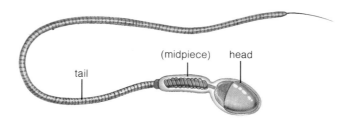

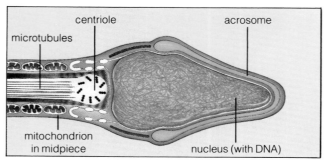

Figure 43.5 Structure of a mature human sperm.

Figure 43.6 (*Right*) Hormonal control of reproductive function in human males. The black dashed line indicates that increased testosterone secretion inhibits LH secretions through its negative effect on hypothalamic GnRH. The blue dashed line indicates that an inhibitory signal from Sertoli cells influences GnRH and FSH secretions. The "signal" is the hormone inhibin.

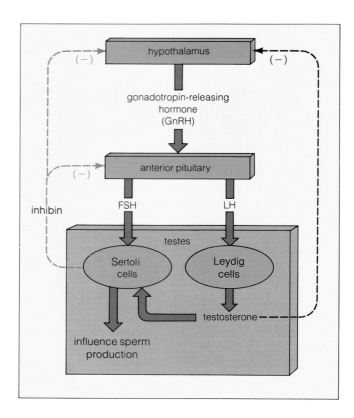

which gradually develop into **sperm**—the male gametes. The entire process takes about nine to ten weeks. All the while, the developing cells receive nourishment and chemical signals from adjacent *Sertoli cells*, which are the only other type of cell inside the tubule.

From puberty onward, sperm are produced continuously, with many millions in different stages of development on any given day. A mature sperm has a tail, a midpiece, and a head with a DNA-packed nucleus. An enzyme-containing cap (acrosome) covers most of the head. Its enzymes help the sperm penetrate an egg at fertilization. Mitochondria in the midpiece supply energy for the tail's whiplike movements (Figure 43.5).

How Semen Forms. Sperm become part of *semen*, a thick fluid that eventually is expelled from the penis. When sperm move out of a testis, they enter a long coiled duct, the epididymis. The sperm are not fully developed at this time, but secretions from the duct walls help them mature. Until sperm leave the body, they are stored in the last part of the epididymis. When they are about to leave, they pass through a thick-walled tube (vas deferens), ejaculatory ducts, then through the urethra—the channel leading outside, to the body's surface.

Secretions from glands along this route become mixed with sperm, forming the semen. Secretions from the seminal vesicles include the sugar fructose, which nourishes sperm. They also include prostaglandins, which may

trigger contractions in the female reproductive tract and assist sperm movement. Secretions from the prostate gland probably help buffer acid conditions in the vagina. Vaginal pH is about 3.5–4, but sperm motility and fertility improve when it is about 6. Bulbourethral glands secrete some mucus-rich fluid into the urethra during sexual arousal. This fluid lubricates the penis, facilitating its penetration into the vagina. It also aids sperm movement.

Hormonal Control of Male Reproductive Functions

Three hormones control male reproductive functions. One is **testosterone**, produced by endocrine cells in the testes. The others are **LH** (luteinizing hormone) and **FSH** (follicle-stimulating hormone). As we have seen, LH and FSH are produced by the anterior lobe of the pituitary gland (page 584). They were named for their effects in females, but we now know that their molecular structure is exactly the same in males.

Testosterone stimulates sperm production and controls the growth, form, and function of all parts of the male reproductive tract. This hormone has roles in normal sexual behavior and may tend to promote aggressive behavior. Growth of facial hair and pubic hair, lowering of the voice, and other secondary sexual traits depend on testosterone secretions.

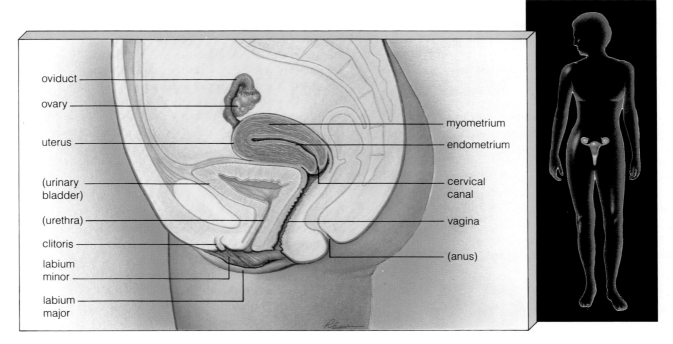

Figure 43.7 Reproductive system of the human female. The uterus has a thick layer of smooth muscle (myometrium) and an inner lining (endometrium).

The hypothalamus governs testosterone secretion. When testosterone levels are low, the hypothalamus secretes **GnRH**, a releasing hormone. GnRH prods the anterior pituitary into secreting LH, which the bloodstream carries to the testes. There, LH triggers testosterone secretion by Leydig cells. The testosterone acts on the Sertoli cells in ways that stimulate sperm production. GnRH also prods the pituitary into secreting FSH, which acts directly on Sertoli cells. FSH stimulates the development of sperm during puberty.

Negative-feedback loops between the hypothalamus and testes control sperm production. These loops are shown in Figure 43.6. As the concentration of blood-borne testosterone increases, it exerts an inhibitory effect on the hypothalamus. GnRH secretion slows down as a result.

Female Reproductive Organs

Figure 43.7 shows the female reproductive system and Table 43.2 lists its components. The two ovaries reside in the abdominal cavity. During a woman's reproductive years, they release eggs on a monthly basis, and they secrete the sex hormones **estrogen** and **progesterone**.

Take a look at Figure 10.12, the generalized picture of how meiosis I and II proceed in an immature egg, or *oocyte*. Even before a female is born, about 2 million oocytes start forming in her ovaries. They enter meio-

Table 43.2	Female Reproductive Organs
Ovaries	Oocyte production, sex hormone production
Oviducts	Conduction of oocyte from ovary to uterus
Uterus	Chamber in which new individual develops
Cervix	Secretion of mucus that enhances sperm movement into uterus and (after fertilization) reduces the embryo's risk of bacterial infection
Vagina	Organ of sexual intercourse; birth canal

sis I, but in this case the division process is arrested! Meiosis will resume in one oocyte at a time—but not until puberty. By then, only 30,000 or 40,000 oocytes will still be around. And only about 400 of those will mature and escape from the ovary on a monthly basis, over the next three decades or so. Even then, meiosis II will not be completed unless fertilization occurs.

An oocyte released from an ovary enters a nearby channel, an **oviduct**. Fingerlike projections from the oviduct extend over part of the ovary, and they sweep the oocyte into the channel. From there, the oocyte moves into a hollow, pear-shaped organ, the **uterus**. Following fertilization, the new individual grows and develops here. The uterus is mostly a thick layer of smooth muscle (the myometrium). Its interior lining, the **endometrium**, consists of connective tissue, glands, and blood vessels. The lower portion of the uterus (the narrow part of the "pear") is the cervix. A muscular tube, the vagina, extends from the cervix to the body surface. This tube receives sperm and functions as part of the birth canal.

At the body surface are the external genitalia (vulva), which include the organs for sexual stimulation. The outermost structures are a pair of skin folds (the labia majora), which contain adipose tissue. Within the cleft formed by these folds are the labia minora—a smaller pair of skin folds that are highly vascularized but have no fatty tissue. At the anterior end of the vulva, the interior folds of skin partly enclose the *clitoris*, a small organ sensitive to sexual stimulation. The opening of the urethra is about midway between the clitoris and the vaginal opening.

Menstrual Cycle

Most mammalian females follow an "estrous" cycle. They can become pregnant only during estrus. At that time, they are said to be in heat, or sexually receptive to males. Estrus occurs only at certain times of year, when oocytes mature and hormone action primes the endometrium to receive a fertilized egg.

The females of humans and other primates follow a **menstrual cycle**. For them, the release of oocytes and priming of the endometrium is cyclic and intermittent. A menstrual cycle differs from estrus, for there is no correspondence between heat and the time of fertility. In other words, all female primates can be physically and behaviorally receptive to sexual activity at any time.

Table 43.3	Events of the Menstrual Cycle	
Phase	Events	Days of the Cycle*
Follicular phase	Menstruation; endometrium breaks down	1–5
	Follicle matures in ovary; endometrium rebuilds	6–13
Ovulation	Secondary oocyte released from ovary	14
Luteal phase	Corpus luteum forms; endometrium thickens and develops	15–28

*Assuming a 28-day cycle.

Human menstrual cycles begin at about age thirteen. As Table 43.3 indicates, it takes about twenty-eight days to complete one cycle. Twenty-eight days simply is the average time span; the cycle runs longer for some women and shorter for others. Menstrual cycles continue until menopause, in the late forties or early fifties. By then, the egg supply is dwindling, hormone secretions slow down, and eventually menstruation (and fertility) is over. The eggs that are still to be released late in a woman's life are at some risk of acquiring chromosome abnormalities when meiosis finally resumes. A newborn with Down syndrome is one of the possible outcomes (page 198).

Ovarian Function. Figure 43.8 provides a closer look at how an oocyte develops in an ovary. Each oocyte becomes surrounded by a single layer of *granulosa cells*. Each "primary" oocyte, together with the surrounding cell layer, is a **follicle**. This is what FSH, the "follicle-stimulating hormone," will act upon. The granulosa cells gradually deposit a layer of material, the *zona pellucida*, around the oocyte.

Usually only one follicle reaches maturity during a menstrual cycle. Within that follicle, meiosis I resumes in the oocyte and two cells form. This division process is shown in Figure 10.12. One cell, the **secondary oocyte**, ends up with nearly all the cytoplasm. The other cell is the first **polar body**. It is a tiny cell that functions as a "dumping ground" for half the diploid number of chromosomes. The distribution of chromosomes between the secondary oocyte and the first polar body assures that both cells will be haploid. Neither cell will complete meiosis II until fertilization.

As the follicle develops, it secretes an estrogen-containing fluid. The fluid accumulates in the follicle and causes it to balloon outward from the ovary's surface, then rupture. The fluid escapes, carrying the secondary oocyte with it. The release of a secondary oocyte from an ovary is called **ovulation**. The granulosa cells left behind in the follicle differentiate into a glandular structure, the **corpus luteum**, which secretes progesterone and some estrogen.

A corpus luteum can persist for about twelve days if fertilization does not follow ovulation. During that time, the hypothalamus signals the anterior pituitary to decrease its FSH secretions. This prevents other follicles from developing until the menstrual cycle is over.

The corpus luteum degenerates during the last days of the cycle if fertilization does not occur. Apparently, it self-destructs by secreting prostaglandins, which interfere with its function. With the corpus luteum gone, progesterone and estrogen levels fall rapidly. Now FSH secretions can increase, another follicle can be stimulated to mature—and the cycle begins anew.

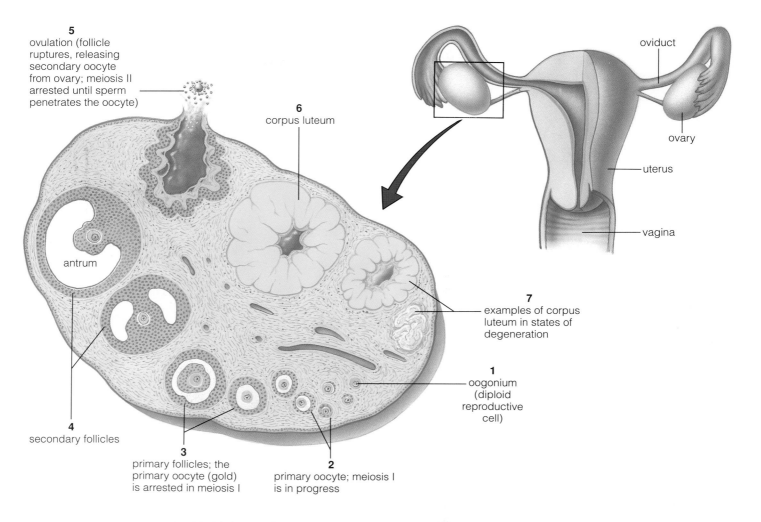

5
ovulation (follicle ruptures, releasing secondary oocyte from ovary; meiosis II arrested until sperm penetrates the oocyte)

6
corpus luteum

oviduct

ovary

uterus

vagina

antrum

7
examples of corpus luteum in states of degeneration

1
oogonium (diploid reproductive cell)

4
secondary follicles

3
primary follicles; the primary oocyte (gold) is arrested in meiosis I

2
primary oocyte; meiosis I is in progress

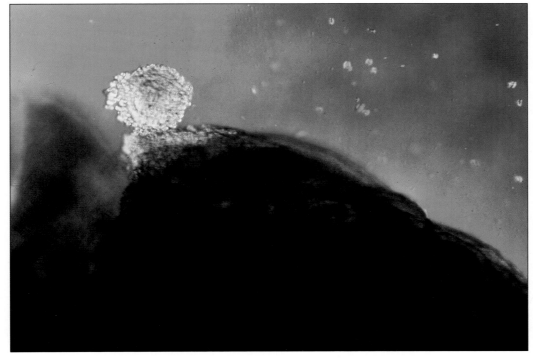

Figure 43.8 A human ovary, drawn as if sliced lengthwise through its midsection. Events in the ovarian cycle proceed from the growth and maturation of follicles, through ovulation (rupturing of a mature follicle with a concurrent release of a secondary oocyte), through the formation and maintenance (or degeneration) of an endocrine structure called the corpus luteum. The positions of the oocyte and corpus luteum are varied for illustrative purposes only. The maturation of an oocyte occurs at the *same* site, from the beginning of the cycle to ovulation. The photograph shows a secondary oocyte at the moment of ovulation.

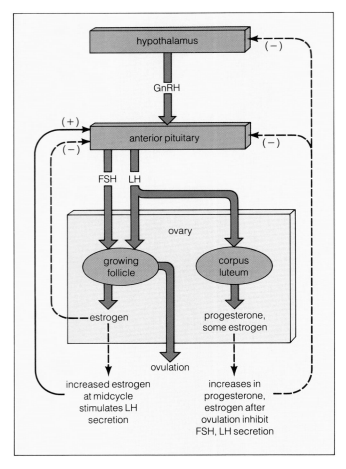

Figure 43.9 Feedback loops among the hypothalamus, anterior pituitary, and ovaries during the menstrual cycle.

Feedback loops among the hypothalamus, pituitary, and ovaries control events in the ovary. Figure 43.9 shows the loops. When the menstrual cycle begins, the hypothalamus signals the anterior pituitary to release LH and FSH, which in turn signal the ovary to secrete estrogen. About midway through the cycle, the increased estrogen level in the blood causes a brief outpouring of LH from the pituitary. *It is this midcycle surge of LH that triggers ovulation.*

Uterine Function. The changing estrogen and progesterone levels just described cause profound changes that prepare the uterus for pregnancy. Estrogen stimulates the growth of the endometrium and its glands in the uterus. Progesterone causes blood vessels to grow rapidly in the thickened endometrium.

At ovulation, estrogen acts on the cervix, the narrow opening to the uterus from the vagina. It causes the cervix to secrete large amounts of a thin, clear mucus—an ideal medium through which sperm can travel. Right after ovulation, progesterone from the corpus luteum

acts on the cervix. The mucus becomes thick and sticky, forming a barrier against vaginal bacteria that might enter the uterus through the cervix and endanger a new zygote.

When fertilization does not occur and the corpus luteum self-destructs, the endometrium starts to break down. Deprived of oxygen and nutrients, its blood vessels constrict, and its tissues die. Blood escapes from the ruptured walls of weakened capillaries. This menstrual flow consists of blood and sloughed endometrial tissues, and its appearance marks the first day of a new cycle. The menstrual sloughing continues for three to six days, until rising estrogen levels stimulate the repair and growth of the endometrium.

Each year, between 4 and 10 million American women are affected by *endometriosis*, the spread and growth of endometrial tissue outside the uterus. Estrogen acts on endometrial tissue wherever it occurs. Its action may lead to sensations of pain during menstruation, sexual relations, or urination. Also, endometrial scar tissue on the ovaries or oviduct can cause infertility. Endometriosis might arise when some menstrual flow backs up through the oviducts and spills into the pelvic cavity. Or perhaps some embryonic cells were positioned in the wrong place before birth and are stimulated to grow at puberty, when sex hormones become active.

Figure 43.10 summarizes the correlations between changing hormone levels and changes in the ovary and uterus during the menstrual cycle.

Sexual Intercourse

Suppose a secondary oocyte happens to be on its way down the oviduct when a female and male are engaged in sexual intercourse, or *coitus*. Within mere seconds of sexual arousal, the penis can undergo changes that will help it penetrate into the vaginal channel. As Figure 43.3 shows, the penis contains three cylinders of spongy tissue. The mushroom-shaped tip of one cylinder (the glans penis) is loaded with sensory receptors that are activated by friction. Between times of sexual arousal, blood vessels leading into the three cylinders are constricted and the penis is limp. Upon arousal, blood flows into the cylinders faster than it flows out and collects in the spongy tissue, so the penis lengthens and stiffens.

During coitus, pelvic thrusts stimulate the penis as well as the female's vaginal walls and clitoral region. The mechanical stimulation causes rhythmic, involuntary contractions in the male reproductive tract. The contractions move sperm from their main storage site in the last portion of the epididymis. The contractions force the contents of seminal vesicles and the prostate into the urethra, then ejaculation of semen into the vagina. (During ejaculation, a sphincter closes and prevents urine from being excreted from the bladder.)

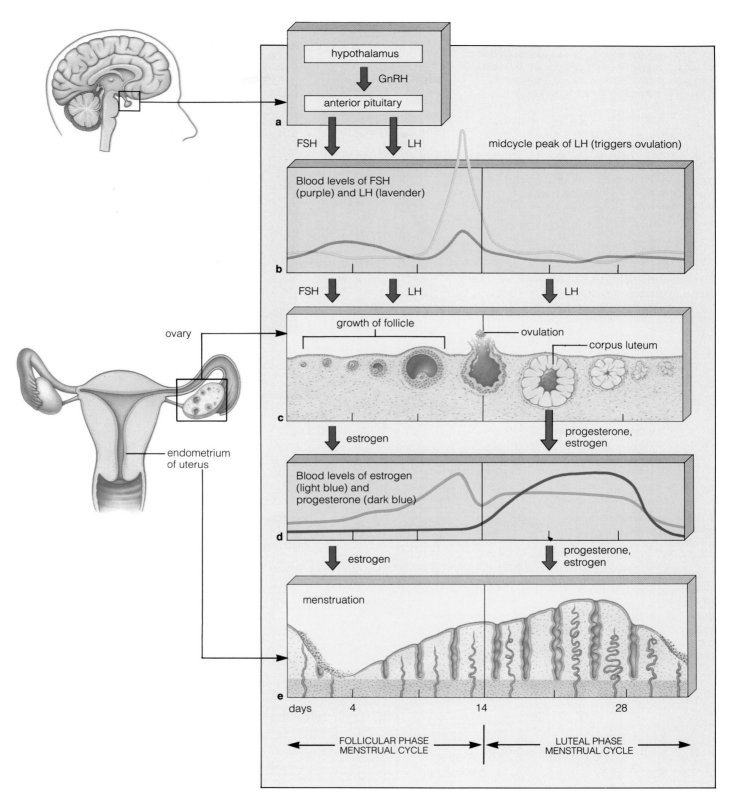

Figure 43.10 Correlation between changes in the ovary and uterus with changing hormone levels during the menstrual cycle. Green arrows indicate which hormones dominate the follicular phase or the luteal phase of the cycle. A releasing hormone (GnRH) from the hypothalamus (**a**) controls the release of FSH and LH from the pituitary. The FSH and LH promote changes in ovarian structure and function (**b**, **c**), then estrogen and progesterone from the ovary promote changes in the endometrium (**d**, **e**).

Figure 43.11 Fertilization and implantation in the uterus. The left photograph shows a human sperm about to penetrate the membranous covering (zona pellucida) around a secondary oocyte. The right photograph shows the three polar bodies above a mature ovum; these products of meiosis will degenerate shortly.

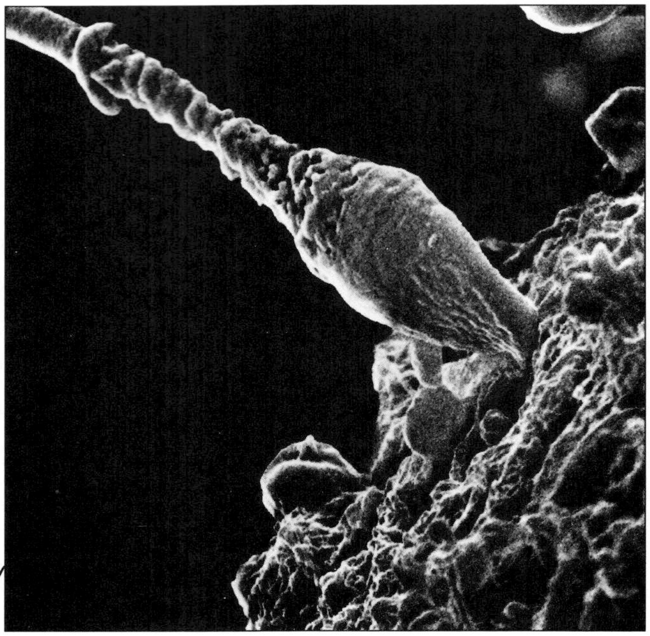

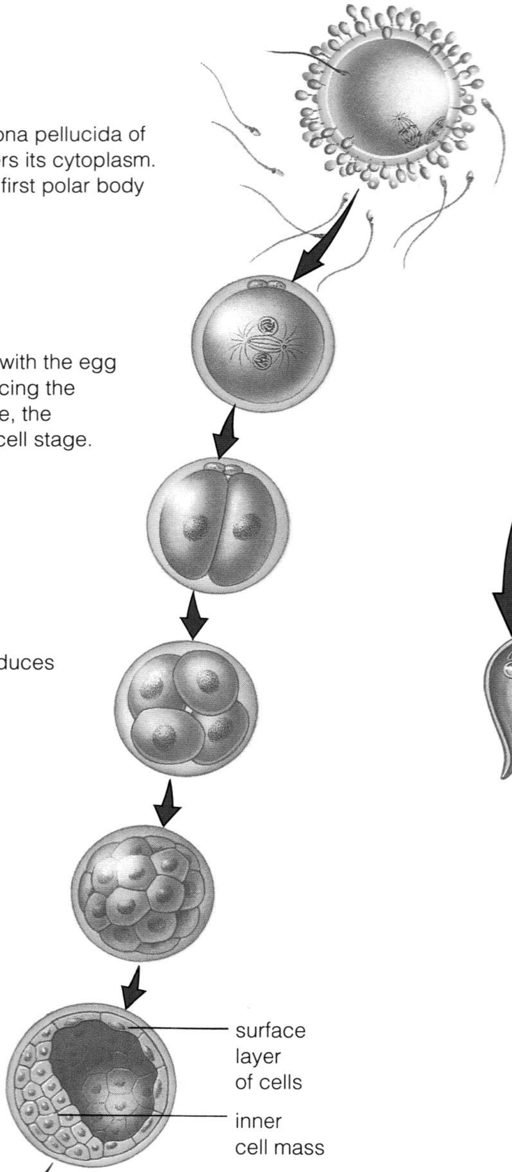

a. A sperm penetrates the zona pellucida of a secondary oocyte and enters its cytoplasm. This triggers meiosis II in the first polar body and in the oocyte.

b. The sperm nucleus fuses with the egg nucleus at fertilization, producing the zygote. With the first cleavage, the fertilized egg enters the two-cell stage.

c. The second cleavage produces the four-cell stage.

d. Successive cleavages produce a solid ball of cells, the morula.

e. Fluid enters the ball and lifts some cells, forming a cavity. This produces the blastocyst, a ball of cells having a surface layer and an inner cell mass.

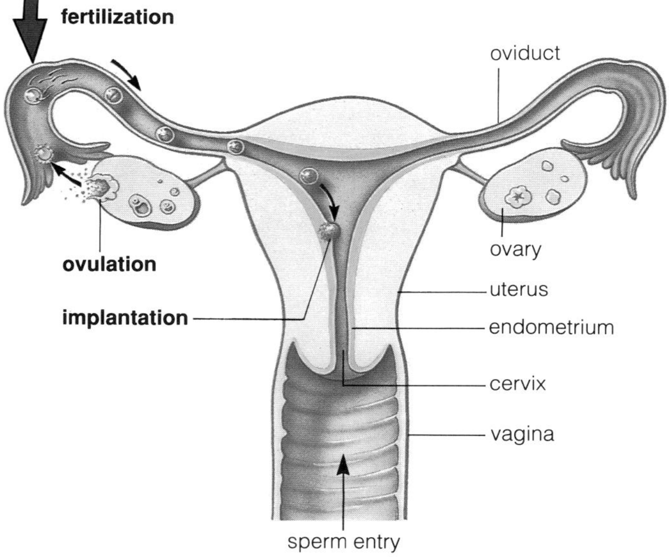

surface layer of cells

inner cell mass

fertilization

oviduct

ovulation

ovary

implantation

uterus

endometrium

cervix

vagina

sperm entry

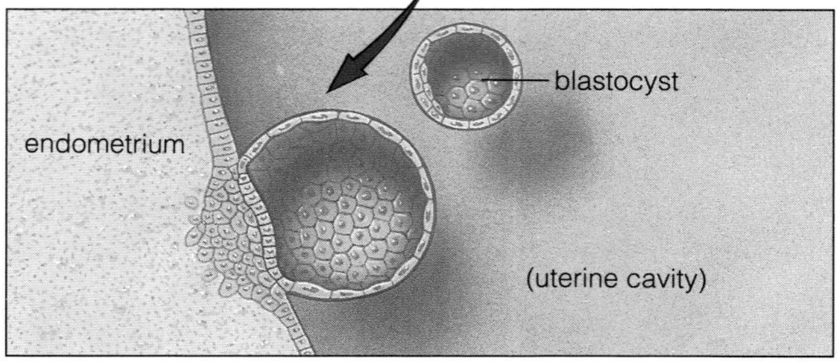

blastocyst

endometrium

(uterine cavity)

proliferating cell mass

maternal blood vessel

embryonic disk

amniotic cavity

f. Implantation begins when the blastocyst attaches to and invades the endometrium. During the second week after fertilization, it slowly embeds itself in the endometrium.

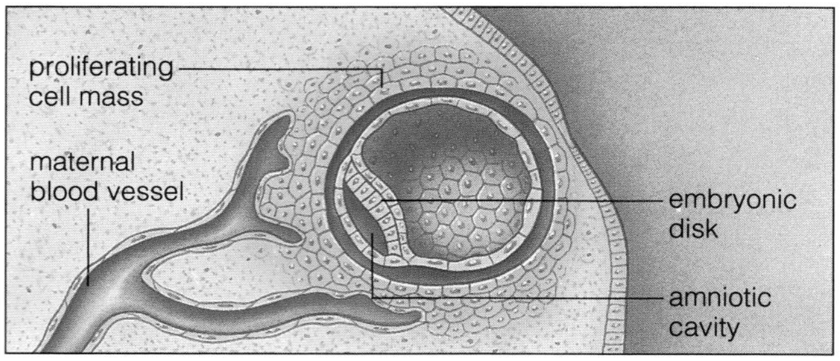

g. During implantation, a slitlike cavity forms between the inner cell mass and the surface layer of the embryo. The inner cell mass is transformed into a flattened, embryonic disk from which the embryo develops.

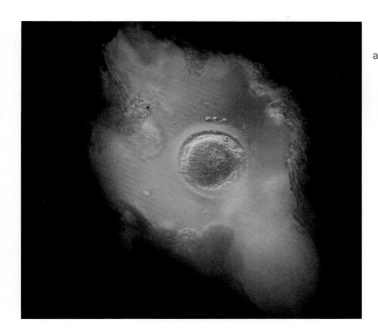

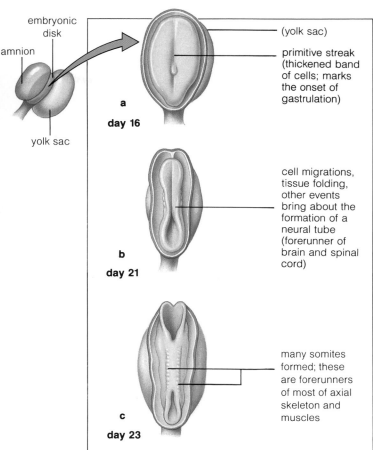

Figure 43.12 Transformation of the pancake-shaped embryonic disk into the early embryo. Shown are three dorsal views (the embryo's back).

Together, the muscular contractions, ejaculation, and associated sensations of release, warmth, and relaxation are called *orgasm*. Female orgasm involves similar events, including an intense vaginal awareness, involuntary uterine and vaginal contractions, and sensations of relaxation and warmth. Even if the female does not reach this state of excitation, she can still get pregnant.

FROM FERTILIZATION TO BIRTH

Fertilization

Fertilization may occur if sperm enter the vagina any time between a few days before ovulation to a few days afterward. Within thirty minutes after ejaculation, muscle contractions move sperm deeper into the female reproductive tract. As many as 150 million to 350 million sperm may be deposited in the vagina during one ejaculation. Yet only a few hundred reach the upper region of the oviduct, where fertilization most commonly takes place.

When a sperm encounters a secondary oocyte, it releases digestive enzymes from its acrosome. Those enzymes clear a path through the zona pellucida (Figure 43.11a). Several sperm can reach the egg, but usually only one enters its cytoplasm. The arrival of that sperm stimulates both the secondary oocyte and the first polar body into completing meiosis II. As Figure 43.11 shows, there are now three polar bodies and a mature egg, or **ovum**. The sperm nucleus fuses with the nucleus of the ovum and their chromosomes intermingle, restoring the diploid number for the zygote.

Implantation

For the first three or four days after fertilization, the zygote travels down the oviduct. It picks up required nutrients from maternal secretions and undergoes the first cleavages, in the manner described on page 746. By the time the cluster of dividing cells reaches the uterus, it is a solid ball of cells (the morula). The ball becomes transformed into a *blastocyst*, an embryonic stage consisting of a surface layer of cells and an inner cell mass (Figure 43.11e).

Implantation occurs before the end of the first week. By this process, the blastocyst adheres to the uterine lining, and some of its cells send out projections that invade the maternal tissues. While the invasion is proceeding, the inner cell mass becomes transformed into an **embryonic disk**. This is the oval, flattened pancake described in the chapter introduction (Figure 43.12). The disk will give rise to the embryo proper during the week following implantation, when all three germ layers will form.

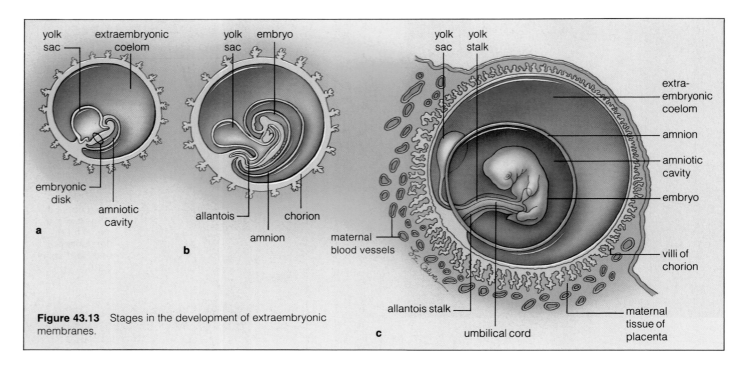

Figure 43.13 Stages in the development of extraembryonic membranes.

Labels for panel a: yolk sac, extraembryonic coelom, embryonic disk, amniotic cavity

Labels for panel b: yolk sac, embryo, allantois, chorion, amnion

Labels for panel c: yolk sac, yolk stalk, maternal blood vessels, allantois stalk, umbilical cord, extra-embryonic coelom, amnion, amniotic cavity, embryo, villi of chorion, maternal tissue of placenta

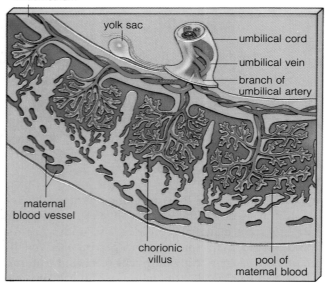

Labels: endometrium, yolk sac, umbilical cord, umbilical vein, branch of umbilical artery, maternal blood vessel, chorionic villus, pool of maternal blood

Figure 43.14 Relationship between fetal and maternal tissues in the placenta. The diagram shows how chorionic villi become progressively developed (from left to right across the illustration).

Membranes Around the Embryo

To understand what happens after implantation, think back to the shelled egg, which figured in the vertebrate invasion of land (page 456). Inside most shelled eggs are four membranes, the **yolk sac**, **allantois**, **amnion**, and **chorion**. These "extraembryonic" membranes protect the embryo and serve in its nutrition, respiration, and excretion.

A human embryo is not housed in a shell or nourished by yolk, but it is still served by a yolk *sac* as part of its vertebrate heritage. The sac forms below the embryonic disk *as if* yolk were still there (Figures 43.12 and 43.13). The sac will play a role in the formation of a digestive tube.

In hard-shelled eggs, the allantois stores wastes from protein metabolism and its blood vessels supply the embryo with oxygen. In humans, the allantois is not involved in waste storage, but its blood vessels still function in oxygen and nutrient transport by the placenta.

The amnion of all land vertebrates is a fluid-filled sac that completely surrounds the embryo and keeps it from drying out. The fluid inside also absorbs shocks. Just before childbirth, "water" flows freely from the vagina. This is amniotic fluid, released when the amnion ruptures.

In time, only a thick cord connects the growing human embryo to parts of the yolk sac, allantois, and amnion. This **umbilical cord** is well endowed with blood vessels (Figure 43.14). The chorion develops as a protective membrane around the embryo and other structures. As you will see shortly, this membrane becomes part of the placenta. The chorion secretes a hormone (chorionic gonadotropin) that maintains the corpus luteum. And progesterone secreted from the corpus luteum in turn maintains the uterine lining during the first three months of pregnancy. After that, the placenta produces sufficient progesterone and estrogen.

gill arches   somites

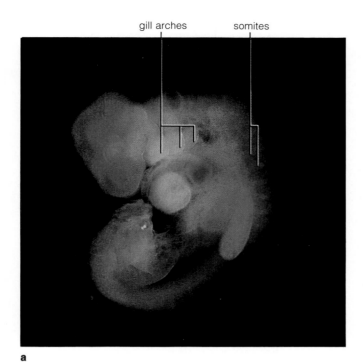

a

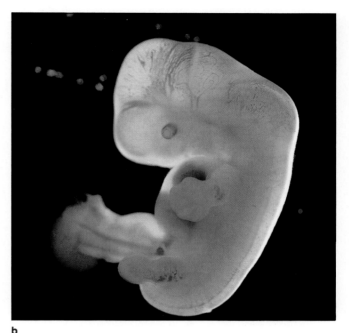

b

umbilical cord   amniotic sac

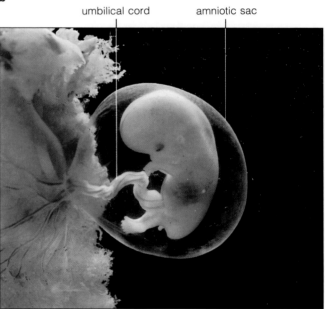

c

The Placenta

Three weeks after fertilization, almost a fourth of the inner surface of the uterus has become a spongy tissue composed of endometrium *and* embryonic membranes, the chorion especially. By this tissue, the **placenta**, the embryo receives nutrients and oxygen from the mother and sends out wastes in return. The embryo's wastes are quickly disposed of through the mother's lungs and kidneys.

The tiny projections sent out from the blastocyst during implantation develop into many chorionic villi, each endowed with small blood vessels (Figure 43.14). When the embryo starts developing, its bloodstream will remain distinct from that of its mother. Substances simply will diffuse out of the mother's blood vessels, across the blood-filled spaces in the uterine lining, then into the embryo's blood vessels. At the same time, they diffuse in the opposite direction from the embryo.

Embryonic and Fetal Development

First Trimester. The "first trimester" of the nine months of human development extends from fertilization to the end of the third month. The introduction to this chapter sketched out a picture of what the embryo looks like during this period. Figure 43.15 gives a more detailed picture of its shape. As indicated in the *Commentary* on the next page, the first trimester is a very critical period of embryonic development.

Figure 43.15 (**a**) Another view of the embryo at four weeks, about 7 millimeters (0.3 inch) long. Notice the tail and gill arches, features that emerge in developing embryos of all vertebrates. Arm and leg buds are visible now. (**b**) Embryo at the end of five weeks, about 12 millimeters long. The head starts to enlarge and the trunk starts to straighten. Finger rays appear in the paddlelike forelimbs. A pigmented retina outlines the forming eyes. (**c**) An embryo poised at the boundary between the end of the first trimester and the start of the second, which will extend from the fourth month through sixth. Once past this boundary, the new individual is called a *fetus*. Here, the embryo floats in fluid within the amniotic sac. The chorion, which covers the amniotic sac, has been opened and pulled aside.

Mother as Protector, Provider, Potential Threat

Many safeguards are built into the female reproductive system. The placenta, for example, is a highly selective filter that prevents many noxious substances in the mother's bloodstream from gaining access to the embryo or fetus. Even so, from fertilization to birth, the developing individual is at the mercy of the mother's diet, health habits, and life-style.

Some Nutritional Considerations

During pregnancy, a balanced intake of carbohydrates, amino acids, and fats or oils usually provides all vitamins and minerals in amounts sufficient for normal development. The mother's vitamin needs are definitely increased, but the developing fetus is more resistant than she is to vitamin and mineral deficiencies. (The placenta preferentially absorbs vitamins and minerals from her blood.) In cases where the diet is marginal, the money spent on vitamin pills and other food supplements would usually do the mother (hence the fetus) more good if spent on wholesome, protein-rich food.

A few years ago, it was accepted medical practice for a pregnant woman to keep her total weight gain to ten or fifteen pounds. It is now clear that if the woman restricts her food intake too severely, especially during the last trimester, fetal development will be affected and the newborn will be underweight. Significantly underweight infants face more post-delivery complications than do infants of normal weight; in fact, they represent nearly half of all newborn deaths. They also will suffer a much higher incidence of mental retardation and other handicaps later in life. In most cases, a woman should gain somewhere between twenty and twenty-five pounds during pregnancy.

As birth approaches, the growing fetus demands more and more nutrients from the mother's body. During this last phase of pregnancy, the mother's diet profoundly influences the course of development. Poor nutrition damages most organs—particularly the brain, which undergoes its greatest growth in the weeks just before and after birth.

Risk of Infections

Throughout pregnancy, antibodies transferred across the placenta protect the developing individual from all but the most severe bacterial infections. However, certain viral diseases can have damaging effects if they are contracted

a Critical periods of embryonic and fetal development. Red indicates periods in which organs are most sensitive to damage from cigarette smoke, alcohol, viral infection, and so on. Numbers signify the week of development.

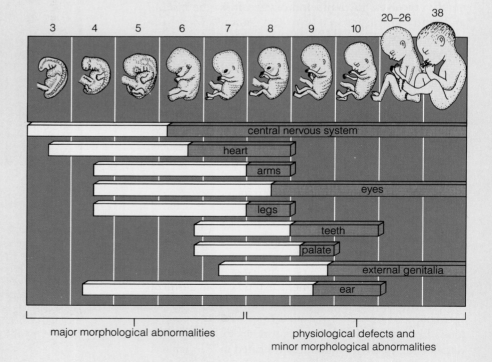

during the first six weeks after fertilization, the critical time of organ formation. For example, if the woman contracts German measles during this period, there is a 50 percent chance that her embryo will become malformed. If she contracts the measles virus when the embryo's ears are forming, her newborn may be deaf. (German measles can be avoided by vaccination *before* pregnancy.) The likelihood of damage to the embryo diminishes after the first six weeks. The same disease, contracted during the fourth month or thereafter, has no discernible effect on the fetus.

Effects of Prescription Drugs

During the first trimester, the embryo is highly sensitive to drugs. A shocking example of drug effects came during the first two years after *thalidomide* was introduced in Europe. Women using this prescription tranquilizer during the first trimester gave birth to infants with missing or severely deformed arms and legs. Once the deformities were traced to thalidomide, the drug was withdrawn from the market. However, there is evidence that other tranquilizers (and sedatives and barbiturates) might cause similar, although less severe, damage. Even certain anti-acne drugs increase the risk of facial and cranial deformities. Tetracycline, a commonly prescribed antibiotic, causes yellowed teeth. Streptomycin causes hearing problems and may affect the nervous system.

At no stage of development is the embryo impervious to drugs in the maternal bloodstream. Clearly, the woman should take no drugs at all during pregnancy unless they are prescribed by a knowledgeable physician.

Effects of Alcohol

As the fetus matures, its physiology becomes increasingly like that of the mother's. Alcohol passes freely across the placenta and has the same kind of effect on the fetus as on the woman who drinks it. *Fetal alcohol syndrome* (FAS) is a constellation of deformities that are thought to result from excessive use of alcohol by the mother during pregnancy. FAS is the third most common cause of mental retardation in the United States. It also is characterized by facial deformities, poor coordination and, sometimes, heart defects (see Figure *b*). Between 60 and 70 percent of alcoholic women give birth to infants with FAS. Some researchers now suspect that drinking any alcohol at all during pregnancy may be dangerous for the fetus. Increasingly, physicians are urging total or near-abstinence during pregnancy.

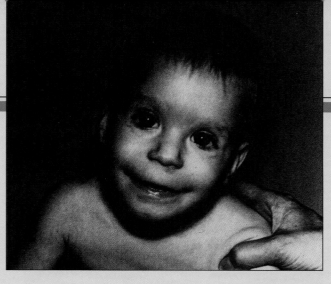

b An infant affected by FAS. Symptoms include a small head, low and prominent ears, poorly developed cheekbones, and a long, smooth upper lip. The child can expect to encounter growth problems and abnormalities of the nervous system. About 1 in 750 newborns in the United States are affected by this disorder.

Effects of Cocaine

Cocaine, particularly crack cocaine, disrupts the function of the fetal nervous system as well as the mother's. The consequences can be devastating, and they extend beyond birth. Here you may wish to read again the introduction to Chapter 33.

Effects of Cigarette Smoke

Cigarette smoking has an adverse effect on fetal growth and development. Newborns of women who have smoked every day throughout pregnancy have a low birth weight. That is true even when the woman's weight, nutritional status, and all other relevant variables are identical with those of pregnant women who do not smoke. Smoking has other effects as well. For example, for seven years in Great Britain, records were kept for all births during a particular week. The newborns of women who had smoked were not only smaller, they also had a 30 percent greater incidence of death shortly after delivery and a 50 percent greater incidence of heart abnormalities. More startling, at age seven, their average "reading age" was nearly half a year behind that of children born to nonsmokers.

In this last study, the critical period was shown to be the last half of pregnancy. Newborns of women who had stopped smoking by the middle of the second trimester were indistinguishable from those born to women who had never smoked. Although the mechanisms by which smoking exerts its effects on the fetus are not known, its demonstrated effects are further evidence that the placenta—marvelous structure that it is—cannot prevent all assaults on the fetus that the human mind can dream up.

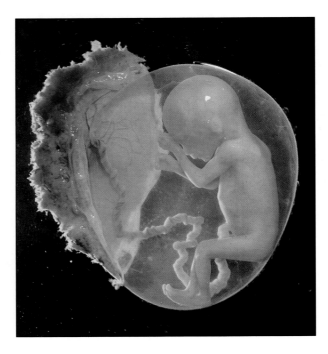

Figure 43.16 The fetus at sixteen weeks.

Figure 43.17 The fetus at eighteen weeks, about 18 centimeters (a little more than 7 inches) long. The sucking reflex begins during the earliest fetal stage, as soon as nerves establish functional connections with developing muscles. Legs kick, arms wave, fingers make grasping motions—all reflexes that will be vital skills in the world outside the uterus.

Table 43.4	Tissues and Organs Derived from the Three Germ Layers in Human Embryos
Germ Layer	Main Derivatives in the Adult
Endoderm	Various epithelia, as in the gut, respiratory tract, urinary bladder and urethra, and parts of the inner ear; also portions of the tonsils, thyroid and parathyroid glands, thymus, liver, and pancreas
Mesoderm	Cartilage, bone, muscle, and various connective tissues; gives rise to cardiovascular system (including blood), lymphatic system, spleen, and adrenal cortex
Ectoderm	Central and peripheral nervous systems; sensory epithelia of the eyes, ears, and nose; epidermis and its derivatives (including hair and nails), mammary glands, pituitary gland, subcutaneous glands, tooth enamel, and adrenal medulla

Data from Keith Moore, *The Developing Human.*

Once the embryonic disk forms, development proceeds along the course described in Chapter 42. Gastrulation starts during the second week. It leads to the formation of three germ layers—ectoderm, mesoderm, and endoderm. The mesoderm forms when certain cells migrate inward from the surface of the embryo. Ectoderm remaining at the surface will give rise to the nervous system as well as to certain glands and other structures (Table 43.4). Later, endoderm will give rise to parts of the respiratory and digestive systems. Mesoderm will develop into the heart, muscles, bone, and many other internal organs. After the third week, an early, tubelike heart is beating.

By the end of the fourth week, the embryo has grown 500 times its original size. Its growth spurt gives way to four weeks in which the main organs develop rather slowly. The nerve cord and the four heart chambers form. Respiratory organs form but are not yet functional.

Finally, the segmentation characteristic of all vertebrate embryos becomes apparent during the first trimester. Figure 43.15 shows some of the paired segments, or *somites*—the start of connective tissues, bones, and muscles. Arms, legs, fingers, and toes now develop, along with the tail that also emerges in all vertebrate embryos. The human tail gradually disappears after the eighth week.

Second Trimester. The "second trimester" extends from the start of the fourth month to the end of the sixth. All major organs have formed, and the growing individual is now called a **fetus**. Figure 43.16 shows what the fetus looks like at the ninth and the sixteenth week of development. Movements of facial muscles produce frowns and squints. The sucking reflex also is evident (Figure 43.17). Before the second trimester draws to a close, the mother already can sense movements of the fetal arms and legs.

When the fetus is five months old, its heart can be heard through a stethoscope on the mother's abdomen. Soft, fuzzy hair (the lanugo) covers its body. Its skin is wrinkled, rather red, and protected from abrasion by a thick, cheesy coating. During the sixth month, eyelids and eyelashes form. During the seventh month, the eyes open.

Third Trimester. The "third trimester" extends from the seventh month until birth. Not until the middle of the third trimester will the fetus be able to survive on its own if born prematurely or removed surgically from the uterus. However, with intensive medical support, fetuses as young as 23–25 weeks have survived early delivery. Although development appears to be relatively complete by the seventh month, few fetuses would be able to breathe normally or maintain a normal body temperature, even with the best medical care. By the ninth month, survival chances increase to about 95 percent.

Birth and Lactation

Birth takes place about thirty-nine weeks after fertilization, give or take a few weeks. The birth process begins when the uterus starts to contract. For the next two to eighteen hours, the contractions become stronger and more frequent. The cervical canal dilates fully and the amniotic sac usually ruptures. Birth typically occurs less than an hour after full dilation. Immediately afterward, uterine contractions force fluid, blood, and the placenta from the body (Figure 43.18). The umbilical cord—the lifeline to the mother—is now severed, and the newborn embarks on its nurtured existence in the outside world.

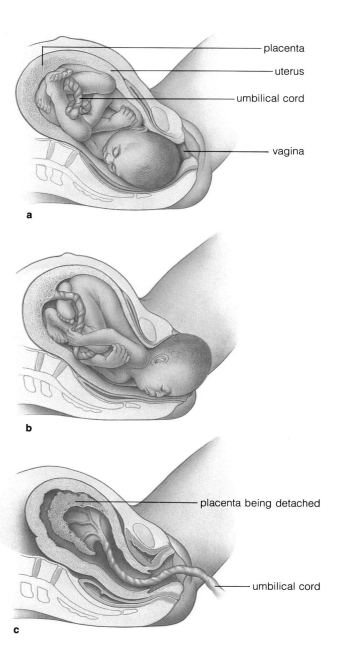

a

b

c

Figure 43.18 Expulsion of the fetus during the birth process. The placenta, fluid, and blood are expelled shortly afterward (this is the "afterbirth").

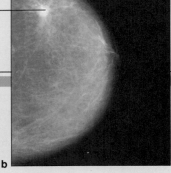

tumor

Commentary

Cancer in the Human Reproductive System

For both men and women, reproductive function depends absolutely on hormonal controls of the sort described in this chapter. Hormonal imbalances contribute to a variety of disorders of the reproductive system, including many forms of cancer. Unless it is eradicated from the body, cancer kills—and it is not often eradicated easily. This makes cancer one of the most feared disorders of modern times. Here we focus on two types: cancer of the breast and of the testis. Both often can be detected through routine self-examination, a habit that may save your life.

Breast Cancer

Of all cancers in women, breast cancer currently is second only to lung cancer in having the highest mortality. Despite intensive medical research, that rate has not been lowered by much over the past fifty years. Each year in the United States, well over 100,000 women develop breast cancer; more than a third die from it. Obesity, high blood cholesterol, and excessively high levels of estrogen and perhaps other hormones play roles in the development of cancer, but how they do this is not clear.

Chances for cure are excellent if breast cancer is detected early and treated promptly. That is why a woman should examine herself once a month, about a week after each menstrual period. The following steps have been recommended by the American Cancer Society:

1. Lie down and put a folded towel or pillow under the right shoulder, then put your right hand behind your head. With the left hand (fingers flat), begin the examination by following the outer circle of arrows shown in the diagram of Figure *a*. Gently press the fingers in small, circular motions to check for any lump, hard knot, or thickening. Next, follow the inner circle of arrows. Continue doing this for at least three more circles, one of which should include the nipple. Then repeat the procedure for the left breast.

2. For a complete examination, repeat the procedure of step 1 while standing in a shower or tub (hands glide more easily over wet skin).

3. Stand before a mirror, lift your arms over your head, and look for any unusual changes in the contour of your breasts, such as a swelling, dimpling, or retraction (inward sinking) of the nipple. Also check for any unusual discharge from the nipple.

If you discover a lump or any other change during a breast self-examination, it's important to see a physician at once.

Most changes are not cancerous, but let the physician make the diagnosis.

Currently, *mammography* (which uses low doses of x-rays) is the only imaging procedure with a proven record for detecting small cancers in the breast; it is 80 percent reliable. Figure *b* shows a mammogram that revealed a tumor, which a biopsy indicated to be cancerous. (The white patches at the front of the breast are milk ducts and fibrous tissue.)

Most often, cancerous tumors are removed through *modified radical mastectomy*. All the breast tissue, the overlying skin, and lymph nodes in adjacent tissues are removed, but muscles of the chest wall are left intact to permit more normal shoulder motions following surgery. In another procedure (*lumpectomy*), which is followed by radiation therapy, some of the breast tissue is left in place. In both cases, the removed lymph nodes are examined to determine the need for further treatment and to predict the prospects of a cure. If tumor cells are present in the nodes, there is a high risk of metastasis (cancer spread). Treatment then may include hormone therapy (aimed at shrinking any tumor masses) and radiation therapy. Treatment is most promising when cancer is detected at an early stage.

Cancer of the Testis

You may be surprised to learn that cancer of the testis is a frequent cause of death in young men. About 5,000 cases are diagnosed in a given year in the United States alone. In its early stages, testicular cancer is painless. If not detected in time, however, it can spread to lymph nodes in the abdomen, chest, neck and, eventually, the lungs. Once it has metastasized, the cancer kills as many as half of its victims.

Once a month, from high school onward, men should examine each testis separately after a warm bath or shower (when the scrotum is relaxed). The testis should be rolled gently between the thumb and forefinger to check for any type of lump, enlargement, or hardening. Changes of that sort may or may not cause discomfort—but they must be reported to a physician, who can make a complete examination. Treatment of testicular cancer has one of the highest rates of success—when the cancer is caught before it can spread.

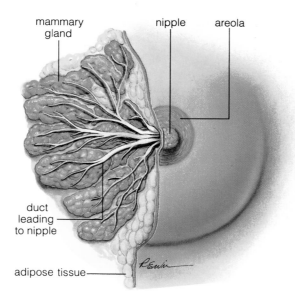

mammary gland   nipple   areola

duct leading to nipple

adipose tissue

Figure 43.19 Breast of a lactating female. This cutaway view shows the mammary glands and ducts. The *Commentary* describes how to examine breast tissues for cancer.

During pregnancy, estrogen and progesterone were stimulating the growth of mammary glands and ducts in the mother's breasts (Figure 43.19). For the first few days after birth, those glands produce a fluid rich in proteins and lactose. Then prolactin secreted by the pituitary stimulates milk production (page 584).

When the newborn suckles, the pituitary also releases oxytocin, which causes breast tissues to contract and so force milk into the ducts. Oxytocin also triggers uterine contractions that "shrink" the uterus back to its normal size.

POSTNATAL DEVELOPMENT, AGING, AND DEATH

Following birth, the new individual follows a prescribed course of further growth and development that leads to the **adult**, the mature form of the species. Table 43.5 summarizes all the prenatal and postnatal stages. (Prenatal means before birth; postnatal means after birth.) Figure 42.18 shows how the human body changes in proportions as the course is followed.

Late in life, the body gradually deteriorates through processes that are collectively called **aging**. Cell structure and function starts to break down, and this is accompanied by structural changes and gradual loss of body functions. As we saw in Chapter 42, all organisms with extensively differentiated cells undergo aging.

Table 43.5 Stages of Human Development: A Summary

Prenatal Period:

1. Zygote	Single cell resulting from fusion of sperm nucleus and egg nucleus at fertilization
2. Morula	Solid ball of cells produced by cleavages
3. Blastocyst	Ball of cells with surface layer and inner cell mass
4. Embryo	All developmental stages from two weeks after fertilization until end of eighth week
5. Fetus	All developmental stages from the ninth week until birth (about thirty-nine weeks after fertilization)

Postnatal Period:

6. Newborn	Individual during the first two weeks after birth
7. Infant	Individual from two weeks to about fifteen months after birth
8. Child	Individual from infancy to about twelve or thirteen years
9. Pubescent	Individual at puberty, when secondary sexual traits develop; girls between twelve and fifteen years, boys between thirteen and sixteen years
10. Adolescent	Individual from puberty until about three or four years later; physical, mental, emotional maturation
11. Adult	Early adulthood (between eighteen and twenty-five years), bone formation and growth completed. Changes proceed very slowly afterward.
12. Old age	Aging follows late in life

CONTROL OF HUMAN FERTILITY

Some Ethical Considerations

The transformation of a zygote into an intricately detailed adult raises profound questions. *When does development begin?* As we have seen, key developmental events occur even before fertilization. *When does life begin?* During her lifetime, a human female can produce as many as four hundred eggs, all of which are alive. During one ejaculation, a human male can release a quarter of a billion sperm, which also are alive. Even before sperm and egg merge by chance and establish the genetic makeup of a new individual, they are as much alive as any other form of life. It is scarcely tenable, then, to say "life begins" when they fuse. *Life began billions of years ago; and each gamete, each zygote, each mature individual is only a fleeting stage in the continuation of that beginning.*

This fact cannot diminish the meaning of conception, for it is no small thing to entrust a new individual with the gift of life, wrapped in the unique evolutionary threads of our species and handed down through an immense sweep of time.

Yet how can we reconcile the marvel of individual birth with the growing awareness of the astounding birth rate for our whole species? While this book is being written, an average of 10,700 newborns enter the world every single hour. By the time you go to bed tonight, there will be 257,000 more people on earth than there were last night at that hour. Within a week, the number will reach 1,800,000—about as many people as there are now in the entire state of Massachusetts. *Within one week.* Worldwide population growth has outstripped resources, and each year millions face the horrors of starvation. Living as we do on one of the most productive continents on earth, few of us can know what it means to give birth to a child, to give it the gift of life, and have no food to keep it alive.

And how can we reconcile the marvel of birth with the confusion surrounding unwanted pregnancies? Even highly developed countries have inadequate educational programs concerning fertility control, and a good number of their members are not inclined to exercise control. Each year in the United States alone, there are more than 100,000 "shotgun" marriages, about 200,000 unwed teenage mothers, and perhaps 1,500,000 abortions. Many parents encourage early boy–girl relationships, at the same time ignoring the risk of premarital intercourse and unplanned pregnancy. Advice is often condensed to a terse, "Don't do it. But if you do it, be careful!"

The motivation to engage in sex has been evolving for more than 500 million years. A few centuries of moral and ecological reasoning that call for its suppression have not prevented unwanted pregnancies. And complex social factors have contributed to a population growth rate that is out of control.

How will we reconcile our biological past and the need for a stabilized cultural present? Whether and how fertility is to be controlled is one of the most volatile issues of our time. We will return to this issue in the next chapter, in the context of principles governing the growth and stability of populations. Here, we can briefly consider some possible control options.

Birth Control Options

The most effective method of birth control is complete *abstinence*, no sexual intercourse whatsoever. It is unrealistic to expect many people to practice it.

A modified form of abstinence is the *rhythm method*. The idea is to avoid intercourse during the woman's fertile period, beginning a few days before ovulation and ending a few days after. Her fertile period is identified and tracked either by keeping records of the length of her menstrual cycles or by taking her temperature each morning when she wakes up. (It rises by one-half to one degree just before the fertile period.) But ovulation can be irregular, and miscalculations are frequent. Also, sperm deposited in the vaginal tract a few days before ovulation may survive until ovulation. The method *is* inexpensive (it costs nothing after you buy the thermometer) and does not require fittings and periodic checkups by a physician. But its practitioners do run a large risk of pregnancy (Figure 43.20).

Withdrawal, or removing the penis from the vagina before ejaculation, dates back at least to biblical times. But withdrawal requires very strong willpower, and the method may fail anyway. Fluid released from the penis just before ejaculation may contain some sperm.

Douching, or rinsing out the vagina with a chemical right after intercourse, is next to useless. Sperm can move past the cervix and out of reach of the douche within ninety seconds after ejaculation.

Other methods involve physical or chemical barriers to prevent sperm from entering the uterus and moving to the ovarian ducts. *Spermicidal foam* and *spermicidal jelly* are toxic to sperm. They are packaged in an applicator and placed in the vagina just before intercourse. These products are not always reliable unless used with another device, such as a diaphragm or condom.

A *diaphragm* is a flexible, dome-shaped device, inserted into the vagina and positioned over the cervix before intercourse. A diaphragm is relatively effective when fitted initially by a doctor, used with foam or jelly before each sexual contact, and inserted correctly with each use.

Condoms are thin, tight-fitting sheaths of rubber or animal skin, worn over the penis during intercourse. They are about 85 to 93 percent reliable, and they help prevent the spread of sexually transmitted diseases (see the *Commentary* on page 780). However, condoms can tear and leak, in which case they are rendered useless.

The most widely used method of fertility control is *the Pill*, an oral contraceptive of synthetic estrogens and progesterones. It suppresses the normal release of these hormones from the pituitary and so stops eggs from maturing and being released at ovulation. The Pill is a prescription drug. Formulations vary and are selected to match each patient's needs. That is why it is not wise for a woman to borrow oral contraceptives from someone else.

When a woman does not forget to take her daily dosage, the Pill is one of the most reliable methods of controlling fertility. It does not interrupt sexual intercourse, and the method is easy to follow. Often the Pill corrects erratic menstrual cycles and decreases cramping. However, the Pill has some side effects for a small number of users. In the first month or so of use, it may cause nausea, weight gain, tissue swelling, and minor headaches. Its continued use may lead to blood clotting in the veins of a few women (3 out of 10,000) predisposed to this disorder. Some cases of elevated blood pressure and some abnormalities in fat metabolism might be linked to a growing number of gallbladder disorders in Pill users.

In *vasectomy*, a tiny incision is made in a man's scrotum, and each vas deferens is severed and tied off. The simple operation can be performed in twenty minutes in a physician's office, with only a local anesthetic. After vasectomy, sperm cannot leave the testes and so will not be present in semen. So far there is no firm evidence that vasectomy disrupts the male hormone system, and there seems to be no noticeable difference in sexual activity. Vasectomies can be reversed, but half the men who have had the surgery develop antibodies against sperm and may not be able to regain fertility.

For females, surgical intervention includes *tubal ligation*, in which the oviducts are cauterized or cut and tied off. Tubal ligation is usually performed in a hospital. A small number of women who have had the operation suffer recurring bouts of pain and inflammation of tissues in the pelvic region where the surgery was performed. The operation can be reversed, although major surgery is required and success is not always assured.

Once conception and implantation have occurred, the only way to terminate a pregnancy is *abortion*, the dislodging and removal of the embryo from the uterus. *RU-486*, the "morning-after Pill," can induce termination of pregnancy. It it administered under a physician's supervision, at least in Europe. Those opposed to abortion are currently fighting its use in the United States.

At one time, abortions were generally forbidden by law in the United States unless the pregnancy endangered the mother's life. The Supreme Court in 1973 ruled that the government does not have the right to forbid abortions during the early stages of pregnancy (typically up to five months). Before this ruling, there were

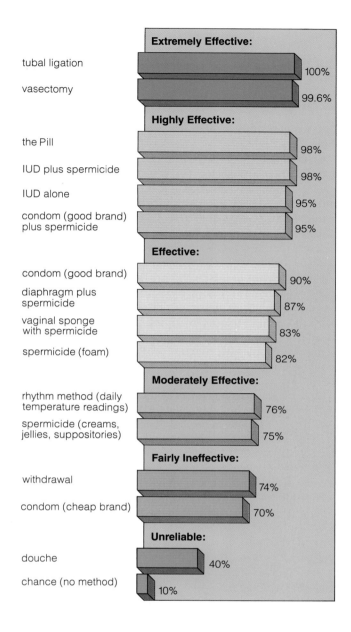

Figure 43.20 Comparison of the effectiveness of some contraceptive methods.

Commentary

Sexually Transmitted Diseases

(This Commentary is based on information from the Centers for Disease Control, Atlanta)

Sexually transmitted diseases (STDs) have reached epidemic proportions, even in countries with the highest medical standards. The disease agents are mostly bacteria and viruses. Usually they are transmitted from infected to uninfected persons during sexual intercourse. In the United States alone, many millions of young adults have reported that they have some form of STD. No one can estimate the number of unreported cases.

The economics of this health problem are staggering. The cost of treatment far exceeds $2 billion a year—and this does not include the accelerating cost of treating AIDS patients. In many developing countries, AIDS alone threatens to overwhelm health-care delivery systems and to unravel decades of economic progress. The social consequences are sobering. Of every twenty babies born in the United States, one will have a chlamydial infection. Type II *Herpes* virus will infect as many as one of every 10,000 newborns, possibly killing half of them and leaving a fourth with serious neurological defects. Each year 1 million female Americans contract pelvic inflammatory disease, usually as a complication of gonorrhea and other STDs. Of every 200,000 who are hospitalized, over 100,000 become permanently sterile, and 900 die.

The examples just given only hint at the alarming complications of many sexually transmitted diseases.

AIDS

Acquired immune deficiency syndrome (AIDS) is a set of chronic disorders that can follow infection by the human immunodeficiency virus (HIV). The virus cripples the immune system, in the manner described in Chapter 39. The body becomes highly vulnerable to illnesses, many of which would not otherwise be life-threatening. (Hence the description, "opportunistic" infections.)

AIDS is mainly a sexually transmitted disease, with most infections occurring through the transfer of bodily fluids during vaginal or anal intercourse. Such fluids include blood, semen, urine, and vaginal secretions. The virus enters the body through cuts or abrasions on the penis, vagina, or rectum. Mucous membranes in the mouth may be another point of entry. Once inside the body, the virus locks onto cells that are capable of sustaining its replication (page 692). Helper T cells (the T4 lymphocytes), macrophages, brain cells, and epithelial cells of the cervix are known targets.

At present there is no effective treatment or vaccine for AIDS. *There is no cure.* Infected persons may be symptom-free at first but as many as half develop AIDS within five to ten years. Others develop ARC (AIDS-related complex), which are milder symptoms than those characterizing AIDS (page 694). Will most or all of those infected by HIV eventually develop AIDS? We do not know enough about the natural history of the virus and the progression rates of the disease to discount that possibility.

HIV apparently existed in some parts of Central Africa for at least several decades. In the 1970s and early 1980s, it spread to different countries and was finally identified in 1981. Today it is spreading through African populations mainly by heterosexual contact. In the United States and other developed countries, HIV spread at first among male homosexuals. The pool of infection now includes a significant portion of the heterosexual population, largely as a result of needle-sharing among intravenous (IV) drug abusers. *During the next decade, as many as 10 million may be infected worldwide.*

As an example of the relative frequencies of the main modes of transmission, the following cases were reported to the Centers for Disease Control through July 1991:

	White Males	Black Males	White Females	Black Females
Homosexual or bisexual contact:	80%	44%	—	—
IV drug abuse (heterosexual):	7%	36%	41%	57%
Both of the above:	7%	8%	—	—
Heterosexual contact:	1%	7%	30%	33%
Blood transfusion:	2%	1%	21%	3%

Free or low-cost, confidential testing for AIDS is available through public health facilities and many physicians' offices. Keep in mind that there may be a time lag from a few weeks to six months or longer before detectable antibodies form in response to infection. The presence of antibodies indicates exposure to the virus, but this in itself does not mean that AIDS will develop. Even so, anyone who tests positive should be considered capable of spreading the virus.

The spread of HIV has led to massive public education programs, including the strongest possible advocacy for safe sex. Yet there is confusion about what constitutes "safe" sex. Proper use of high-quality, latex condoms, together with a spermicide that contains nonoxynol-9, is assumed to be highly effective in stopping transmission—but there is still a small risk of irreversible infection. Open-mouthed, intimate kissing with a person who tests positive for the virus should be avoided. Caressing carries no risk—if there are no lesions or cuts through which the virus can enter the body. Such lesions commonly accompany other sexually transmitted diseases, and they apparently are correlated with increased susceptibility to HIV infection.

In sum, AIDS has reached epidemic proportions mainly for three reasons. First, we did not know that the virus is transmitted by semen, blood, and vaginal fluid and that *behavioral* controls can limit its spread. Second, we did not have tests that could be used to identify symptom-free carriers who could unwittingly infect others; we do now. Third, we thought AIDS was a threat associated only with homosexual behavior. The medical, social, and economic consequences of its rapid spread throughout the world make it everyone's problem.

Gonorrhea

Unlike AIDS, gonorrhea is a sexually transmitted disease that can be cured by prompt diagnosis and treatment. Gonorrhea ranks first among the reported communicable diseases in the United States, with 1 million new cases reported each year. There may be anywhere from 3 million to 10 million unreported cases.

Gonorrhea is caused by *Neisseria gonorrhoeae*. This bacterium can infect epithelial cells of the genital tract, eye membranes, and the throat. Since 1960 its incidence in the population has been rising at an alarming rate. The increase has coincided with the use of birth control pills and increased sexual permissivity.

Males have a greater chance than females do of detecting the disease in early stages. Within a week, yellow pus is discharged from the penis. Urination becomes more frequent and painful. Females may or may not experience a burning sensation while urinating. They may or may not have a slight vaginal discharge; even if they do, the

(a) *Neisseria gonorrhoeae*, a bacterium that typically is seen as paired cells, as shown here. The threadlike structures (pili) evident in this electron micrograph help the bacterium attach to its host, upon which it bestows gonorrhea.

discharge may not be perceived as abnormal. Thus, in the absence of worrisome symptoms, gonorrhea often goes untreated. The bacteria may spread into the oviducts. Eventually, there may be violent cramps, fever, vomiting and, often, sterility due to scarring and blocking.

Complications arising from gonorrheal infection can be avoided with prompt treatment. As a preventive measure, males who have multiple sexual partners can wear condoms to help prevent the spread of infection. Part of the problem is that the initial stages of the disease are so uneventful that the dangers are masked. Also, many infected persons wrongly believe that once cured of gonorrhea, they are safe from reinfection—which simply is not true. Multiple reinfections can and do occur.

Syphilis

Syphilis is caused by a motile, corkscrew-shaped bacterium, *Treponema pallidum*. As many as 300,000 humans may become infected in a given year in the United States, but only about 30,000 are reported. In the past five years, its incidence has nearly doubled among females between ages fifteen and twenty-four. The bacterium is transmitted by sexual contact. After it has penetrated exposed tissues, it produces a chancre (localized ulcer) that teems with treponeme offspring. Usually the chancre is flat, not bumpy, and it is not painful. It becomes visible between one and eight weeks following infection—and it is a symptom of the primary stage of syphilis. By then, treponemes have already moved into the lymph vascular system and bloodstream.

During the second stage of infection, lesions can occur in mucous membranes, the eyes, bones, and the central nervous system. Afterward, the infection enters a latent stage, and there are no outward symptoms. Syphilis can be detected only by laboratory tests during the latent stage, which can last many years. All the while, the immune system works against the bacterium. Sometimes the body does cure itself, but this is not the usual outcome.

If untreated, syphilis in its tertiary stage can produce lesions of the skin and internal organs, including the liver, bones, and aorta. Scars form; the walls of the aorta can weaken. Treponemes also damage the brain and spinal cord in ways that lead to various forms of insanity and paralysis. Women who have been infected typically have miscarriages, stillbirths, or sickly and syphilitic infants.

Chlamydial Infections

An intracellular parasite, *Chlamydia trachomatis*, is the culprit behind a variety of sexually transmitted diseases. Each year, anywhere from 3 million to 10 million Americans—college students particularly—are affected.

Among other things, the parasite infects cells of the genitals and urinary tract. Infected men may have a discharge from the penis and have a burning sensation when they urinate. Women may have a vaginal discharge as well as burning and itching sensations. Sometimes, however, there may be no apparent evidence of infection—yet the parasite can still be passed on to others.

Following infection, the parasites migrate to lymph nodes, which become enlarged and tender. The enlargement can impair lymph drainage and lead to pronounced tissue swelling. Chlamydial infections can be treated with tetracycline and sulfonamides. Most of the infections have

no long-term complications. However, in some females the infection leads to pelvic inflammatory disease.

Pelvic Inflammatory Disease

A condition called pelvic inflammatory disease (PID) affects about 14 million women each year. It is one of the serious complications of gonorrhea, chlamydial infections, and other STDs. It also can arise when microorganisms that normally inhabit the vagina ascend into the pelvic region. Most often, the uterus, oviducts, and ovaries are affected. Pain may be so severe, infected women often think they are having an attack of acute appendicitis. The oviducts may become scarred, and scarring can lead to abnormal pregnancies as well as to sterility.

Genital Herpes

Genital herpes is an extremely contagious viral infection of the genitals. It is transmitted when any part of a person's body comes into direct contact with active *Herpes* viruses or sores that contain them. Mucous membranes of the mouth or genital area are especially susceptible to invasion, as is broken or damaged skin. Transmission seems to require direct contact; the virus does not survive for long outside the human body.

There are an estimated 5 million to 20 million persons with genital herpes in the United States alone. From 1965 to 1979, the number of reported cases increased by 830 percent. And 200,000 to 500,000 cases are still being reported annually.

Newborns of infected mothers are among those cases. Contact with the mother's active lesions during normal vaginal delivery can lead to a form of herpes that is often fatal. Lesions arising in the infant's eyes can cause blindness. Chronic herpes infection of the cervix also increases the risk of cervical cancer.

The many strains of *Herpes* viruses are classified as types I and II. The type I strains infect mainly the lips, tongue, mouth, and eyes. Type II strains cause most of the genital infections. Disease symptoms occur two to ten days after exposure to the virus, although sometimes symptoms are mild or absent. Among infected women, small, painful blisters appear on the vulva, cervix, urethra, or anal tissues. Among men, the blisters occur on the penis and anal tissues. Within three weeks, the sores crust over and heal without leaving scars.

After the first sores disappear, sporadic reactivation of the virus can produce new, painful sores at or near the original site of infection. Recurrent infections may be triggered by sexual intercourse, emotional stress, menstruation, or other infections. At present there is no cure for genital herpes. Acyclovir, an antiviral drug, decreases the healing time and often decreases the pain and viral shedding.

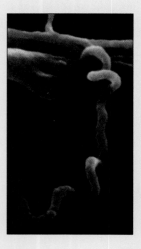

(**b**) *Treponema pallidum*, a bacterium that causes syphilis

dangerous, traumatic, and often fatal attempts to abort embryos, either by pregnant women themselves or by quacks.

Newer methods have made abortion relatively rapid, painless, and free of complications when performed during the first trimester. Abortions in the second and third trimesters will probably remain extremely controversial unless the mother's life is clearly threatened. For both medical and humanitarian reasons, however, it is generally agreed in this country that the preferred route to birth control is not through abortion but through control of conception in the first place.

In Vitro Fertilization

Controls over fertility also extend in the other direction—to help childless couples who are desperate to conceive a child. In the United States, about 15 percent of all couples cannot do so because of sterility or infertility. For example, hormonal imbalances may prevent ovulation in females, or the sperm count in the male may be too low to assure fertilization.

With *in vitro fertilization*, external conception is possible, provided sperm and oocytes obtained from the couple are normal. A hormone is administered that prepares the ovaries for ovulation. Then a physician locates and removes the preovulatory oocyte with a suction device. Before the oocyte is removed, sperm from the male is placed in a solution that simulates the fluid in oviducts. When the suctioned oocyte is placed with the sperm, fertilization may occur a few hours later. About twelve hours later, the newly dividing zygote is transferred to a solution that will support further development, and about two to four days after that, it is transferred to the female's uterus. Implantation occurs in about 20 percent of the cases, and each attempt costs several thousand dollars.

SUMMARY

1. Most animals reproduce sexually. Separation into male and female sexes involves specialized reproductive structures and forms of behavior that help assure successful fertilization and that lend initial nutritional support to offspring.

2. Humans have a pair of primary reproductive organs (sperm-producing testes in males, egg-producing ovaries in females), accessory ducts, and glands. Testes and ·ovaries also produce hormones that influence reproductive functions and secondary sexual traits.

3. The hormones testosterone, LH, and FSH control sperm formation. They are part of feedback loops among the hypothalamus, anterior pituitary, and testes.

4. The hormones estrogen, progesterone, FSH, and LH control egg maturation and release, as well as changes in the lining of the uterus (endometrium). They are part of feedback loops involving the hypothalamus, anterior pituitary, and ovaries.

5. The following events occur during a menstrual cycle:
   a. A follicle, which is an oocyte surrounded by a cell layer, matures in an ovary, and the endometrium starts to rebuild. (It breaks down at the end of each menstrual cycle when pregnancy does not occur.)
   b. A midcycle peak of LH triggers the release of a secondary oocyte from the ovary. This event is called ovulation.
   c. A corpus luteum forms from the remainder of the follicle. Its secretions prime the endometrium for fertilization. When fertilization occurs, the corpus luteum is maintained, and its secretions help maintain the endometrium.

6. Human development proceeds through gamete formation, fertilization, cleavage, gastrulation, organ formation, and growth and tissue specialization.

7. All tissues and organs in the developing embryo arise from three germ layers: the endoderm, ectoderm, and mesoderm of the early embryo.

8. Embryonic development depends on the formation of four extraembryonic membranes:
   a. Yolk sac: parts give rise to the embryo's digestive tube.
   b. Allantois: its blood vessels function in oxygen transport.
   c. Amnion: a fluid-filled sac that surrounds and protects the embryo from mechanical shocks and keeps it from drying out.
   d. Chorion: a protective membrane around the embryo and the other membranes; a primary component of the placenta.

9. The embryo and the mother exchange substances by way of the placenta (a spongy tissue of endometrium and extraembryonic membranes).

10. The placental barrier provides some protection for the fetus, but it may suffer harmful effects from the mother's nutritional deficiencies, infections, intake of prescription drugs, illegal drugs, alcohol, and smoking.

11. At delivery, contractions of the uterus dilate the cervical canal and expel the fetus and afterbirth. Estrogen and progesterone stimulate growth of the mammary glands. After delivery, nursing causes the release of hormones that stimulate milk production and release.

12. Control of human fertility raises important ethical questions. These questions extend to the physical, chemical, surgical, or behavioral interventions used in the control of unwanted pregnancies.

Review Questions

1. Study Table 43.1. Then list the main organs of the human male reproductive tract and identify their functions. *761*

2. Which hormones influence male reproductive function? *762*

3. Label the component parts of the female reproductive tract: *765*

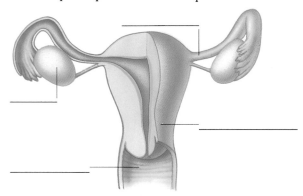

4. What is the menstrual cycle? Which four hormones influence this cycle? *764–766*

5. List four events that are triggered by the surge of LH at the midpoint of the menstrual cycle. *766–767*

6. What changes occur in the endometrium during the menstrual cycle? *766*

Self-Quiz *(Answers in Appendix IV)*

1. Besides producing gametes, human male and female gonads also produce sex hormones. The _____ and the pituitary gland control secretion of both.

2. _____ production is continuous from puberty onward in males; _____ production is cyclic and intermittent in females.
   a. egg; sperm   c. testosterone; sperm
   b. sperm; egg   d. estrogen; egg

3. Sperm formation is controlled through _____ secretions.
   a. testosterone   c. FSH
   b. LH   d. all of the above are correct

4. During the menstrual cycle, a midcycle surge of _____ triggers ovulation.
   a. estrogen   c. LH
   b. progesterone   d. FHS

5. Which is the correct order for one turn of the menstrual cycle?
   a. corpus luteum forms, ovulation, follicle forms
   b. follicle forms, ovulation, corpus luteum forms

6. Parts of the _____, an extraembryonic membrane, give rise to the embryo's digestive tube.
   a. yolk sac   c. amnion
   b. allantois   d. chorion

7. The _____, a fluid-filled sac, surrounds and protects the embryo from mechanical shocks and keeps it from drying out.
   a. yolk sac   c. amnion
   b. allantois   d. chorion

8. Blood vessels of the _____, an extraembryonic membrane, transport oxygen and nutrients to the embryo.
   a. yolk sac   c. amnion
   b. allantois   d. chorion

9. Substances are exchanged between the embryo and mother through the _____, which is composed of maternal and embryonic tissues.
   a. yolk sac   d. amnion
   b. allantois   e. chorion
   c. placenta

10. Match the reproduction and development concepts.
   _____ extraembryonic   a. endoderm, ectoderm,
         membranes   mesoderm
   _____ corpus luteum   b. immature oocyte and
   _____ follicle   surrounding cell layer
   _____ egg   c. yolk sac, allantois, amnion,
   _____ germ layers   chorion
                             d. mature ovum
                             e. its secretions help prepare
                                endometrium for fertilization

Selected Key Terms

adult *777*	FSH *762*	polar body *764*
aging *777*	GnRH *763*	progesterone *763*
allantois *770*	granulosa cell *764*	secondary oocyte *764*
amnion *770*	implantation *769*	semen *762*
blastocyst *769*	Leydig cell *761*	seminiferous tubule *761*
chorion *770*	LH *762*	Sertoli cell *762*
clitoris *764*	menstrual cycle *764*	somite *775*
coitus *766*	oocyte *763*	sperm *762*
corpus luteum *764*	orgasm *769*	testis *759*
embryonic disk *769*	ovary *759*	testosterone *762*
endometrium *764*	oviduct *763*	umbilical cord *770*
estrogen *763*	ovulation *764*	uterus *763*
fetus *775*	ovum *769*	yolk sac *770*
follicle *764*	placenta *771*	zona pellucida *764*

Readings

Carlson, B. 1988. *Patten's Foundations of Embryology.* Fifth edition. New York: McGraw-Hill.

Gilbert, S. 1991. *Developmental Biology.* Sunderland, Massachusetts: Sinauer. Third edition. Acclaimed reference text.

Nilsson, L. et al. 1986. *A Child Is Born.* New York: Delacorte Press/Seymour Lawrence. Extraordinary photographs of embryonic development.

Saunders, J. W. 1982. *Developmental Biology: Patterns, Problems, Principles.* New York: Macmillan.

Schatten, G. 1983. "Motility During Fertilization." *Endeavor* 7(4): 173–182.

FACING PAGE: *Two organisms—a fox in the shadows cast by a snow-dusted spruce tree. What are the nature and consequences of their interactions with one another, with other organisms, and with their environment? By the end of this last unit of the book, you possibly will see worlds within worlds in such photographs.*

APPENDIX I
Units of Measure

Metric-English Conversions

Length

English		Metric
inch	=	2.54 centimeters
foot	=	0.30 meter
yard	=	0.91 meter
mile (5,280 feet)	=	1.61 kilometer

To convert	multiply by	to obtain
inches	2.54	centimeters
feet	30.00	centimeters
centimeters	0.39	inches
millimeters	0.039	inches

Weight

English		Metric
grain	=	64.80 milligrams
ounce	=	28.35 grams
pound	=	453.60 grams
ton (short) (2,000 pounds)	=	0.91 metric ton

To convert	multiply by	to obtain
ounces	28.3	grams
pounds	453.6	grams
pounds	0.45	kilograms
grams	0.035	ounces
kilograms	2.2	pounds

Volume

English		Metric
cubic inch	=	16.39 cubic centimeters
cubic foot	=	0.03 cubic meter
cubic yard	=	0.765 cubic meters
ounce	=	0.03 liter
pint	=	0.47 liter
quart	=	0.95 liter
gallon	=	3.79 liters

To convert	multiply by	to obtain
fluid ounces	30.00	milliliters
quart	0.95	liters
milliliters	0.03	fluid ounces
liters	1.06	quarts

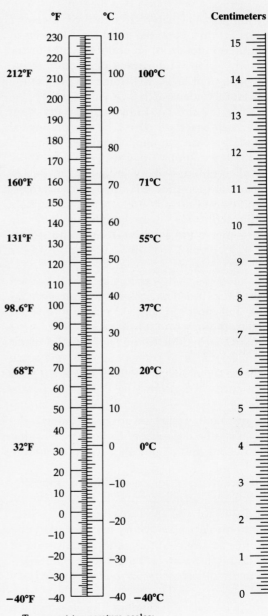

To convert temperature scales:

Fahrenheit to Celsius: $°C = 5/9 (°F - 32)$

Celsius to Fahrenheit: $°F = 9/5 (°C) + 32$

APPENDIX II
Brief Classification Scheme

This classification scheme is a composite of several used in microbiology, botany, and zoology. Although major groupings are more or less agreed upon, what to call them and (sometimes) where to place them in the overall hierarchy are not. As Chapter 20 indicated, there are several reasons for this. First, the fossil record varies in its quality and completeness, so certain evolutionary relationships are open to interpretation. Comparative studies at the molecular level are firming up the picture, but this work is still under way.

Second, since the time of Linnaeus, classification schemes have been based on perceived morphological similarities and differences among organisms. Although some original interpretations are now open to question, we are so used to thinking about organisms in certain ways that reclassification proceeds slowly. For example, birds and reptiles traditionally are considered separate classes (Reptilia and Aves)—even though there now are compelling arguments for grouping lizards and snakes as one class, and crocodilians, dinosaurs, and birds as another.

Finally, botanists as well as zoologists have inherited a wealth of literature based on schemes that are peculiar to their fields; and most see no good reason to give up established terminology and so disrupt access to the past. Thus botanists continue to use Division as a major taxon in the hierarchical schemes and zoologists use Phylum in theirs. Opinions are notably divergent with respect to an entire Kingdom (the Protista), certain members of which could just as easily be called single-celled forms of plants, fungi, or animals. Indeed, the term protozoan is a holdover from earlier schemes that ranked the amoebas and some other forms as simple animals.

Given the problems, why do we bother imposing hierarchical schemes on the natural history of life on earth? We do this for the same reason that a writer might decide to break up the history of civilization into several volumes, many chapters, and a multitude of paragraphs. Both efforts are attempts to impart structure to what might otherwise be an overwhelming body of information.

One more point to keep in mind: The classification scheme in this Appendix is primarily for reference purposes, and it is by no means complete (numerous phyla of existing and extinct organisms are not represented). Our strategy is to focus mainly on the organisms mentioned in the text, with numerals referring to some of the pages on which representatives are illustrated or described. A few examples of organisms are also listed under the entries.

SUPERKINGDOM PROKARYOTA. Prokaryotes (single-celled organisms with no nucleus or other membrane-bound organelles in the cytoplasm).

KINGDOM MONERA. Bacteria, either single cells or simple associations of cells; autotrophic and heterotrophic forms. *Bergey's Manual of Systematic Bacteriology,* the authoritative reference in the field, calls this "a time of taxonomic transition" and groups bacteria mainly on the basis of form, physiology, and behavior, not on phylogeny (Table 22.4 gives examples). The scheme presented here does reflect the growing evidence of evolutionary relationships for at least some bacterial groups.

SUBKINGDOM ARCHAEBACTERIA. Methanogens, halophiles, thermophiles. Strict anaerobes, distinct from other bacteria in their cell wall, membrane lipids, ribosomes, and RNA sequences. 356, 357

SUBKINGDOM EUBACTERIA. Gram-negative and Gram-positive forms. Peptidoglycan in cell walls. Photosynthetic autotrophs, chemosynthetic autotrophs, and heterotrophs. 358–359

DIVISION GRACILICUTES. Typical Gram-negative, thin wall. Autotrophs (photosynthetic and chemosynthetic) and heterotrophs. *Anabaena, Chlorobium; Escherichia, Shigella, Desulfovibrio, Agrobacterium, Pseudomonas, Neisseria.* 57, 354, 356, 781
DIVISION FIRMICUTES. Typical Gram-positive, thick wall. Heterotrophs. *Staphylococcus, Streptococcus, Clostridium, Bacillus, Actinomyces.* 346, 359, 557, 680
DIVISION TENERICUTES. Gram-negative, wall absent. Heterotrophs (saprobes, parasites, pathogens). *Mycoplasma.* 356

SUPERKINGDOM EUKARYOTA. Eukaryotes (single-celled and multicelled organisms; cells typically have a nucleus and other organelles).

KINGDOM PROTISTA. Mostly single-celled eukaryotes. Some colonial forms.

PHYLUM GYMNOMYCOTA. Heterotrophs.
  Class Acrasiomycota. Cellular slime molds. *Dictyostelium.* 362, 363
  Class Myxomycota. Plasmodial slime molds. *Physarum.* 362
PHYLUM EUGLENOPHYTA. Euglenoids. Mostly heterotrophic, some photosynthetic. Flagellated. *Euglena.* 364
PHYLUM CHRYSOPHYTA. Golden algae, yellow-green algae, diatoms. Photosynthetic. Some flagellated, others not. *Vaucheria.* 50, 364–368
PHYLUM PYRRHOPHYTA. Dinoflagellates. Mostly photosynthetic, some heterotrophs. *Gonyaulax.* 365
PHYLUM MASTIGOPHORA. Flagellated protozoans. Heterotrophs. *Trypanosoma, Trichomonas.* 366–367

APPENDIX III
Answers to Self-Quizzes

CHAPTER 31
1. epithelial, connective, muscle, nervous
2. e
3. d
4. c
5. b
6. a
7. d
8. c
9. a
10. Receptors, integrator, effectors
11. d, e, c, b, a

CHAPTER 32
1. neurons
2. sensory neurons, interneurons, motor neurons
3. a
4. c
5. a
6. d
7. b
8. d
9. d
10. c, d, a, b, e

CHAPTER 33
1. nerve nets
2. cephalization; bilateral
3. a
4. b
5. c
6. a
7. d
8. c
9. a
10. b
11. e, d, b, c, a

CHAPTER 34
1. Hormones; transmitter substances; local signaling molecules; pheromones
2. hypothalamus; pituitary
3. negative feedback
4. receptors
5. a
6. d
7. e
8. b
9. d
10. d, b, e, c, a

CHAPTER 35
1. stimulus
2. sensation
3. perception
4. d
5. b
6. d
7. c
8. b
9. a
10. f, d, c, e, a, b

CHAPTER 36
1. integumentary
2. Skeletal; muscular
3. smooth; cardiac; skeletal
4. d
5. d
6. c
7. b
8. c
9. d, e, b, a, c

CHAPTER 37
1. digestive; circulatory; respiratory and urinary
2. digesting; absorbing; eliminating
3. caloric; energy
4. carbohydrates
5. essential amino acids; essential fatty acids
6. b
7. c
8. d
9. c
10. e, d, a, c, b

CHAPTER 38
1. circulatory; lymphatic
2. Arteries; veins; capillaries; venules; arterioles
3. d
4. d
5. c
6. b
7. c
8. b
9. e, b, a, d, c
10. d, e, a, f, c, b

CHAPTER 39
1. Antigens
2. antigen that is circulating or attached to pathogen; infected, cancerous, or mutated cells
3. e
4. d
5. b
6. b
7. b
8. a
9. c
10. b, a, e, d, c

CHAPTER 40
1. Oxygen; carbon dioxide
2. gas; partial pressure gradient
3. epithelium; diffuse
4. adaptations; air flow
5. d
6. e
7. b
8. d
9. d
10. e, a, d, b, c

CHAPTER 41
1. filtration; absorption; secretion
2. c
3. c
4. d
5. d
6. a
7. a
8. c
9. d
10. b, d, e, c, a

CHAPTER 42
1. d
2. a
3. a
4. a
5. c
6. morphogenesis
7. b
8. c
9. e
10. d, a, f, e, c, b

CHAPTER 43
1. hypothalamus
2. b
3. d
4. c
5. b
6. a
7. c
8. b
9. c
10. c, e, b, d, a

GLOSSARY OF BIOLOGICAL TERMS

ABO blood typing Method of characterizing blood according to particular proteins at the surface of red blood cells.

abortion Spontaneous or induced expulsion of the embryo or fetus from the uterus.

abscisic acid (ab-SISS-ik) Plant hormone that promotes stomatal closure, bud dormancy, and seed dormancy.

abscission (ab-SIH-zhun) [L. *abscissus*, to cut off] The dropping of leaves, flowers, fruits, or other plant parts due to hormonal action.

absorption For complex animals, the movement of nutrients, fluid, and ions across the gut lining and into the internal environment.

acid [L. *acidus*, sour] A substance that releases hydrogen ions (H⁺) in solution.

acid deposition, dry The falling to earth of airborne particles of sulfur and nitrogen oxides.

acid deposition, wet The falling to earth of snow or rain that contains sulfur and nitrogen oxides.

acoelomate (ay-SEE-la-mate) Type of animal that has no fluid-filled cavity between the gut and body wall.

actin (AK-tin) A globular contractile protein. Within sarcomeres, the basic units of muscle contraction, actin molecules form two beaded strands, twisted together into filaments, that are pulled by myosin molecules during contraction.

action potential An abrupt but brief reversal in the steady voltage difference across the plasma membrane (that is, the resting membrane potential) of a neuron and some other cells.

activation energy The minimum amount of collision energy required to bring reactant molecules to an activated condition (the transition state) at which a reaction will proceed spontaneously. Enzymes enhance reaction rates by lowering the activation energy.

active site A crevice on the surface of an enzyme molecule where a specific reaction is catalyzed.

active transport The pumping of specific solutes through transport proteins that span the lipid bilayer of a plasma membrane, most often against their concentration gradient. The proteins act when they receive an energy boost, as from ATP.

adaptation [L. *adaptare*, to fit] In evolutionary biology, the process of becoming adapted (or more adapted) to a given set of environmental conditions. Of sensory neurons, a decrease in the frequency of action potentials (or their cessation) even when a stimulus is being maintained at constant strength.

adaptive radiation A burst of speciation events, with lineages branching away from one another as they partition the existing environment or invade new ones.

adaptive trait Any aspect of form, function, or behavior that helps an organism survive and reproduce under a given set of environmental conditions.

adaptive zone A way of life, such as "catching insects in the air at night." A lineage must have physical, ecological, and evolutionary access to an adaptive zone to become a successful occupant of it.

adenine (AH-de-neen) A purine; a nitrogen-containing base found in nucleotides.

adenosine diphosphate (ah-DEN-uh-seen die-FOSS-fate) ADP, a molecule involved in cellular energy transfers; typically formed by hydrolysis of ATP.

adenosine phosphate Any of several relatively small molecules, some of which function as chemical messengers within and between cells, and others as energy carriers.

adenosine triphosphate *See* ATP.

ADH Antidiuretic hormone, produced by the hypothalamus and released from the posterior pituitary; stimulates reabsorption in the kidneys and so reduces urine volume.

adrenal cortex (ah-DREE-nul) Outer portion of either of two adrenal glands; its hormones have roles in metabolism, inflammation, maintaining extracellular fluid volume, and other functions.

adrenal medulla Inner region of the adrenal gland; its hormones help control blood circulation and carbohydrate metabolism.

aerobic respiration (air-OH-bik) [Gk. *aer*, air, + *bios*, life] Degradative, oxygen-requiring pathway of ATP formation. Oxygen serves as the final acceptor of electrons stripped from glucose or some other organic compound. The pathway proceeds from glycolysis, then through the Krebs cycle and electron transport phosphorylation. Of all degradative pathways, aerobic respiration has the greatest energy yield, with 36 ATP typically formed for each glucose molecule.

agglutination (ah-glue-tin-AY-shun) Clumping of foreign cells, induced by the cross-linking of antigen-antibody complexes at their surface.

aging A range of processes, including the breakdown of cell structure and function, by which the body gradually deteriorates. Characteristic of all organisms showing extensive cell differentiation.

AIDS Acquired immune deficiency syndrome, a set of chronic disorders following infection by the human immunodeficiency virus (HIV), which destroys key cells of the immune system.

alcoholic fermentation Anaerobic pathway of ATP formation in which pyruvate from glycolysis is broken down to acetaldehyde; the acetaldehyde then accepts electrons from NADH to become ethanol.

aldosterone (al-DOSS-tuh-rohn) Hormone secreted by the adrenal cortex that helps regulate sodium reabsorption.

allantois (ah-LAN-twahz) [Gk. *allas*, sausage] A vascularized extraembryonic membrane; in reptiles and birds, it functions in excretion and respiration; in placental mammals, it functions in oxygen transport by way of the umbilical cord.

allele (uh-LEEL) One of two or more alternative forms of a gene at a given locus on a chromosome.

allele frequency The relative abundance of each kind of allele that can occur at a given gene locus for all individuals in a population.

allergy An abnormal, secondary immune response to a normally harmless substance.

allopatric speciation [Gk. *allos*, different, + *patria*, native land] Formation of new species as a result of geographic isolation.

allosteric control (AL-oh-STARE-ik) Control of enzyme functioning through the binding of a specific substance at a control site on the enzyme molecule.

altruistic behavior (al-true-ISS-tik) Self-sacrificing behavior; the individual behaves in a way that helps others but, in so doing, decreases its own chance to produce offspring.

alveolar sac (al-VEE-uh-lar) Any of the pouch-like clusters of alveoli in lungs; the major sites of gas exchange.

alveolus (ahl-VEE-uh-lus), plural **alveoli** [L. *alveus*, small cavity] Any of the many cup-shaped, thin-walled outpouchings of the respiratory bronchioles; a site where oxygen from air in the lungs diffuses into the bloodstream and where carbon dioxide from the bloodstream diffuses into the lungs.

amino acid (uh-MEE-no) A small organic molecule having a hydrogen atom, an amino group, an acid group, and an R group covalently bonded to a central carbon atom; amino acids strung together as polypeptide chains represent the primary structure of proteins.

ammonification (uh-moan-ih-fih-KAY-shun) A decomposition process by which certain bacteria and fungi break down nitrogen-containing wastes and remains of other organisms.

amnion (AM-nee-on) In land vertebrates, an extraembryonic membrane in the form of a fluid-filled sac around the embryo; it absorbs shocks and keeps the embryo from drying out.

anaerobic pathway (an-uh-ROW-bok) [Gk. *an*, without, + *aer*, air] Degradative metabolic pathway in which a substance other than oxygen serves as the final electron acceptor for the reactions.

analogous structures Occurrence of similar body parts being used for similar functions in evolutionarily remote lineages; evolutionary outcome of morphological convergence.

anaphase (AN-uh-faze) The stage of mitosis when sister chromatids of each chromosome separate and move to opposite poles of the spindle.

anaphase I and II Stages of meiosis when each chromosome separates from its homologous partner (anaphase I) and, later, when sister chromatids of each chromosome separate and move to opposite poles of the spindle (anaphase II).

angiosperm (AN-gee-oh-spurm) [Gk. *angeion*, vessel, + *spermia*, seed] A flowering plant.

animal A heterotroph that eats or absorbs nutrients from other organisms; is multicelled, usually with tissues arranged in organs and organ systems; is usually motile during at least part of the life cycle; and goes through a period of embryonic development.

Animalia The kingdom of animals.

annual plant Vascular plant that completes its life cycle in one growing season.

anther [Gk. *anthos*, flower] In flowering plants, the pollen-bearing part of the male reproductive structure (stamen).

antibody [Gk. *anti*, against] Any of a variety of Y-shaped receptor molecules with binding sites for a specific antigen (molecule that triggers an immune response); produced by B cells of the immune system.

anticodon In a tRNA molecule, a sequence of three nucleotide bases that can pair with an mRNA codon.

antigen (AN-tih-jen) [Gk. *anti*, against, + *genos*, race, kind] Any large molecule (usually a protein or polysaccharide) with a distinct configuration that triggers an immune response.

aorta (ay-OR-tah) [Gk. *airein*, to lift, heave] Main artery of systemic circulation; carries oxygenated blood away from the heart.

apical dominance The influence exerted by a terminal bud in inhibiting the growth of lateral buds.

apical meristem (AY-pih-kul MARE-ih-stem) [L. *apex*, top, + Gk. *meristos*, divisible] In most plants, a mass of self-perpetuating cells at a root or shoot tip that is responsible for primary growth, or elongation, of plant parts.

appendicular skeleton (ap-en-DIK-you-lahr) In vertebrates, bones of the limbs, pelvic girdle (at the hips), and pectoral girdle (at the shoulders).

arteriole (ar-TEER-ee-ole) Any of the blood vessels between arteries and capillaries; arterioles serve as control points where the volume of blood delivered to different body regions can be adjusted.

artery Any of the large-diameter blood vessels that conduct oxygen-poor blood to the lungs and oxygen-enriched blood to all body tissues; with their thick, muscular wall, arter-ies are pressure reservoirs that smooth out pulsations in blood pressure caused by heart contractions.

asexual reproduction Mode of reproduction in which offspring arise from a single parent, and inherit the genes of that parent only.

atmosphere A region of gases, airborne particles, and water vapor enveloping the earth; 80 percent of its mass is distributed within seventeen miles of the earth's surface.

atmospheric cycle A biogeochemical cycle in which a large portion of the element being cycled between the physical environment and ecosystems occurs in a gaseous phase in the atmosphere; examples are the carbon cycle and nitrogen cycle.

atom The smallest unit of matter that is unique to a particular kind of element.

atomic number The number of protons in the nucleus of each atom of an element; differs for each element.

ATP Adenosine triphosphate, an energy carrier composed of adenine, ribose, and three phosphate groups; directly or indirectly transfers energy to or from nearly all metabolic pathways; produced during photosynthesis, aerobic respiration, fermentation, and other pathways.

australopith (OHSS-trah-low-pith) [L. *australis*, southern, + Gk. *pithekos*, ape] Any of the earliest known species of hominids; that is, the first species on the evolutionary branch leading to humans.

autoimmune response Abnormal immune response in which lymphocytes mount an attack against the body's own cells.

autonomic nervous system (auto-NOM-ik) Those nerves leading from the central nervous system to the smooth muscle, cardiac muscle, and glands of internal organs and structures; that is, to the visceral portion of the body.

autosomal dominant inheritance Condition in which a dominant allele on an autosome (not a sex chromosome) is always expressed to some extent.

autosomal recessive inheritance Condition in which a mutation produces a recessive allele on an autosome (not a sex chromosome); only recessive homozygotes show the resulting phenotype.

autosome Any of those chromosomes that are of the same number and kind in both males and females of the species.

autotroph (AH-toe-trofe) [Gk. *autos*, self, + *trophos*, feeder] An organism able to build all the organic molecules it requires using carbon dioxide (present in air and in water) and energy from the physical environment. Photosynthetic autotrophs use sunlight energy; chemosynthetic autotrophs extract energy from chemical reactions involving inorganic substances. Compare *heterotroph*.

auxin (AWK-sin) Any of a class of plant growth-regulating hormones; auxins promote stem elongation as one effect.

axial skeleton In vertebrates, the skull, backbone, ribs, and breastbone (sternum).

axon A long, cylindrical extension of a neuron with finely branched endings. Action potentials move rapidly, without alteration, along an axon; their arrival at the axon endings can trigger the release of transmitter substances that may affect the activity of an adjacent cell.

bacterial flagellum Whiplike motile structure of many bacterial cells; unlike other flagella, it does not contain a core of microtubules.

bacteriophage (bak-TEER-ee-oh-fahj) [Gk. *baktērion*, small staff, rod, + *phagein*, to eat] Category of viruses that infect bacterial cells.

balanced polymorphism Of a population, the maintenance of two or more forms of a trait in fairly stable proportions over the generations.

Barr body In the cells of female mammals, a condensed X chromosome that was inactivated during embryonic development.

basal body A centriole which, after having given rise to the microtubules of a flagellum or cilium, remains attached to the base of either motile structure.

base A substance that, in solution, releases ions that can combine with hydrogen ions.

base pair A pair of hydrogen-bonded nucleotide bases; in the two strands of a DNA double helix, either A–T (adenine with thymine), or G–C (guanine with cytosine). When an mRNA strand forms on a DNA strand during transcription, uracil (U) base-pairs with the DNA's adenine.

behavior, animal A response to external and internal stimuli, following integration of sensory, neural, endocrine, and effector components. Behavior has a genetic basis, hence is subject to natural selection, and it commonly can be modified through experience.

benthic province All of the sediments and rocky formations of the ocean bottom; begins with the continental shelf and extends down through deep-sea trenches.

bilateral symmetry Of animals, a body plan with left and right halves that are basically mirror-images of each other.

biennial Flowering plant that lives through two growing seasons.

binary fission Of bacteria, a mode of asexual reproduction in which the parent cell replicates its single chromosome, then divides into two genetically identical daughter cells.

biogeochemical cycle The movement of carbon, oxygen, hydrogen, or some mineral element necessary for life from the environment to organisms, then back to the environment.

biogeographic realm [Gk. *bios*, life, + *geographein*, to describe the surface of the earth] In one scheme, one of six major land regions, each having distinguishing types of plants and animals.

biological clocks In many organisms, internal time-measuring mechanisms that have roles in adjusting daily and often seasonal activities in response to environmental cues.

biological magnification The increasing concentration of a nondegradable or slowly degradable substance in body tissues as it is passed along food chains.

biological systematics Branch of biology that assesses patterns of diversity based on information from taxonomy, phylogenetic reconstruction, and classification.

biomass The combined weight of all the organisms at a particular trophic (feeding) level in an ecosystem.

biome A broad, vegetational subdivision of some biogeographic realm, shaped by climate, topography, and composition of regional soils.

biosphere [Gk. *bios*, life, + *sphaira*, globe] the entire realm in which organisms exist; the lower regions of the atmosphere, the earth's waters, and the surface rocks, soils, and sediments of the earth's crust; the most inclusive level of biological organization.

biosynthetic pathway A metabolic pathway in which small molecules are assembled into lipids, proteins, and other large organic molecules.

biotic potential Of a population, the maximum rate of increase, per individual, under ideal conditions.

bipedalism A habitual standing and walking on two feet. Humans and ostriches are examples of bipedal animals.

blastocyst (BLASS-tuh-sist) [Gk. *blastos*, sprout, + *kystis*, pouch] In mammalian development, a modified blastula stage consisting of a hollow ball of surface cells and an inner cell mass.

blastula (BLASS-chew-lah) Among animals, an embryonic stage consisting of a ball of cells produced by cleavage.

blood A fluid connective tissue composed of water, solutes, and formed elements (blood cells and platelets).

blood-brain barrier Set of mechanisms that help control which blood-borne substances reach neurons in the brain.

blood pressure Fluid pressure, generated by heart contractions, that keeps blood circulating.

brainstem The vertebrate midbrain, pons, and medulla oblongata, the core of which contains the reticular formation that helps govern activity of the nervous system as a whole.

bronchus, plural **bronchi** (BRONG-CUSS, BRONG-kee) [Gk. *bronchos*, windpipe] Tubelike branchings of the trachea that lead to the lungs.

bud An undeveloped shoot of mostly meristematic tissue; often protected by a covering of modified leaves.

buffer A substance that can combine with hydrogen ions, release them, or both, in response to changes in pH.

bulk flow In response to a pressure gradient, a movement of more than one kind of molecule in the same direction in the same medium (as in blood, sap, or air).

C4 pathway Alternative light-independent reactions of photosynthesis in which carbon dioxide is fixed twice, in two different cell types. Carbon dioxide accumulates in the leaf and helps counter photorespiration (a "wasteful" process that reduces synthesis of sugar phosphates). The first compound formed is the 4-carbon oxaloacetate.

calorie (KAL-uh-ree) [L. *calor*, heat] The amount of heat needed to raise the temperature of 1 gram of water by 1°C. Nutritionists sometimes use "calorie" to mean kilocalorie (1,000 calories), which is a source of confusion.

Calvin-Benson cycle Cyclic, light-independent reactions, the "synthesis" part of photosynthesis. For every six carbon dioxide molecules that become affixed to six RuBP molecules, twelve 3-carbon PGA molecules form; ten are used to regenerate RuBP and the other two to form a sugar phosphate. The cycle runs on ATP and NADPH formed in the light-dependent reactions.

CAM plant A plant that conserves water by opening stomata only at night, when carbon dioxide is fixed by way of the C4 pathway.

cambium, plural **cambia** (KAM-bee-um) In vascular plants, one of two types of meristems that are responsible for secondary growth (increase in stem or root diameter). Vascular cambium gives rise to secondary xylem and phloem; cork cambium gives rise to periderm.

camouflage An outcome of form, patterning, color, or behavior that helps an organism blend with its surroundings and escape detection.

cancer A type of malignant tumor, the cells of which show profound abnormalities in the plasma membrane and cytoplasm, abnormal growth and division, and weakened capacity for adhesion within the parent tissue (leading to metastasis), and, unless eradicated, lethality.

capillary [L. *capillus*, hair] A thin-walled blood vessel; component of capillary beds, the diffusion zones for exchanges of gases and materials between blood and interstitial fluid.

carbohydrate [L. *carbo*, charcoal, + *hydro*, water] A simple sugar or a polymer composed of sugar units, and used universally by cells for energy and as structural materials. A monosaccharide, oligosaccharide, or polysaccharide.

carbon cycle Biogeochemical cycle in which carbon moves from reservoirs in the land, atmosphere, and oceans, through organisms, then back to the reservoirs.

carbon dioxide fixation Initial step of the light-independent reactions of photosynthesis. Carbon dioxide becomes affixed to a specific carbon compound (such as RuBP) that undergoes rearrangements leading to regeneration of that carbon compound *and* to a sugar phosphate.

carcinogen (kar-SIN-uh-jen) Ultraviolet radiation and many other agents that can trigger cancer.

cardiac cycle [Gk. *kardia*, heart, + *kyklos*, circle] The sequence of muscle contractions and relaxation constituting one heartbeat.

cardiovascular system Of animals, an organ system composed of blood, one or more hearts, and blood vessels.

carnivore [L. *caro*, *carnis*, flesh, + *vovare*, to devour] An animal that eats other animals; a type of heterotroph.

carotenoid (kare-OTT-en-oyd) A light-sensitive pigment that absorbs violet and blue wavelengths but reflects yellow, orange, and red.

carpel (KAR-pul) One or more closed vessels that serve as the female reproductive parts of a flower. The chamber within a carpel is the ovary where eggs develop and are fertilized, and seeds mature.

carrier protein Type of transport protein that binds specific substances and changes shape in ways that shunt the substances across a plasma membrane. Some carrier proteins function passively, others require an energy input.

carrying capacity The number of individuals of a given species that can be sustained indefinitely by the available resources in a given environment; births are balanced by deaths in a population at its carrying capacity.

Casparian strip In plant roots, a waxy band that acts as an impermeable barrier between the walls of abutting cells of the endodermis or exodermis.

cDNA Any DNA molecule copied from a mature mRNA transcript by way of reverse transcription.

cell [L. *cella*, small room] The basic *living* unit. A cell has the capacity to maintain itself as an independent unit and to reproduce, given appropriate conditions and resources.

cell cycle A sequence of events by which a cell increases in mass, roughly doubles its number of cytoplasmic components, duplicates its DNA, then undergoes nuclear and cytoplasmic division. The cycle extends from the time the cell forms until its own daughter cells form.

cell junction Of multicelled organisms, a point of contact that physically links two cells or that provides functional links between their cytoplasm.

cell plate Of plant cells undergoing mitotic cell division, a partition that forms at the spindle equator, between the two newly forming daughter cells.

cell theory A theory in biology, the key points of which are that (1) all organisms are composed of one or more cells, (2) the cell is the smallest unit that still retains a capacity for independent life, and (3) all cells arise from preexisting cells.

cell wall A rigid or semirigid supportive wall outside the plasma membrane; a cellular feature of plants, fungi, protistans, and most bacteria.

central nervous system Of vertebrates, the brain and spinal cord.

central vacuole Of living plant cells, a fluid-filled organelle that stores nutrients, ions, and wastes; its enlargement during cell growth has the effect of improving the cell's surface-to-volume ratio.

centriole (SEN-tree-ohl) A small cylinder of triplet microtubules near the nucleus in most animal cells. Centrioles occur in pairs; some give rise to the microtubular core of flagella and cilia, and some may govern the plane of cell division.

centromere (SEN-troh-meer) [Gk. *kentron*, center, + *meros*, a part] A small, constricted region of a chromosome having attachment sites for microtubules that help move the chromosome during nuclear division.

cephalization (sef-ah-lah-ZAY-shun) [Gk. *kephalikos*, head] Of an animal body, having sensory structures and nerve cells concentrated in the head.

cerebellum (ser-ah-BELL-um) [L. diminutive of *cerebrum*, brain] Hindbrain region that coordinates motor activity for refined limb movements, appropriate posture, and spatial orientation.

cerebral cortex Thin surface layer of the cerebral hemispheres. Some regions of the cortex receive sensory input, others integrate information and coordinate appropriate motor responses.

cerebrospinal fluid Clear extracellular fluid that surrounds and cushions the brain and spinal cord. The bloodstream exchanges substances with the fluid, which exchanges substances with neurons.

cerebrum (suh-REE-bruhm) Part of the vertebrate forebrain governing responses to olfactory input and, in mammals, the most complex information-encoding and information-processing center. Divided into two cerebral hemispheres; in humans, the left hemisphere deals generally with spoken language skills and the right, with abstract, nonverbal skills.

channel protein Type of transport protein that serves as a pore through which ions or other water-soluble substances move across the plasma membrane. Some channels remain open; others are gated, and open and close in controlled ways.

chemical bond A union between the electron structures of two or more atoms or ions.

chemical synapse (SIN-aps) [Gk. *synapsis*, union] A junction where a small gap, the synaptic cleft, separates two neurons (or a neuron and a muscle cell or gland cell). The presynaptic neuron releases a transmitter substance into the cleft, and this may have an excitatory or inhibitory effect on the postsynaptic cell.

chemiosmotic theory (kim-ee-OZ-MOT-ik) Theory that an electrochemical gradient across a cell membrane drives ATP synthesis. Metabolic reactions cause hydrogen ions (H^+) to accumulate in some type of membrane-bound compartment. The combined force of the resulting concentration and electric gradients propels hydrogen ions down the gradient, through channel proteins. Enzyme action at those proteins links ADP with inorganic phosphate to form ATP.

chemoreceptor (KEE-moe-ree-sep-tur) Sensory receptor that detects chemical energy (ions or molecules) dissolved in body fluids next to the cell.

chemosynthetic autotroph (KEE-moe-sin-THET-ik) One of a few kinds of bacteria able to synthesize all the organic molecules it requires using carbon dioxide as the carbon source and certain inorganic substances (such as sulfur) as the energy source.

chlorophyll (KLOR-uh-fill) [Gk. *chloros*, green, + *phyllon*, leaf] Photosynthetic pigment molecule that absorbs light of blue and red wavelengths and transmits green. Special chlorophyll pigments of photosystems give up electrons used in photosynthesis.

chloroplast (KLOR-uh-plast) Of plants and certain protistans, a membrane-bound organelle that specializes in photosynthesis.

chordate An animal having a notochord, a dorsal hollow nerve cord, a pharynx, and gill

slits in the pharynx wall for at least part of its life cycle.

chorion (CORE-ee-on) Of land vertebrates, a protective membrane surrounding an embryo and other extraembryonic membranes; in placental mammals it becomes part of the placenta.

chromatid The name applied to each of the two parts of a duplicated eukaryotic chromosome for as long as the two parts remain attached at the centromere. Each chromatid consists of a DNA double helix and associated proteins, and it has the same gene sequence as its "sister" chromatid.

chromosome (CROW-moe-some) [Gk. *chroma*, color, + *soma*, body] In eukaryotes, a DNA molecule and many associated proteins. A chromosome that has undergone duplication prior to nuclear division consists of two DNA molecules and associated proteins; the two are called *sister chromatids*. A bacterial chromosome does not have a comparable profusion of proteins associated with the DNA.

cilium (SILL-ee-um), plural **cilia** [L. *cilium*, eyelid] Short, hairlike process extending from the plasma membrane and containing a regular array of microtubules. Cilia are typically more profuse than flagella. Cilia serve as motile structures, help create currents of fluids, or are part of sensory structures.

circadian rhythm (ser-KAYD-ee-un) [L. *circa*, about, + *dies*, day] Of many organisms, a cycle of physiological events that is completed every 24 hours or so, even when environmental conditions remain constant.

circulatory system Of multicelled animals, an organ system consisting of a muscular pump (heart, most often), blood vessels, and blood; the system transports materials to and from cells and often helps stabilize body temperature and pH.

cladistics An approach to biological systematics in which organisms are grouped according to similarities that are derived from a common ancestor.

cladogram Branching diagram that represents patterns of relative relationships between organisms based on discrete morphological, physiological, and behavioral traits that vary among taxa being studied.

cleavage Stage of animal development when mitotic cell divisions convert a zygote to a ball of cells, the *blastula*. Different cytoplasmic components end up in different daughter cells, and this *cytoplasmic localization* helps seal the developmental fate of their descendants.

cleavage furrow Of animal cells undergoing cytokinesis, a depression that forms at the cell surface as contractile microfilaments pull the plasma membrane inward; defines where the cell will be cut in two.

climate Prevailing weather conditions for an ecosystem, including temperature, humidity, wind speed, cloud cover, and rainfall.

climax community Of an ecosystem, a more or less stable array of species that results from the process of succession.

clonal selection theory Theory of immune system function stating that lymphocytes activated by a specific antigen rapidly multiply, giving rise to descendants (clones) that

all retain the parent cell's specificity against that antigen.

cloned DNA Multiple, identical copies of DNA fragments contained within plasmids or some other cloning vector.

codominance Condition in which two alleles of a pair are not identical yet the expression of both can be discerned in heterozygotes. Each gives rise to a different phenotype.

codon One of a series of base triplets in an mRNA molecule that code for a series of amino acids that will be strung together during protein synthesis. Different codons specify different amino acids; a few serve as a stop signal and one type as a start signal.

coelom (SEE-lum) [Gk. *koilos*, hollow] Of many animals, a type of body cavity that occurs between the gut and body wall and that has a lining, the *peritoneum*.

coenzyme An organic molecule that serves as a carrier of electrons or atoms in metabolic reactions and that is necessary for proper functioning of many enzymes. NAD^+ is an example.

coevolution The joint evolution of two or more closely interacting species; when one evolves, the change affects selection pressures operating between the two species so the other also evolves.

cofactor A metal ion or coenzyme that either helps catalyze a reaction or serves briefly as an agent that transfers electrons, atoms, or functional groups from one substrate to another.

cohesion Condition in which molecular bonds resist rupturing when under tension.

cohesion theory of water transport Theory that water moves up through vascular plants due to hydrogen bonding among water molecules confined inside the xylem pipelines. The collective cohesive strength of those bonds allows water to be pulled up as columns in response to transpiration (evaporation from leaves).

collenchyma Of vascular plants, a ground tissue that helps strengthen the plant body.

colon (CO-lun) The large intestine.

commensalism [L. *com*, together, + *mensa*, table] Two-species interaction in which one species benefits significantly while the other is neither helped nor harmed to any notable extent.

communication signal Of social animals, an evolved action or cue that transfers information, to the benefit of both the member of the species sending the signal *and* the member receiving it.

community The populations of all species occupying a habitat; also applies to groups of organisms with similar life-styles in a habitat (such as the bird community).

comparative morphology [Gk. *morph*, form] Detailed study of differences and similarities in body form and structural patterns among major taxa.

competition, exploitation Interaction in which both species have equal access to a required resource, but differ in how fast or how efficiently they exploit it.

competition, interference Interaction in which one species may limit another species' access to some resource regardless of whether the resource is abundant or scarce.

competition, interspecific Two-species interaction in which both species can be harmed due to overlapping niches.

competition, intraspecific Interaction among individuals of the same species that are competing for the same resources.

competitive exclusion The theory that populations of two species competing for a limited resource cannot coexist indefinitely in the same habitat; the population better adapted to exploit the resource will enjoy a competitive (hence reproductive) edge and will eventually exclude the other species from the habitat.

complement system A group of about twenty proteins circulating in blood plasma that are activated during both general responses and immune responses to a foreign agent in the body; part of the *inflammatory response*.

concentration gradient A difference in the number of molecules or ions of a substance between one region and another, as in a given volume of fluid. In the absence of other forces, the molecules tend to move down their gradient.

condensation reaction Enzyme-mediated reaction leading to the covalent linkage of small molecules and, often, the formation of water.

cone cell In the vertebrate eye, a type of photoreceptor that responds to intense light and contributes to sharp daytime vision and color perception.

conjugation [L. *conjugatio*, a joining] Of some bacterial species, the transfer of DNA between two different mating strains that have made cell-to-cell contact.

consumer [L. *consumere*, to take completely] A heterotrophic organism that obtains energy and raw materials by feeding on the tissues of other organisms. Herbivores, carnivores, omnivores, and parasites are examples.

continuous variation For many traits, small degrees of phenotypic variation that occur over a more or less continuous range.

contractile vacuole (kun-TRAK-till VAK-you-ohl) [L. *contractus*, to draw together] In some protistans, a membranous chamber that takes up excess water in the cell body, then contracts, expelling the water through a pore to the outside.

control group In a scientific experiment, a group used to evaluate possible side effects of the manipulation of the experimental group. Ideally, the experimental group differs from the control group only with respect to the key factor, or variable, being studied.

convergence, morphological Resemblance of body parts between dissimilar and only distantly related species, an evolutionary outcome of their ancestors having adopted a similar way of life and having used those body parts for similar functions.

cork cambium Of woody plants, a type of lateral meristem that produces a tough, corky replacement for the epidermis on older plant parts.

corpus callosum (CORE-pus ka-LOW-sum) In the human brain, a band of 200 million axons that functionally links the two cerebral hemispheres.

corpus luteum (CORE-pus LOO-tee-um) A glandular structure; it develops from cells of a ruptured ovarian follicle and secretes progesterone and some estrogen, both of which maintain the lining of the uterus (endometrium).

cortex [L. *cortex*, bark] In general, a rindlike layer; the kidney cortex is an example. In vascular plants, ground tissue that makes up most of the primary plant body, supports plant parts, and stores food.

cotyledon A so-called seed leaf that develops as part of a plant embryo; cotyledons provide nourishment for the germinating seedling.

covalent bond (koe-VAY-lunt) [L. *con*, together, + *valere*, to be strong] A sharing of one or more electrons between atoms or groups of atoms. When electrons are shared equally, the bond is *nonpolar*. When electrons are shared unequally, the bond is *polar*—slightly positive at one end and slightly negative at the other.

crossing over During prophase I of meiosis, an event in which nonsister chromatids of a pair of homologous chromosomes break at one or more sites along their length and exchange corresponding segments at the breakage points. As a result, new combinations of alleles replace old ones in a chromosome.

culture The sum total of behavior patterns of a social group, passed between generations by learning and by symbolic behavior, especially language.

cuticle (KEW-tih-kull) A body covering. In plants, a cuticle consisting of waxes and lipid-rich cutin is deposited on the outer surface of epidermal cell walls. Annelids have a thin, flexible cuticle. Arthropods have a thick, protein- and chitin-containing cuticle that is flexible, lightweight, and protective.

cyclic AMP, cyclic adenosine monophosphate (SIK-lik ah-DEN-uh-seen mon-oh-FOSS-fate) A nucleotide present in cytoplasm that serves as a mediator of the cell's response to hormonal signals; a type of second messenger.

cyclic photophosphorylation (SIK-lik foe-toe-FOSS-for-ih-LAY-shun) Photosynthetic pathway in which electrons excited by sunlight energy move from a photosystem to a transport chain, then back to the photosystem. Operation of the transport chain helps produce electrochemical gradients that lead to ATP formation. (Compare *chemiosmotic theory*.)

cytochrome (SIGH-toe-krome) [Gk. *kytos*, hollow vessel, + *chrōma*, color] Iron-containing protein molecule present in the electron transport systems used in photosynthesis and aerobic respiration.

cytokinesis (SIGH-toe-kih-NEE-sis) [Gk. *kinesis*, motion] The actual splitting of a parental cell into two daughter cells; also called cytoplasmic division.

cytokinin (SIGH-tow-KY-nun) Any of the class of plant hormones that stimulate cell division, promote leaf expansion, and retard leaf aging.

cytomembrane system [Gk. *kytos*, hollow vessel] The membranous system in the cytoplasm in which proteins and lipids take on their final form and are distributed. Components of the system include the endoplasmic reticulum, Golgi bodies, lysosomes, and a variety of vesicles.

cytoplasm (SIGH-toe-plaz-um) [Gk. *plassein*, to mold] All cellular parts, particles, and semifluid substances enclosed by the plasma membrane *except* the nucleus (in eukaryotes) or the nucleoid (in prokaryotes).

cytosine (SIGH-toe-seen) A pyrimidine; one of the nitrogen-containing bases in nucleotides.

cytoskeleton Of eukaryotic cells, an internal "skeleton" that structurally supports the cell, organizes its components, and often moves components about. Microtubules, microfilaments, and intermediate filaments are the most common cytoskeletal elements.

decomposer [L. *de-*, down, away, + *companere*, to put together] Generally, any of the heterotrophic bacteria or fungi that obtain energy by chemically breaking down the remains, products, or wastes of other organisms. Their activities help cycle nutrients back to producers.

degradative pathway A metabolic pathway by which molecules are broken down in stepwise reactions that lead to products of lower energy.

deletion Loss of a chromosome segment, nearly always resulting in a genetic disorder.

demographic transition model Model of human population growth in which changes in the growth pattern correspond to different stages of economic development. These are a preindustrial stage, when birth and death rates are both high, a transitional stage, an industrial stage, and a postindustrial stage, when the death rate exceeds the birth rate.

denaturation (deh-NAY-chur-AY-shun) Disruption of bonds holding a protein in its three-dimensional form, such that its polypeptide chain(s) unfolds partially or completely.

dendrite (DEN-drite) [Gk. *dendron*, tree] A short, slender extension from the cell body of a neuron.

denitrification (DEE-nite-rih-fih-KAY-shun) The conversion of nitrate or nitrite, by certain bacteria, to gaseous nitrogen (N_2) and a small amount of nitrous oxide (N_2O).

density-dependent controls Factors such as predation, parasitism, disease, and competition for resources, which limit population growth by reducing the birth rate, increasing the rates of death and dispersal, or all of these.

density-independent controls Factors such as storms or floods that increase a population's death rate more or less independently of its density.

dentition (den-TIH-shun) The type, size, and number of an animal's teeth.

dermis The layer of skin underlying the epidermis, consisting mostly of dense connective tissue.

desertification (dez-urt-ih-fih-KAY-shun) Conversion of grasslands, rain-fed cropland, or irrigated cropland to desertlike conditions, with a drop of agricultural productivity of 10 percent or more.

detrital food web A network of interlinked food chains in which energy flows from plants through decomposers and detritivores.

detritivore (dih-TRY-tih-vore) [L. *detritus*; after *deterere*, to wear down] An earthworm, crab, nematode, or other heterotroph that obtains energy by feeding on partly decomposed particles of organic matter.

deuterostome (DUE-ter-oh-stome) [Gk. *deuteros*, second, + *stoma*, mouth] Any of the bilateral animals, including echinoderms and chordates, in which the first indentation in the early embryo develops into the anus.

diaphragm (DIE-uh-fram) [Gk. *diaphragma*, to partition] Muscular partition between the thoracic and abdominal cavities, the contraction and relaxation of which contribute to breathing. Also, a contraceptive device used temporarily to prevent sperm from entering the uterus during sexual intercourse.

dicot (DIE-kot) [Gk. *di*, two, + *kotylēdōn*, cup-shaped vessel] Short for dicotyledon; class of flowering plants characterized generally by seeds having embryos with two cotyledons (seed leaves); net-veined leaves; and floral parts arranged in fours, fives, or multiples of these.

differentiation Of the cells of multicelled organisms, differences in composition, structure, and function that arise through selective gene expression. All the cells inherit the same genes but become specialized by activating or suppressing some fraction of those genes in different ways.

diffusion Tendency of molecules or ions of the same substance to move from a region of greater concentration to a region where they are less concentrated. See *concentration gradient*.

digestive system Of most animals, an internal tube or cavity in which ingested food is reduced to molecules small enough to be absorbed into the internal environment; often divided into regions specialized for food transport, processing, and storage.

dihybrid cross An experimental cross between two organisms, each of which breeds true (is homozygous) for forms of *two* traits that are distinctly different from those displayed by the other organism. For each trait, the first-generation offspring inherit a pair of nonidentical alleles.

diploid (DIP-loyd) Of sexually reproducing species, having two chromosomes of each type (that is, homologous chromosomes) in somatic cells. Except for sex chromosomes, the two homologues of a pair resemble each other in length, shape, and which genes they carry. Compare *haploid*.

directional selection A shift in allele frequencies in a population in a steady, consistent direction in response to a new environment (or a directional change in the old one), so that forms of traits at one end of a range of phenotypic variation become more common than the intermediate forms.

disaccharide (die-SAK-uh-ride) [Gk. *di*, two, + *sakcharon*, sugar] A type of simple carbohydrate, of the class called oligosaccharides; two monosaccharides covalently bonded.

disruptive selection Selection that favors forms of traits at both ends of a range of phenotypic variation in a population, and operates against intermediate forms.

distal tubule The tubular section of a nephron most distant from the glomerulus; a major site of water and sodium reabsorption.

divergence Accumulation of differences in allele frequencies between reproductively isolated populations of a species.

divergence, morphological Among similar and evolutionarily related species, decreased resemblance in one or more aspects of body patterning or function, usually corresponding to divergences in life-styles.

diversity, organismic Sum total of variations in form, function, and behavior that have accumulated in different lineages. Those variations generally are adaptive to prevailing conditions or were adaptive to conditions that existed in the past.

DNA Deoxyribonucleic acid (dee-OX-ee-RYE-bow-new-CLAY-ik) Usually, two strands of nucleotides twisted together in the shape of a double helix. The nucleotides differ only in their nitrogen-containing bases (adenine, thymine, guanine, cytosine), but *which* ones follow others in a DNA strand represents instructions for assembling proteins, and, ultimately, new organisms.

DNA library A collection of DNA fragments produced by restriction enzymes and incorporated into plasmids.

DNA ligase (LYE-gaze) Enzyme that links together short stretches of nucleotides on a parent DNA strand during replication; also used by recombinant DNA technologists to join base-paired DNA fragments to cut plasmid DNA.

DNA polymerase (poe-LIM-uh-raze) Enzyme that assembles a new strand on a parent DNA strand during replication; also "proofreads" for mismatched base pairs, which are replaced with correct bases.

dominance hierarchy Form of social organization in which some members of the group are subordinate to other members, which in turn are dominated by others.

dominant allele In a diploid cell, an allele whose expression masks the expression of its partner on the homologous chromosome.

dormancy [L. *dormire*, to sleep] Of plants, the temporary, hormone-mediated cessation of growth under conditions that might appear to be quite suitable for growth.

double fertilization Of flowering plants only, the fusion of one sperm nucleus with the egg nucleus (to produce a zygote), *and* fusion of a second sperm nucleus with the nuclei of the endosperm mother cell, which gives rise to nutritive tissue.

duplication Type of chromosome rearrangement in which a gene sequence occurs in excess of its normal amount in a chromosome.

ecology [Gk. *oikos*, home, + *logos*, reason] Study of the interactions of organisms with one another and with their physical and chemical environment.

ecosystem [Gk. *oikos*, home] A whole complex of organisms and their environment, all of which interact through a one-way flow of energy and a cycling of materials.

ecosystem modeling Method of identifying pieces of information about different components of an ecosystem, then combining that information with computer programs and models in order to predict the outcome of a disturbance to the system.

ectoderm [Gk. *ecto*, outside, + *derma*, skin] Of animal embryos, the outermost primary tissue layer, or *germ layer*, that gives rise to the outer portion of the integument and to tissues of the nervous system.

effector A muscle (or gland) that responds to nerve signals by producing movement (or chemical change) that helps adjust the body to changing conditions.

egg A female gamete; of complex animals, a mature ovum.

electron Negatively charged particle occupying one of the orbitals around the nucleus of an atom.

electron transport phosphorylation (FOSS-for-ih-LAY-shun) Final stage of aerobic respiration, in which ATP forms after hydrogen ions and electrons (from the Krebs cycle) are sent through a transport system that gives up the electrons to oxygen. (Compare *chemiosmotic theory* and *electron transport system*).

electron transport system An organized array of membrane-bound enzymes and cofactors that accept and donate electrons in sequence. Operation of such systems leads to the flow of hydrogen ions (H^+) across a cell membrane, and this flow results in ATP formation and other reactions.

element Any substance that cannot be decomposed into substances with different properties.

embryo (EM-bree-oh) [Gk. *en*, in, + probably *bryein*, to swell] Of animals generally, the early stages of development (cleavage, gastrulation, organogenesis, and morphogenesis). In most plants, a young sporophyte, from the first cell divisions after fertilization until germination.

embryo sac In flowering plants, the female gametophyte.

endergonic reaction (en-dur-GONE-ik) Chemical reaction showing a net gain in energy.

endocrine gland Ductless gland that secretes hormones into interstitial fluid, after which they are distributed by way of the bloodstream.

endocrine system System of cells, tissues, and organs that is functionally linked to the nervous system and that helps control body functions with its hormones and other chemical secretions.

endocytosis (EN-doe-sigh-TOE-sis) A process by which part of the plasma membrane encloses substances (or cells, in the case of phagocytes) at or near the cell surface, then pinches off to form a vesicle that transports the substance into the cytoplasm.

endoderm [Gk. *endon*, within, + *derma*, skin] Of animal embryos, the inner primary tissue layer, or *germ layer*, that gives rise to the inner lining of the gut and organs derived from it.

endodermis In roots, a sheetlike wrapping of single cells around the vascular cylinder; functions in controlling the uptake of water and dissolved nutrients. An impermeable barrier (*Casparian strip*) prevents water from passing between the walls of abutting endodermal cells.

endometrium (EN-doh-MEET-ree-um) [Gk. *metrios*, of the womb] Inner lining of the uterus, consisting of connective tissues, glands, and blood vessels.

endoplasmic reticulum or **ER** (EN-doe-PLAZ-mik reh-TIK-yoo-lum) System of membranous channels, tubes, and sacs in the cytoplasm in which many newly formed proteins become modified and the protein and lipid components of most organelles are manufactured. Rough ER has ribosomes on the surface facing the cytoplasm; smooth ER does not.

endoskeleton [Gk. *endon*, within, + *skleros*, hard, stiff] In chordates, the internal framework of bone, cartilage, or both. Together with skeletal muscle, supports and protects other body parts, helps maintain posture, and moves the body.

endosperm (EN-doe-sperm) Nutritive tissue that surrounds and serves as food for a flowering plant embryo and, later, for the germinating seedling.

endospore Of certain bacteria, a resistant body that forms around the DNA and some cytoplasm under unfavorable conditions; it germinates and gives rise to new bacterial cells when conditions become favorable.

energy The capacity to make things happen, to do work.

energy pyramid A pyramid-shaped representation of the trophic structure of an ecosystem, based on the decreasing energy flow at each upward transfer to a different trophic level.

entropy (EN-trow-pee) A measure of the degree of disorder in a system—that is, how much energy in the system has become so dispersed (usually as low-quality heat) that it is no longer available to do work.

enzyme (EN-zime) One of a special class of proteins that greatly speed up (catalyze) reactions involving specific substrates.

epidermis The outermost tissue layer of a multicelled plant or animal.

epistasis (eh-PISS-tih-sis) An absence of an expected phenotype owing to a masking of the expression of one gene pair by another gene pair.

epithelium (EP-ih-THEE-lee-um) Of multicelled animals, one or more layers of adhering cells having one free surface; the opposite surface rests on a basement membrane that intervenes between it and an underlying connective tissue. Epithelia cover external body surfaces and line internal cavities and tubes.

equilibrium, dynamic [Gk. *aequus*, equal, + *libra*, balance] The point at which a chemical reaction runs forward as fast as it runs in reverse, so that there is no net change in the concentrations of products or reactants.

erythrocyte (eh-RITH-row-site) [Gk. *erythros*, red, + *kytos*, vessel] Red blood cell.

esophagus (ee-SOF-uh-gus) Tubular portion of a digestive system that receives swallowed food and leads to the stomach.

essential amino acid Any of eight amino acids that human cells cannot synthesize and must obtain from food.

essential fatty acid Any of the fatty acids that the human body cannot synthesize and must obtain from food.

estrogen (ESS-trow-jun) A sex hormone required in egg formation, preparing the uterine lining for pregnancy, and maintaining secondary sexual traits; also influences growth and development.

estrus (ESS-truss) [Gk. *oistrus*, frenzy] For mammals generally, the cyclic period of a female's sexual receptivity to the male.

estuary (EST-you-ary) A partly enclosed coastal region where seawater mixes with freshwater from rivers or streams and runoff from the land.

ethylene (ETH-il-een) Plant hormone that stimulates fruit ripening and triggers abscission.

eukaryotic cell (yoo-CARRY-oht-ik) [Gk. *eu*, good, + *karyon*, kernel] A cell that has a "true nucleus" and many other membrane-bound organelles; any cell except bacteria.

evaporation [L. *e-*, out, + *vapor*, steam] Changes by which a substance is converted from a liquid state into vapor.

evolution [L. *evolutio*, act of unrolling] Change within a line of descent over time; entails successive changes in allele frequencies in a population as brought about by mutation, natural selection, genetic drift, and gene flow.

excitatory postsynaptic potential or **EPSP** One of two competing signals at an input zone of a neuron; a graded potential that brings the neuron's plasma membrane closer to threshold.

excretion Any of several processes by which excess water, excess or harmful solutes, or waste materials are passed out of the body. Compare *secretion*.

exergonic reaction (EX-ur-GONE-ik) A chemical reaction that shows a net loss in energy.

exocrine gland (EK-suh-krin) [Gk. *es*, out of, + *krinein*, to separate] Glandular structure that secretes products, usually through ducts or tubes, to a free epithelial surface.

exocytosis (EK-so-sigh-TOE-sis) A process by which substances are moved out of cells. A vesicle forms inside the cytoplasm, moves to the plasma membrane, and fuses with it, so that the vesicle's contents are released outside.

exodermis Layer of cells just inside the root epidermis in most flowering plants; functions in controlling the uptake of water and dissolved nutrients.

exon Any of the portions of a newly formed mRNA molecule that are spliced together to form the mature mRNA transcript and that are ultimately translated into protein.

exoskeleton [Gk. *exo*, out, + *skleros*, hard, stiff] An external skeleton, as in arthropods.

exponential growth (EX-po-NEN-shul) Pattern of population growth that occurs when *r* (the net reproduction per individual) holds constant; then, the number of individuals increases in doubling increments (from 2 to 4, then 8, 16, 32, 64, 128, and so on).

extinction, background A steady rate of species turnover that characterizes lineages through most of their histories.

extinction, mass An abrupt increase in the rate at which major taxa disappear, with several taxa being wiped out simultaneously.

extracellular fluid In animals generally, all the fluid not inside cells; includes blood plasma and interstitial fluid, which occupies the spaces between cells and tissues.

extracellular matrix Of animals, a meshwork of fibrous proteins and other components in a ground substance that helps hold many tissues together in certain shapes and that influences cell metabolism by virtue of its composition.

facilitated diffusion The passive transport of specific solutes through the inside of a channel protein or carrier protein that spans the lipid bilayer of a cell membrane; the solutes simply move in the direction that diffusion would take them.

family pedigree A chart of the genetic relationships of the individuals in a family through a number of generations.

fat A lipid with one, two, or three fatty acid tails attached to a glycerol backbone.

fatty acid A compound having a long, unbranched carbon backbone (a hydrocarbon) with a —COOH group at the end.

feedback inhibition A mechanism of enzyme control in which the output of the reaction (such as a particular molecule) works in a way that inhibits further output.

fermentation [L. *fermentum*, yeast] Type of anaerobic pathway of ATP formation that begins with glycolysis and ends with electrons being transferred back to one of the breakdown products or intermediates. Glycolysis yields two ATP; the rest of the pathway serves to regenerate NAD$^+$.

fertilization [L. *fertilis*, to carry, to bear] Fusion of sperm nucleus with egg nucleus. See also *double fertilization*.

fibrous root system Adventitious roots and their branchings.

filtration In urine formation, the process by which blood pressure forces water and solutes out of glomerular capillaries and into the cupped portion of a nephron wall (Bowman's capsule).

first law of thermodynamics [Gk. *therme*, heat, + *dynamikos*, powerful] Law stating that the total amount of energy in the universe remains constant. Energy cannot be created or destroyed, but can only be converted from one form to another.

flagellum (fluh-JELL-um) plural **flagella**, [L. whip] Motile structure of many free-living eukaryotic cells; has a 9 + 2 microtubule array.

flower The often showy reproductive structure that distinguishes angiosperms from other seed plants.

fluid mosaic model Model of membrane structure in which proteins are embedded in a lipid bilayer or attached to one of its surfaces. Lipids impart structure to the membrane as well as impermeability to water-soluble molecules. Packing variations and movements of lipids impart fluidity to the membrane. Proteins carry out most membrane functions, such as transport, enzyme action, and reception of chemical signals.

follicle (FOLL-ih-kul) In a mammalian ovary, a primary oocyte (immature egg) together with the surrounding layer of cells.

food chain A linear sequence of who eats whom in an ecosystem.

food web A network of crossing, interlinked food chains, encompassing primary producers and an array of consumers, detritivores, and decomposers.

forebrain Brain region that includes the cerebrum and cerebral cortex, the olfactory lobes, and the hypothalamus.

fossil Recognizable evidence of an organism that lived in the distant past. Most fossils are skeletons, shells, leaves, seeds, and tracks that were buried in rock layers before they could be decomposed.

fossil fuel Coal, petroleum, or natural gas; formed in sediments by the compression of carbon-containing plant remains over hundreds of millions of years.

founder effect An extreme case of genetic drift in which a few individuals leave a population and establish a new one. Simply by chance, allele frequencies for many traits may differ from those in the original population.

fruit [L. after *frui*, to enjoy] In flowering plants, the ripened ovary of one or more carpels, sometimes with accessory structures incorporated.

functional group An atom or group of atoms covalently bonded to the carbon backbone of an organic compound, contributing to its structure and properties.

Fungi The kingdom of fungi.

fungus A heterotroph that secretes enzymes able to break down an external food source into molecules small enough to be absorbed by cells (extracellular digestion and absorption). Saprobic types feed on nonliving organic matter; parasitic types feed on living organisms. Fungi as a group are major decomposers.

gall bladder Organ of the digestive system that stores bile secreted from the liver.

gamete (GAM-eet) [Gk. *gametēs*, husband, and *gametē*, wife] Haploid cell (sperm or egg) that functions in sexual reproduction.

gametophyte (gam-EET-oh-fite) [Gk. *phyton*, plant] The haploid, multicelled, gamete-producing phase in the life cycle of most plants.

ganglion (GANG-lee-un), plural **ganglia** [Gk. *ganglion*, a swelling] A clustering of cell bodies of neurons into a distinct structure in regions other than the brain or spinal cord.

gastrulation (gas-tru-LAY-shun) Stage of embryonic development in which cells become arranged into two or three primary tissue layers (germ layers); in humans, the layers are an inner endoderm, an intermediate mesoderm, and a surface ectoderm.

gene [short for German *pangan*, after Gk. *pan*, all + *genes*, to be born] Any of the units of instruction for heritable traits. Each gene is a linear sequence of nucleotides that calls for the assembly of a sequence of specific amino acids into a polypeptide chain.

gene flow Microevolutionary process whereby allele frequencies in a population change due to immigration, emigration, or both.

gene frequency More precisely, allele frequency: the relative abundances of different alleles carried by the individuals of a population.

gene locus Particular location on a chromosome for a given gene.

gene mutation [L. *mutatus*, a change] Change in DNA due to the deletion, addition, or substitution of one to several bases in the nucleotide sequence.

gene pair In diploid cells, the two alleles at a given gene locus on a pair of homologous chromosomes.

gene pool Sum total of all genotypes in a population. More accurately, allele pool.

gene therapy Inserting one or more normal genes into existing cells of an organism as a way to correct some genetic defect.

genetic code [After L. *genesis*, to be born] The correspondence between nucleotide triplets in DNA (then in mRNA) and a specific sequence of amino acids in the resulting polypeptide chains; the basic language of protein synthesis.

genetic drift Microevolutionary process whereby allele frequencies in a population change randomly over time, as a result of chance events.

genetic engineering Altering the information content of DNA through use of recombinant DNA technology.

genetic equilibrium Hypothetical state in a population in which allele frequencies for a trait remain stable through the generations; a reference point for measuring rates of evolutionary change.

genetic recombination Presence of a new combination of alleles in a DNA molecule compared to the parental genotype; the result of processes such as crossing over at meiosis, chromosome rearrangements, gene mutation, and recombinant DNA technology.

genome All the DNA in a haploid number of chromosomes of a species.

genotype (JEEN-oh-type) Genetic constitution of an individual. Can mean a single gene pair or the sum total of the individual's genes. Compare *phenotype*.

genus, plural **genera** (JEEN-us, JEN-er-ah) [L. *genus*, race, origin] A taxon into which all species exhibiting certain phenotypic similarities and evolutionary relationship are grouped.

germ cell Animal cell that may give rise to gametes. Compare *somatic cell*.

germ layer Of animal embryos, one of two or three primary tissue layers that form during gastrulation and that gives rise to certain tissues of the adult body. Compare *ectoderm*, *mesoderm*, and *endoderm*.

germination (jur-min-AY-shun) Of plants, the time at which an embryo sporophyte breaks through its seed coat and resumes growth.

gibberellin (JIB-er-ELL-un) Any of a class of plant hormones that promote stem elongation.

gill A respiratory organ, typically with a moist, thin, vascularized layer of epidermis that functions in gas exchange.

glomerulus (glow-MARE-you-luss) [L. *glomus*, ball] Region where a nephron wall balloons around a cluster of capillaries and where water and solutes are filtered from blood.

glucagon (GLUE-kuh-gone) Type of hormone, secreted by alpha cells of the pancreas, that stimulates conversion of glycogen and amino acids to glucose.

glyceride (GLISS-er-eyed) A molecule having one, two, or three fatty acid tails attached to a backbone of glycerol. Glycerides—fats and oils—are the body's most abundant lipids and its richest source of energy.

glycerol (GLISS-er-ol) [Gk. *glykys*, sweet, + L. *oleum*, oil] A three-carbon molecule with three hydroxyl groups attached; combines with fatty acids to form fat or oil.

glycogen (GLY-kuh-jen) In animals, a storage polysaccharide that is a main food reserve; can be readily broken down into glucose subunits.

glycolysis (gly-CALL-ih-sis) [Gk. *glykys*, sweet, + *lysis*, loosening or breaking apart] Initial stage of both aerobic and anaerobic pathways by which glucose (or some other organic compound) is partially broken down to pyruvate, with a net yield of two ATP.

Golgi body (GOHL-gee) Organelle in which many newly forming proteins and lipids undergo final processing, then are sorted and packaged in vesicles.

gonad (GO-nad) Primary reproductive organ in which gametes are produced.

graded potential Of neurons, a local signal that slightly changes the voltage difference across a small patch of the plasma membrane. Such signals vary in magnitude, depending on the stimulus. With prolonged or intense stimulation, graded potentials may spread to a trigger zone of the membrane and initiate an *action potential*.

granum, plural **grana** Within many chloroplasts, any of the stacks of flattened membranous compartments that incorporate chlorophyll and other light-trapping pigments and reaction sites for ATP formation.

gravitropism (GRAV-i-TROPE-izm) [L. *gravis*, heavy, + Gk. *trepein*, to turn] The tendency of a plant to grow directionally in response to the earth's gravitational force.

gray matter Of vertebrates, the dendrites, neuron cell bodies, and neuroglial cells of the spinal cord and cerebral cortex.

grazing food web A network of interlinked food chains in which energy flows from plants to herbivores, then through some array of carnivores.

greenhouse effect Warming of the lower atmosphere due to the buildup of so-called greenhouse gases—carbon dioxide, methane, nitrous oxide, ozone, water vapor, and chlorofluorocarbons.

green revolution In developing countries, the use of improved crop varieties, modern agricultural practices (including massive inputs of fertilizers and pesticides), and equipment to increase crop yields.

ground meristem (MARE-ih-stem) [Gk. *meristos*, divisible] Of vascular plants, a primary meristem that produces ground tissue, hence the bulk of the plant body.

guard cell Either of two adjacent cells having roles in the movement of gases and water vapor across leaf or stem epidermis. An opening (stoma) forms when both cells swell with water and move apart; it closes when they lose water and collapse against each other.

gut A body region where food is digested and absorbed; of complete digestive systems, the portions from the stomach onward.

gymnosperm (JIM-noe-sperm) [Gk. *gymnos*, naked, + *sperma*, seed] A plant that bears seeds at exposed surfaces of reproductive structures, such as cone scales. Pine trees are examples.

habitat [L. *habitare*, to live in] The type of place where an organism normally lives, characterized by physical features, chemical features, and the presence of certain other species.

hair cell Type of mechanoreceptor that may give rise to action potentials when bent or tilted.

haploid (HAP-loyd) Of sexually reproducing species, having only one of each pair of homologous chromosomes that were present in the nucleus of a parent cell; an outcome of meiosis. Compare *diploid*.

heart Muscular pump that keeps blood circulating through the animal body.

hemoglobin (HEEM-oh-glow-bin) [Gk. *haima*, blood, + L. *globus*, ball] Iron-containing, oxygen-transporting protein that gives red blood cells their color.

hemostasis (HEE-mow-STAY-iss) [Gk. *haima*, blood, + *stasis*, standing] Stopping of blood loss from a damaged blood vessel through coagulation, blood vessel spasm, platelet plug formation, and other mechanisms.

herbivore [L. *herba*, grass, + *vovare*, to devour] Plant-eating animal.

heterocyst (HET-er-oh-sist) Of some filamentous cyanobacteria, a type of thick-walled, nitrogen-fixing cell that forms when nitrogen is scarce.

heterotroph (HET-er-oh-trofe) [Gk. *heteros*, other, + *trophos*, feeder] Organism that cannot synthesize its own organic compounds and must obtain them by feeding on plants or other autotrophs. Animals, fungi, many protistans, and most bacteria are heterotrophs.

heterozygous condition (HET-er-oh-ZYE-guss) [Gk. *zygoun*, join together] For a given trait, having nonidentical alleles at a particular locus on homologous chromosomes.

hindbrain Of the vertebrate brain, the medulla oblongata, cerebellum, and pons; includes reflex centers for respiration, blood circulation, and other basic functions; also coordinates motor responses and many complex reflexes.

histone Of eukaryotic chromosomes, any of a class of structural proteins intimately associated with the DNA.

homeostasis (HOE-me-oh-STAY-sis) [Gk. *homo*, same, + *stasis*, standing] Of multicelled organisms, a physiological state in which the physical and chemical conditions of the internal environment are stabilized within tolerable ranges.

hominid [L. *homo*, man] All species on the evolutionary branch leading to modern humans. *Homo sapiens* is the only living representative.

hominoid Apes, humans, and their recent ancestors.

homologous chromosome (huh-MOLL-uh-gus) [Gk. *homologia*, correspondence] One of a pair of chromosomes that resemble each other in length, shape, and the genes they carry, and that pair with each other at meiosis. X and Y chromosomes differ in these respects but still function as homologues.

homologous structures Similarity in some aspect of body form or patterning between different species; evolutionary outcome of descent from a common ancestor.

homozygous condition (HOE-moe-ZYE-guss) For a given trait, having two identical alleles at a particular locus on homologous chromosomes.

homozygous dominant An individual having two dominant alleles at a given gene locus (on a pair of homologous chromosomes).

homozygous recessive An individual having two recessive alleles at a given gene locus (on a pair of homologous chromosomes).

hormone [Gk. *hormon*, to stir up, set in motion] Any of the signaling molecules secreted from endocrine glands, endocrine cells, and some neurons and that travel the bloodstream to nonadjacent target cells.

hydrogen bond Type of chemical bond in which an atom of a molecule interacts weakly with a hydrogen atom already taking part in a polar covalent bond.

hydrogen ion A hydrogen atom that has lost its electron and so bears a positive charge (H^+); a "naked" proton.

hydrologic cycle A biogeochemical cycle in which hydrogen and oxygen move, in the form of water molecules, through the atmosphere, on or through the uppermost layers of land masses, to the oceans, and back again; driven by solar energy.

hydrolysis (high-DRAWL-ih-sis) [L. *hydro*, water, + Gk. *lysis*, loosening or breaking apart] Enzyme-mediated reaction that breaks covalent bonds in a molecule, which splits into two or more parts; at the same time, H^+ and OH^- (derived from a water molecule) become attached to the exposed bonding sites.

hydrophilic substance [Gk. *philos*, loving] A polar substance that is attracted to water molecules and so dissolves easily in water.

hydrophobic substance [Gk. *phobos*, dreading] A nonpolar substance that is repelled by water molecules and so does not readily dissolve in water. Oil is an example.

hydrosphere All liquid or frozen water on or near the earth's surface.

hypha, plural **hyphae** (HIGH-fuh) [Gk. *hyphe*, web] Of fungi, a generally tube-shaped filament with chitin-reinforced walls and, often, reinforcing cross-walls; component of the mycelium.

hypodermis A subcutaneous layer having stored fat that helps insulate the body; although not part of skin, it anchors skin while allowing it some freedom of movement.

hypothalamus [Gk. *hypo*, under, + *thalamos*, inner chamber or possibly *tholos*, rotunda] Of vertebrate forebrains, a brain center that monitors visceral activities (such as salt-water balance, temperature control, and reproduction) and that influences related forms of behavior (as in hunger, thirst, and sex).

hypothesis A plausible answer, or "educated guess," concerning a question or problem. In science, predictions drawn from hypotheses are tested by making observations, developing models, and performing repeatable experiments.

immune system White blood cells (macrophages, T lymphocytes, and B lymphocytes) and their interactions and products; the system shows specificity in response to a particular foreign agent, and memory—the ability to mount a more rapid attack if that specific agent returns.

immunization Deliberate introduction into the body of an antigen that can provoke an immune response and the production of memory lymphocytes.

imprinting Category of learning in which an animal that has been exposed to specific key stimuli early in its behavioral development forms an association with the object.

incomplete dominance Of heterozygotes, a condition in which one allele of a pair only partially dominates expression of its partner.

independent assortment Mendelian principle that each gene pair tends to assort into gametes independently of other gene pairs located on nonhomologous chromosomes.

indirect selection A theory in evolutionary biology that self-sacrificing individuals can indirectly pass on their genes by helping relatives survive and reproduce.

induced-fit model Model of enzyme action whereby a bound substrate induces changes in the shape of the enzyme's active site, resulting in a more precise molecular fit between the enzyme and its substrate.

inflammatory response A series of events involving many cells, complement proteins, and other substances that destroy foreign agents in the body and that restore tissues and internal operating conditions to normal. Occurs during both nonspecific defense responses and immune responses.

inheritance The transmission, from parents to offspring, of structural and functional patterns that have a genetic basis and are characteristic of each species.

inhibiting hormone A signaling molecule produced and secreted by the hypothalamus that controls secretions by the anterior lobe of the pituitary gland.

inhibitor A substance that can bind with an enzyme and interfere with its functioning.

inhibitory postsynaptic potential, or **IPSP** Of neurons, one of two competing types of graded potentials at an input zone; tends to drive the resting membrane potential away from threshold.

instinctive behavior The capacity of an animal to complete fairly complex, stereotyped responses to particular environmental cues without having had prior experience with those cues.

insulin Hormone that lowers the glucose level in blood; it is secreted from beta cells of the pancreas and stimulates cells to take up glucose; also promotes protein and fat synthesis and inhibits protein conversion to glucose.

integration, neural [L. *integrare*, to coordinate] Moment-by-moment summation of all excitatory and inhibitory synapses acting on a neuron; occurs at each level of synapsing in a nervous system.

integument Of animals, a protective body covering such as skin. Of flowering plants, a protective layer around the developing ovule; when the ovule becomes a seed, its integument(s) harden and thicken into a seed coat.

integumentary exchange (in-teg-you-MEN-tuh-ree) Of some animals, a mode of respiration in which oxygen and carbon dioxide diffuse across a thin, vascularized layer of moist epidermis at the body surface.

interleukin A type of communication signal that sensitizes specific cells of the immune system to the presence of a foreign agent and that stimulates them into action.

interneuron Any of the neurons in the vertebrate brain and spinal cord that integrate information arriving from sensory neurons and that influence other neurons in turn.

internode In vascular plants, the stem region between two successive nodes.

interphase Of cell cycles, the time interval between nuclear divisions in which a cell increases its mass, roughly doubles the number of its structures and organelles, and replicates its DNA. The interval is different for different species.

interstitial fluid (IN-ter-STISH-ul) [L. *interstitus*, to stand in the middle of something] In multicelled animals, that portion of the extracellular fluid occupying spaces between cells and tissues.

intertidal zone Generally, the area on a rocky or sandy shoreline that is above the low water mark and below the high water mark; organisms inhabiting it are alternately submerged, then exposed, by tides.

intervertebral disk One of a number of disk-shaped structures containing cartilage that serve as shock absorbers and flex points between bony segments of the vertebral column.

intron A noncoding portion of a newly formed mRNA molecule.

inversion Type of chromosome rearrangement in which a segment that has become separated from the chromosome is reinserted at the same place but in reverse, so the position and sequence of genes are altered.

invertebrate Animal without a backbone.

ion, negatively charged (EYE-on) An atom or a compound that has gained one or more electrons, hence has acquired an overall negative charge.

ion, positively charged An atom or a compound that has lost one or more electrons, hence has acquired an overall positive charge.

ionic bond An association between ions of opposite charge.

isotonic condition Equality in the relative concentrations of solutes in two fluids; for two fluids separated by a cell membrane, there is no net osmotic (water) movement across the membrane.

isotope (EYE-so-tope) An atom that contains the same number of protons as other atoms of the same element, but that has a different number of neutrons.

karyotype (CARRY-oh-type) Cut-and-paste micrograph of a cell's metaphase chromosomes, arranged according to length, shape, banding patterns, and other features.

keratin A tough, water-insoluble protein manufactured by most epidermal cells.

kidney In vertebrates, one of a pair of organs that filter mineral ions, organic wastes, and other substances from the blood, and help regulate the volume and solute concentrations of extracellular fluid.

kinetochore At the centromere of a chromosome, a specialized group of proteins and DNA that serves as an attachment point for several spindle microtubules during mitosis or meiosis. Each chromatid of a duplicated chromosome has its own kinetochore.

Krebs cycle Stage of aerobic respiration in which pyruvate is completely broken down to carbon dioxide and water. Resulting hydrogen ions and electrons are shunted to the next stage, which yields most of the ATP produced in aerobic respiration.

lactate fermentation Anaerobic pathway of ATP formation in which pyruvate from glycolysis is converted to the three-carbon compound lactate, with a net yield of two ATP.

large intestine The colon; a region of the gut that receives unabsorbed food residues from the small intestine and concentrates and stores feces until they are expelled from the body.

larva, plural **larvae** Of animals, a sexually immature, free-living stage between the embryo and the adult.

larynx (LARE-inks) A tubular airway that leads to the lungs. In humans, contains vocal cords, where sound waves used in speech are produced.

lateral meristem Of vascular plants, a type of meristem responsible for secondary growth; either vascular cambium or cork cambium.

leaf For most vascular plants, a structure having chlorophyll-containing tissue that is the major region of photosynthesis.

learning The adaptive modification of behavior in response to neural processing of information that has been gained from specific experiences.

lichen (LY-kun) A symbiotic association between a fungus and a captive photosynthetic partner such as a green alga.

life cycle A recurring, genetically programmed frame of events in which individuals grow, develop, maintain themselves, and reproduce.

light-dependent reactions First stage of photosynthesis, in which sunlight energy is absorbed and converted to the chemical energy of ATP alone (by the cyclic pathway) or to ATP and NADPH (by the noncyclic pathway).

light-independent reactions Second stage of photosynthesis, in which sugar phosphates are assembled with the help of the ATP and NADPH produced during the first stage.

lignification Of mature land plants, a process by which lignin is deposited in secondary cell walls. The deposits impart strength and rigidity, stabilize and protect other wall components, and form a waterproof barrier around the cellulose. Probably a key factor in the evolution of vascular plants.

lignin An inert substance containing different sugar alcohols in amounts that vary among plant species.

limbic system Brain regions that, along with the cerebral cortex, collectively govern emotions.

lineage (LIN-ee-age) A line of descent.

linkage Tendency of genes located on the same chromosome to stay together during meiosis and to end up together in the same gamete.

lipid A compound of mostly carbon and hydrogen that generally does not dissolve in water, but that does dissolve in nonpolar substances. Some lipids serve as energy reserves; others are components of membranes and other cell structures.

lipid bilayer Of cell membranes, two layers of mostly phospholipid molecules, with all the fatty acid tails sandwiched between the hydrophilic heads as a result of hydrophobic interactions.

liver Glandular organ with roles in storing and interconverting carbohydrates, lipids, and proteins absorbed from the gut; maintaining blood; disposing of nitrogen-containing wastes; and other tasks.

local signaling molecules Secretions from cells in many different tissues that alter chemical conditions in the immediate vicinity where they are secreted, then are swiftly degraded.

locus (LOW-cuss) The specific location of a particular gene on a chromosome.

logistic growth (low-JIS-tik) Pattern of population growth in which the growth rate of a

low-density population goes through a rapid growth phase and then levels off.

loop of Henle The hairpin-shaped, tubular region of a nephron that functions in reabsorption of water and solutes.

lung An internal respiratory surface in the shape of a cavity or sac.

lymph (LIMF) [L. *lympha*, water] Tissue fluid that has moved into the vessels of the lymphatic system.

lymphatic system System of lymphoid organs (which function in defense responses) and lymph vessels (which return excess tissue fluid to the bloodstream and also transport absorbed fats to it).

lymphocyte Any of various white blood cells that take part in vertebrate immune responses.

lymphoid organs The lymph nodes, spleen, thymus, tonsils, adenoids, and patches of tissue in the small intestine and appendix.

lysis [Gk. *lysis*, a loosening] Gross induced leakage across a plasma membrane that leads to cell death. Examples are lysis of a virus-infected cell or of a pathogen under chemical attack by complement proteins.

lysosome (LYE-so-sohm) In eukaryotic cells, an organelle containing digestive enzymes that can break down polysaccharides, proteins, nucleic acids, and some lipids.

lytic pathway A mode of viral replication; the virus quickly takes over a host cell's metabolic machinery, the viral genetic material is replicated, and new virus particles are produced and then released as the cell undergoes lysis.

macroevolution The large-scale patterns, trends, and rates of change among major taxa.

macrophage A phagocytic white blood cell that develops from circulating monocytes and defends tissues; takes part in inflammatory responses and in both cell-mediated and antibody-mediated immune responses.

mass extinction An abrupt rise in extinction rates above the background level; a catastrophic, global event in which major taxa are wiped out simultaneously.

mass number The total number of protons and neutrons in an atom's nucleus. The relative masses of atoms are also called atomic weights.

mechanoreceptor Sensory cell or cell part that detects mechanical energy associated with changes in pressure, position, or acceleration.

medulla oblongata Part of the vertebrate brainstem with reflex centers for respiration, blood circulation, and other vital functions.

medusa (meh-DOO-sah) [Gk. *Medusa*, one of three sisters in Greek mythology having snake-entwined hair; this image probably evoked by the tentacles and oral arms extending from the medusa] Free-swimming, bell-shaped stage in cnidarian life cycles.

megaspore Of seed-bearing plants, a type of spore that develops into a female gametophyte.

meiosis (my-OH-sis) [Gk. *meioun*, to diminish] Two-stage nuclear division process in which the parental number of chromosomes in each daughter nucleus becomes haploid (with one of each type of chromosome that was present in the parent nucleus). Basis of gamete formation, also of spore formation in plants. Compare *mitosis*.

membrane excitability A membrane property of any cell that can produce action potentials in response to appropriate stimulation.

memory The storage and retrieval of information about previous experiences; underlies the capacity for learning.

memory lymphocyte Any of the various B or T lymphocytes of the immune system that are formed in response to invasion by a foreign agent and that circulate for some period, available to mount a rapid attack if the same type of invader reappears.

menopause (MEN-uh-pozz) [L. *mensis*, month, + *pausa*, stop] Physiological changes that mark the end of a human female's potential to bear children.

menstrual cycle The cyclic release of oocytes and priming of the endometrium (lining of the uterus) to receive a fertilized egg; the complete cycle averages about 28 days in female humans.

menstruation Periodic sloughing of the blood-enriched lining of the uterus when pregnancy does not occur.

mesoderm (MEH-so-derm) [Gk. *mesos*, middle, + *derm*, skin] In most animal embryos, a primary tissue layer (germ layer) between ectoderm and endoderm; gives rise to muscle, organs of circulation, reproduction, and excretion, most of the internal skeleton (when present), and connective tissue layers of the gut and body covering.

messenger RNA A linear sequence of ribonucleotides transcribed from DNA and translated into a polypeptide chain; the only type of RNA that carries protein-building instructions.

metabolic pathway A linear or cyclic series of breakdown or synthesis reactions in cells, the steps of which are catalyzed by the action of specific enzymes.

metabolism (meh-TAB-oh-lizm) [Gk. *meta*, change] All those chemical reactions by which cells acquire and use energy as they synthesize, accumulate, break apart, and eliminate substances in ways that contribute to growth, maintenance, and reproduction.

metamorphosis (met-uh-MOR-foe-sis) [Gk. *meta*, change, + *morphe*, form] Transformation of a larva into an adult form by way of major tissue reorganization.

metaphase Stage of mitosis when spindle has fully formed, sister chromatids of each chromosome become attached to opposite spindle poles, and all chromosomes lie at the spindle equator.

metaphase I Stage of meiosis when all pairs of homologous chromosomes are aligned with their partners at the spindle equator.

metaphase II Stage of meiosis when the chromosomes, already separated from their homologous partner but still in the duplicated state, are aligned at the spindle equator.

metazoan Any multicelled animal.

MHC marker Any of the surface receptors that mark an individual's cells as self; except for identical twins, the markers are unique to each individual.

microevolution Changes in allele frequencies brought about by mutation, genetic drift, gene flow, and natural selection.

microfilament [Gk. *mikros*, small, + L. *filum*, thread] Component of the cytoskeleton; involved in cell shape, motion, and growth.

microspore Of seed-bearing plants, a type of spore that develops into an immature male gametophyte called a pollen grain.

microtubular spindle Of eukaryotic cells, a bipolar structure composed of organized arrays of microtubules; it forms during mitosis or meiosis and moves the chromosomes.

microtubule Hollow cylinder of mainly tubulin subunits; a cytoskeletal element with roles in cell shape, motion, and growth and in the structure of cilia and flagella.

microtubule organizing center, or **MTOC** Small mass of proteins and other substances in the cytoplasm; the number, type, and location of MTOCs determine the organization and orientation of microtubules.

microvillus (MY-crow-VILL-us) [L. *villus*, shaggy hair] A slender, cylindrical extension of the animal cell surface that functions in absorption or secretion.

midbrain Of vertebrates, a brain region that evolved as a coordinating center for reflex responses to visual and auditory input; together with the pons and medulla oblongata, part of the brainstem, which includes the reticular formation.

migration Of certain animals, a cyclic movement between two distant regions at times of year corresponding to seasonal change.

mimicry (MIM-ik-ree) Situation in which one species (the mimic) bears deceptive resemblance in color, form, and/or behavior to another species (the model) that enjoys some survival advantage.

mineral Any of a number of small inorganic substances required for the normal functioning of body cells.

mitochondrion, plural **mitochondria** (MY-toe-KON-dree-on) Organelle in which the second and third stages of aerobic respiration occur; those stages are the Krebs cycle and preparatory conversions for it, as well as electron transport phosphorylation.

mitosis (my-TOE-sis) [Gk. *mitos*, thread] Type of nuclear division that maintains the parental number of chromosomes for daughter cells. It is the basis of bodily growth and, in some cases, of asexual reproduction of eukaryotes.

molecule A unit of two or more atoms of the same or different elements, bonded together.

molting The shedding of hair, feathers, horns, epidermis, or exoskeleton in a process of growth or periodic renewal.

Monera The kingdom of single-celled prokaryotes; bacteria.

monocot (MON-oh-kot) Short for monocotyledon; a flowering plant in which seeds have only one cotyledon, whose floral parts generally occur in threes (or multiples of threes), and whose leaves typically are parallel-veined. Compare *dicot*.

monohybrid cross [Gk. *monos*, alone] An experimental cross between two parent organisms that breed true for distinctly different forms of a single trait; heterozygous offspring result.

monomer A simple sugar or some other small organic compound that can serve as one of the individual units of *polymers*.

monophyletic group A set of independently evolving lineages that share a common evolutionary heritage.

monosaccharide (MON-oh-SAK-ah-ride) [Gk. *monos*, along, single, + *sakharon*, sugar] A sugar monomer; the simplest carbohydrate. Glucose is an example.

monosomy Abnormal condition in which one chromosome of diploid cells has no homologue.

morphogenesis (MORE-foe-JEN-ih-sis) [Gk. *morphe*, form, + *genesis*, origin] Processes by which differentiated cells in an embryo become organized into tissues and organs, under genetic controls and environmental influences.

motor neuron Nerve cell that relays information away from the brain and spinal cord to the body's effectors (muscles, glands, or both), which carry out responses.

mouth An oral cavity; in human digestion, the site where polysaccharide breakdown begins.

multicelled organism An organism that has differentiated cells arranged into tissues, and often into organs and organ systems.

multiple allele system Three or more alternative molecular forms of a gene (alleles), any of which may occur at the gene's locus on a chromosome.

muscle tissue Tissue having cells able to contract in response to stimulation, then passively lengthen and so return to their resting state.

mutagen (MEW-tuh-jen) An environmental agent that can permanently modify the structure of a DNA molecule. Certain viruses and ultraviolet radiation are examples.

mutation [L. *mutatus*, a change, + *-ion*, result of a process or an act] A heritable change in the molecular structure of DNA.

mutualism [L. *mutuus*, reciprocal] A type of community interaction in which members of two species each receive benefits from the association. When the interaction is intimate and involves a permanent dependency, it is called *symbiosis*.

mycelium (my-SEE-lee-um), plural **mycelia** [Gk. *mykes*, fungus, mushroom, + *helos*, callus] A mesh of tiny, branching filaments (hyphae) that is the food-absorbing part of a multicelled fungus.

mycorrhiza (MY-coe-RISE-uh) "Fungus-root," a symbiotic arrangement between fungal hyphae and young roots of many vascular plants, in which the fungus obtains carbohydrates from the plant and in turn releases dissolved mineral ions to the plant roots.

myelin sheath Of many sensory and motor neurons, an axonal sheath that affects how fast action potentials travel; formed from the plasma membranes of Schwann cells that are wrapped repeatedly around the axon and are separated from each other by a small node.

myofibril (MY-oh-FY-brill) One of many threadlike structures inside a muscle cell; composed of actin and myosin molecules arranged as sarcomeres, the fundamental units of contraction.

myosin (MY-uh-sin) One of two types of protein filaments that make up sarcomeres, the contractile units of a muscle cell; the other is actin.

NAD$^+$ Nicotinamide adenine dinucleotide, a large organic molecule that serves as a cofactor in enzyme reactions. When carrying electrons and protons (H$^+$) from one reaction site to another, it is abbreviated NADH.

NADP$^+$ Nicotinamide adenine dinucleotide phosphate. When carrying electrons and protons (H$^+$) from one reaction site to another, it is abbreviated NADPH.

natural selection A microevolutionary process; a measure of the differences in survival and reproduction that have occurred among individuals of a population that differ from one another in one or more traits.

negative feedback mechanism A homeostatic feedback mechanism in which an activity changes some condition in the internal environment and so triggers a response that reverses the changed condition.

nematocyst (NEM-ad-uh-sist) [Gk. *nema*, thread, + *kystis*, pouch] Of cnidarians only, a stinging capsule that assists in prey capture and possibly protection.

nephridium, plural, **nephridia** (neh-FRID-ee-um) Of earthworms and some other invertebrates, a system of regulating water and solute levels.

nephron (NEFF-ron) [Gk. *nephros*, kidney] Of the vertebrate kidney, a slender tubule in which water and solutes filtered from blood are selectively reabsorbed and in which urine forms.

nerve Cordlike communication line of the peripheral nervous system, composed of axons of sensory neurons, motor neurons, or both packed within connective tissue. In the brain and spinal cord, similar cordlike bundles are called nerve pathways or tracts.

nerve cord Of many animals, a cordlike communication line consisting of axons of neurons.

nerve impulse See *action potential*.

nerve net Cnidarian nervous system.

nervous system System of neurons oriented relative to one another in precise message-conducting and information-processing pathways.

neuroendocrine control center Those portions of the hypothalamus and pituitary gland that interact to control many body functions.

neuroglial cell (NUR-oh-GLEE-uhl) Of vertebrates, one of the cells that provide structural and metabolic support for neurons and that collectively represent about half the volume of the nervous system.

neuromodulator Type of signaling molecule that influences the effects of transmitter substances by enhancing or reducing membrane responses in target neurons.

neuromuscular junction Chemical synapses between the axon terminals of a motor neuron and a muscle cell.

neuron A nerve cell; the basic unit of communication in nervous systems. Neurons collectively sense environmental change, integrate sensory inputs, then activate muscles or glands that initiate or carry out responses.

neutral mutation Mutation in which the altered allele has no more measurable effect on survival and reproduction than do other alleles for the trait.

neutron Subatomic particle of about the same size and mass as a proton but having no electric charge.

niche (NITCH) [L. *nidas*, nest] Of a species, the full range of physical and biological conditions under which its members can live and reproduce.

nitrification (nye-trih-fih-KAY-shun) Process by which certain soil bacteria strip electrons from ammonia or ammonium, releasing nitrite (NO$_2$) that other soil bacteria break down, releasing nitrate (NO$_3$).

nitrogen cycle Cycling of nitrogen atoms between living organisms and the environment, through nitrogen fixation, assimilation and biosynthesis of nitrogen-containing compounds, decomposition, ammonification, nitrification, and denitrification.

nitrogen fixation Process by which a few kinds of bacteria convert gaseous nitrogen (N$_2$) to ammonia, which dissolves rapidly in water to produce ammonium. Other organisms as well as the bacteria use the fixed nitrogen.

nociceptor A sensory receptor, such as a free nerve ending, that detects any stimulus causing tissue damage.

node In vascular plants, a point on a stem where one or more leaves are attached.

noncyclic photophosphorylation (non-SIK-lik foe-toe-FOSS-for-ih-LAY-shun) [L. *non*, not, + Gk. *kylos*, circle] Photosynthetic pathway in which new electrons derived from water molecules flow through two photosystems and two transport chains, and ATP and NADPH form.

nondisjunction Failure of one or more chromosomes to separate during meiosis.

notochord (KNOW-toe-kord) Of chordates, a rod of stiffened tissue (not cartilage or bone) that serves as a supporting structure for the body.

nuclear envelope A double membrane (two lipid bilayers and associated proteins) that is the outermost portion of a cell nucleus.

nucleic acid (new-CLAY-ik) A single- or double-stranded chain of nucleotide units; DNA and RNA are examples.

nucleoid Of bacteria, a region in which DNA is physically organized apart from other cytoplasmic components.

nucleolus (new-KLEE-oh-lus) [L. *nucleolus*, a little kernel] Within the nucleus of a nondividing cell, a mass of proteins, RNA, and other material used in ribosome synthesis.

nucleosome (NEW-klee-oh-sohm) Of eukaryotic chromosomes, an organizational unit consisting of a segment of DNA looped twice around a core of histone molecules.

nucleotide (NEW-klee-oh-tide) A small organic compound having a five-carbon sugar (deoxyribose), nitrogen-containing base, and phosphate group. Nucleotides are the structural units of adenosine phosphates, nucleotide coenzymes, and nucleic acids.

nucleotide coenzyme A protein that transports hydrogen atoms (free protons) and electrons from one reaction site to another in cells.

nucleus (NEW-klee-us) [L. *nucleus*, a kernel] In atoms, the central core of one or more positively charged protons and (in all but hydrogen) electrically neutral neutrons. In eukaryotic cells, a membranous organelle containing the DNA.

obesity An excess of fat in the body's adipose tissues, caused by imbalances between caloric intake and energy output.

oligosaccharide A carbohydrate consisting of a small number of covalently linked sugar monomers. One subclass, disaccharides, consists of two sugar monomers. Compare *monosaccharide* and *polysaccharide*.

omnivore [L. *omnis*, all, + *vovare*, to devour] An organism able to obtain energy from more than one source rather than being limited to one trophic level.

oncogene (ON-coe-jeen) Any gene having the potential to induce cancerous transformations in a cell.

oogenesis (oo-oh-JEN-uh-sis) Formation of a female gamete, from a germ cell to a mature haploid ovum (egg).

operator A short base sequence between a promoter and the start of a gene; interacts with regulatory proteins to control transcription.

operon Of transcription, a promoter-operator sequence that services more than a single gene. The lactose operon of *E. coli* is an example.

orbitals Volumes of space around the nucleus of an atom in which electrons are likely to be at any instant.

organ A structure of definite form and function that is composed of more than one tissue.

organ formation Stage of development in which primary tissue layers (germ layers) split into subpopulations of cells, and different lines of cells become unique in structure and function; foundation for growth and tissue specialization, when organs acquire specialized chemical and physical properties.

organ system Two or more organs that interact chemically, physically, or both in performing a common task.

organelle Any of various membranous sacs, envelopes, and other compartmental portions of cytoplasm. Organelles separate different, often incompatible metabolic reactions in the space of the cytoplasm and in time (by allowing certain reactions to proceed only in controlled sequences).

organic compound In biology, a compound assembled in cells and having a carbon backbone, often with carbon atoms arranged as a chain or ring structure.

osmosis (oss-MOE-sis) [Gk. *osmos*, act of pushing] Of cell membranes, the passive movement of water through the interior of membrane-spanning proteins in response to solute concentration gradients, a pressure gradient, or both.

ovary (OH-vuh-ree) In female animals, the primary reproductive organ in which eggs form. In seed-bearing plants, the portion of the carpel where eggs develop, fertilization takes place, and seeds mature. A mature ovary and sometimes other plant parts is a fruit.

oviduct (OH-vih-dukt) Duct through which eggs travel from the ovary to the uterus. Formerly called Fallopian tube.

ovulation (AHV-you-LAY-shun) During each turn of the menstrual cycle, the release of a secondary oocyte (immature egg) from an ovary.

ovule (OHV-youl) [L. *ovum*, egg] Any of one or more structures that form on the inner wall of the ovary of seed-bearing plants and that, at maturity, are seeds. An ovule contains the female gametophyte with its egg, surrounded by nutritive and protective tissues.

ovum (OH-vum) A mature female gamete (egg).

oxidation-reduction reaction An electron transfer from one atom or molecule to another. Often hydrogen is transferred along with the electron or electrons.

pancreas (PAN-cree-us) Glandular organ that secretes enzymes and bicarbonate into the small intestine during digestion, and that also secretes the hormones insulin and glucagon.

pancreatic islets Any of the 2 million clusters of endocrine cells in the pancreas, including alpha cells, beta cells, and delta cells.

parasite [Gk. *para*, alongside, + *sitos*, food] An organism that obtains nutrients directly from the tissues of a living host, which it lives on or in and may or may not kill.

parasitoid An insect larva that grows and develops inside a host organism (usually another insect), eventually consuming the soft tissues and killing it.

parasympathetic nerve Of the autonomic nervous system, any of the nerves carrying signals that tend to slow the body down overall and divert energy to basic tasks; such nerves also work continually in opposition with *sympathetic nerves* to bring about minor adjustments in internal organs.

parathyroid glands (PARE-uh-THY-royd) In vertebrates, endocrine glands embedded in the thyroid gland that secrete parathyroid hormone, which helps restore blood calcium levels.

parenchyma Most abundant ground tissue of root and shoot systems. Its cells function in photosynthesis, storage, secretion, and other tasks.

parthenogenesis Development of an embryo from an unfertilized egg.

passive immunity Temporary immunity conferred by deliberately introducing antibodies into the body.

passive transport Movement of a solute across a cell membrane in response to its concentration gradient, through the interior of

proteins that span the membrane. No energy expenditure is required.

pathogen (PATH-oh-jen) [Gk. *pathos*, suffering, + *-genēs*, origin] Disease-causing organism.

pattern formation Of animals, mechanisms responsible for specialization and positioning of tissues during embryonic development.

PCR *See* polymerase chain reaction.

pelagic province The entire volume of ocean water; subdivided into *neritic zone* (relatively shallow waters overlying continental shelves) and *oceanic zone* (water over ocean basins).

penis A male organ that deposits sperm into a female reproductive tract.

perennial [L. *per-*, throughout, + *annus*, year] A plant that lives for three or more growing seasons.

pericycle (PARE-ih-sigh-kul) [Gk. *peri-*, around, + *kyklos*, circle] Of a root vascular cylinder, one or more layers just inside the endodermis that give rise to lateral roots and contribute to secondary growth.

periderm Of vascular plants showing secondary growth, a protective covering that replaces epidermis.

peripheral nervous system (per-IF-ur-uhl) [Gk. *peripherein*, to carry around] Of vertebrates, the nerves leading into and out from the spinal cord and brain, and the ganglia along those communication lines.

peristalsis (pare-ih-STAL-sis) A rhythmic contraction of muscles that moves food forward through the animal gut.

peritoneum A lining of the coelom that also covers and helps maintain the position of internal organs.

permafrost A permanently frozen, water-impenetrable layer beneath the soil surface in arctic tundra.

PGA Phosphoglycerate (FOSS-foe-GLISS-er-ate); a key intermediate in glycolysis as well as the Calvin-Benson cycle.

PGAL Phosphoglyceraldehyde; a key intermediate in glycolysis as well as the Calvin-Benson cycle.

pH scale A scale used in measuring the concentration of free (unbound) hydrogen ions in solutions; pH 0 is the most acidic, 14 the most basic, and 7, neutral.

phagocyte (FAG-uh-sight) [Gk. *phagein*, to eat, + *kytos*, hollow vessel] A macrophage or certain other white blood cells that engulf and destroy foreign agents in body tissues.

phagocytosis (FAG-uh-sigh-TOE-sis) [Gk. *phagein*, to eat, + *kytos*, hollow vessel] Engulfment of foreign cells or substances by amoebas and some white blood cells, by means of endocytosis.

pharynx (FARE-inks) A muscular tube by which food enters the gut and, in land vertebrates, the windpipe (trachea).

phenotype (FEE-no-type) [Gk. *phainein*, to show, + *typos*, image] Observable trait or traits of an individual; arises from interactions between genes, and between genes and the environment.

pheromone (FARE-oh-moan) [Gk. *phero*, to carry, + *-mone*, as in hormone] A type of signaling molecule secreted by exocrine glands that serves as a communication signal between individuals of the same species.

phloem (FLOW-um) Of vascular plants, a tissue with living cells that interconnect and form the tubes through which sugars and other solutes are conducted.

phospholipid A type of lipid with a glycerol backbone, two fatty acid tails, and a phosphate group to which an alcohol is attached; the main lipid of plant and animal cell membranes.

phosphorylation (FOSS-for-ih-LAY-shun) The attachment of inorganic phosphate to a molecule; also the transfer of a phosphate group from one molecule to another, as when ATP phosphorylates glucose.

photolysis (foe-TALL-ih-sis) [Gk. *photos*, light, + *-lysis*, breaking apart] First step in noncyclic photophosphorylation, when water is split into oxygen, hydrogen, and associated electrons; photon energy indirectly drives the reaction.

photoreceptor Light-sensitive sensory cell.

photosynthesis The trapping and conversion of sunlight energy to chemical energy (ATP, NADPH, or both), followed by synthesis of sugar phosphates that become converted to sucrose, cellulose, starch, and other end products.

photosynthetic autotroph An organism able to synthesize all organic molecules it requires using carbon dioxide as the carbon source and sunlight as the energy source. All plants, some protistans, and a few bacteria are photosynthetic autotrophs.

photosystem Of photosynthetic membranes, a light-trapping unit having organized arrays of pigment molecules and enzymes.

photosystem I A type of photosystem that operates during the cyclic pathway of photosynthesis.

photosystem II A type of photosystem that operates during both the cyclic and noncyclic pathways of photosynthesis.

phototropism [Gk. *photos*, light, + *trope*, turning, direction] Adjustment in the direction and rate of plant growth in response to light.

phylogeny Evolutionary relationships among species, starting with the most ancestral forms and including all the branches leading to all their descendants.

phytochrome Light-sensitive pigment molecule, the activation and inactivation of which trigger plant hormone activities governing leaf expansion, stem branching, stem lengthening, and often seed germination and flowering.

phytoplankton (FIE-toe-PLANK-tun) [Gk. *phyton*, plant, + *planktos*, wandering] A freshwater or marine community of floating or weakly swimming photosynthetic autotrophs, such as diatoms, green algae, and cyanobacteria.

pineal gland (py-NEEL) Of vertebrates, a light-sensitive endocrine gland that secretes melatonin, a hormone that influences the development of reproductive organs and reproductive cycles.

pioneer species Typically small plants with short life cycles that are adapted to growing in exposed, often windy areas with intense sunlight, wide swings in air temperature, and soils deficient in nitrogen and other nutrients. By improving conditions in areas they colonize, pioneers invite their replacement by other species.

pituitary gland Of vertebrate endocrine systems, a gland that interacts with the hypothalamus to coordinate and control many physiological functions, including the activity of many other endocrine glands. Its *posterior lobe* stores and secretes hypothalamic hormones; the *anterior lobe* produces and secretes its own hormones.

placenta (play-SEN-tuh) Of a uterus, an organ composed of maternal tissues and extraembryonic membranes (chorion especially); delivers nutrients to and carries away wastes from the embryo.

plankton [Gk. *planktos*, wandering] Any community of floating or weakly swimming organisms, mostly microscopic, in freshwater and saltwater environments. See *phytoplankton* and *zooplankton*.

plant Most often, multicelled autotroph able to build its own food molecules through photosynthesis.

Plantae The kingdom of plants.

plasma (PLAZ-muh) Liquid component of blood; consists of water, various proteins, ions, sugars, dissolved gases, and other substances.

plasma cell Of immune systems, any of the antibody-secreting daughter cells of a rapidly dividing population of B cells.

plasma membrane Of cells, the outermost membrane that separates internal metabolic events from the environment but selectively permits passage of various substances. Composed of a lipid bilayer and proteins that carry out most functions, including transport across the membrane and reception of outside signals.

plasmid Of many bacteria, a small, circular DNA molecule that carries some genes and replicates independently of the bacterial chromosome.

plasmodesma (PLAZ-moe-DEZ-muh) Of multicelled plants, a junction between linked walls of adjacent cells through which nutrients and other substances flow.

plasticity Of the human species, the ability to remain flexible and adapt to a wide range of environments, rather than becoming narrowly adapted to one specific environment.

plate tectonics Arrangement of the earth's outer layer (lithosphere) in slablike plates, all in motion and floating on a hot, plastic layer of the underlying mantle.

platelet (PLAYT-let) Any of the cell fragments in blood that release substances necesary for clot formation.

pleiotropy (PLEE-oh-troh-pee) [Gk. *pleon*, more, + *trope*, direction] Form of gene expression in which a single gene exerts multiple effects on seemingly unrelated aspects of an individual's phenotype.

pollen grain [L. *pollen*, fine dust] Of gymnosperms and flowering plants, an immature male gametophyte (gamete-producing body).

pollen sac In anthers of flowers, any of the chambers in which pollen grains develop.

pollen tube A tube formed after a pollen grain germinates; grows down through carpel tissues and carries sperm to the ovule.

pollination Of flowering plants, the arrival of a pollen grain on the landing platform (stigma) of a carpel.

pollutant Any substance with which an ecosystem has had no prior evolutionary experience, in terms of kinds or amounts, and that can accumulate to disruptive or harmful levels. Can be naturally occurring or synthetic.

polymer (POH-lih-mur) [Gk. *polus*, many, + *meris*, part] A molecule composed of three to millions of small subunits that may or may not be identical.

polymerase chain reaction DNA amplification method; DNA having a gene of interest is split into single strands, which enzymes (polymerases) copy; the enzymes also act on the accumulating copies, multiplying the gene sequence by the millions.

polymorphism (poly-MORE-fizz-um) [Gk. *polus*, many, + *morphe*, form] Of a population, the persistence through the generations of two or more forms of a trait, at a frequency greater than can be maintained by new mutations alone.

polyp (POH-lip) Vase-shaped, sedentary stage of cnidarian life cycles.

polypeptide chain Three or more amino acids joined by peptide bonds.

polyploidy (POL-ee-PLOYD-ee) A condition in which offspring end up with three or more of each type of chromosome characteristic of the parental stock.

polysaccharide [Gk. *polus*, many, + *sakcharon*, sugar] A straight or branched chain of hundreds of thousands of covalently linked sugar monomers, of the same or different kinds. The most common polysaccharides are starch, cellulose, and glycogen.

polysome Of protein synthesis, several ribosomes all translating the same messenger RNA molecule, one after the other.

population A group of individuals of the same species occupying a given area.

positive feedback mechanism Homeostatic mechanism by which a chain of events is set in motion and intensifies an original condition.

post-translational controls Of eukaryotes, controls that govern modification of newly formed polypeptide chains into functional enzymes and other proteins.

predator [L. *prehendere*, to grasp, seize] An organism that feeds on and may or may not kill other living organisms (its *prey*); unlike parasites, predators do not live on or in their prey.

pressure flow theory Of vascular plants, a theory that organic compounds move through phloem because of gradients in solute concentrations and pressure between source regions (such as photosynthetically active leaves) and sink regions (such as growing plant parts).

primary growth Plant growth originating at root tips and shoot tips.

primary immune response Actions by white blood cells and their products, elicited by a first-time encounter with an antigen; includes both antibody-mediated and cell-mediated responses.

primary productivity, gross Of an ecosystem, the total rate at which the producers capture and store a given amount of energy, as by photosynthesis, during a specified interval.

primary productivity, net Of an ecosystem, the rate of energy storage in the tissues of producers in excess of their rate of aerobic respiration.

procambium (pro-KAM-bee-um) Of vascular plants, a primary meristem that gives rise to the primary vascular tissues.

producer, primary An organism such as a plant that directly or indirectly nourishes consumers, decomposers, and detritivores.

progesterone (pro-JESS-tuh-rown) Female sex hormone secreted by the ovaries.

prokaryote (pro-CARRY-oht) [L. *pro*, before, + Gk. *karyon*, kernel] Single-celled organism that has no nucleus or other membrane-bound organelles; only bacteria are prokaryotes.

promoter Of transcription, a base sequence that signals the start of a gene; the site where RNA polymerase initially binds.

prophase First stage of mitosis, when each duplicated chromosome becomes condensed into a thicker, rodlike form.

prophase I Stage of meiosis when each duplicated chromosome condenses and pairs with its homologous partner, followed by crossing over and genetic recombination among nonsister chromatids.

prophase II Brief stage of meiosis after interkinesis during which each chromosome still consists of two chromatids.

protein Organic compound composed of one or more chains of amino acids (polypeptide chains).

Protista The kingdom of protistans.

protistan (pro-TISS-tun) [Gk. *prōtistos*, primal, very first] Single-celled eukaryote.

proton Positively charged unit of energy in the atomic nucleus.

proto-oncogene A gene sequence similar to an oncogene but that codes for a protein required in normal cell function; may trigger cancer, generally when specific mutations alter its structure or function.

protostome (PRO-toe-stome) [Gk. *proto*, first, + *stoma*, mouth] A bilateral animal in which the first indentation in the early embryo develops into the mouth. Includes mollusks, annelids, and arthropods.

proximal tubule Of a nephron, the tubular region that receives water and solutes filtered from the blood.

pulmonary circuit Blood circulation route leading to and from the lungs.

Punnett-square method A diagramming technique for predicting the possible outcome of a mating or an experimental cross.

purine Nucleotide base having a double ring structure. Adenine and guanine are two examples.

pyrimidine (pih-RIM-ih-deen) Nucleotide base having a single ring structure. Cytosine and thymine are examples.

pyruvate (PIE-roo-vate) Three-carbon compound produced by the initial breakdown of a glucose molecule during glycolysis.

radial symmetry Body plan having four or more roughly equivalent parts arranged around a central axis.

rain shadow A reduction in rainfall on the leeward side of high mountains, resulting in arid or semiarid conditions.

reabsorption Of urine formation, the diffusion or active transport of water and usable solutes out of a nephron and into capillaries leading back to the general circulation; regulated by ADH and aldosterone.

receptor Of cells, a molecule at the cell surface or within the cytoplasm that may be activated by a specific hormone, virus, or some other outside agent. Of nervous systems, a sensory cell or cell part that may be activated by a specific stimulus.

receptor protein Protein that binds a signaling molecule such as a hormone, then triggers alterations in cell behavior or metabolism.

recessive allele [L. *recedere*, to recede] In heterozygotes, an allele whose expression is fully or partially masked by expression of its partner; fully expressed only in the homozygous recessive condition.

recognition protein Protein at cell surface recognized by cells of like type; helps guide the ordering of cells into tissues during development and functions in cell-to-cell interactions.

recombinant technology Procedures by which DNA (genes) from different species may be isolated, cut, spliced together, and the new recombinant molecules multiplied in quantity.

red blood cell Erythrocyte; an oxygen-transporting cell in blood.

red marrow A substance in the spongy tissue of many bones that serves as a major site of blood cell formation.

reflex [L. *reflectere*, to bend back] A simple, stereotyped, and repeatable movement elicited by a sensory stimulus.

reflex arc [L. *reflectere*, to bend back] Type of neural pathway in which signals from sensory neurons can be sent directly to motor neurons, without intervention by an interneuron.

refractory period Of neurons, the period following an action potential at a given patch of membrane when sodium gates are shut and potassium gates are open, so that the patch is insensitive to stimulation.

releasing hormone A hypothalamic signaling molecule that stimulates or slows down secretion by target cells in the anterior lobe of the pituitary gland.

repressor protein Regulatory protein that provides negative control of gene activity by preventing RNA polymerase from binding to DNA.

reproduction, asexual Production of new individuals by any mode that does not involve gametes.

reproduction, sexual Mode of reproduction that begins with meiosis, proceeds through gamete formation, and ends at fertilization.

reproductive isolating mechanism Any aspect of structure, functioning, or behavior that prevents successful interbreeding (hence gene flow) between populations or between local breeding units within a population.

resource partitioning A community pattern in which similar species generally share the same kind of resource in different ways, in different areas, or at different times.

respiration [L. *respirare*, to breathe] In most animals, the overall exchange of oxygen from the environment and carbon dioxide wastes from cells by way of circulating blood. Compare *aerobic respiration*.

resting membrane potential Of neurons and other excitable cells that are not being stimulated, the steady voltage difference across the plasma membrane.

restriction enzymes Class of enzymes that cut apart foreign DNA that enters a cell, as by viral infection; also used in recombinant DNA technology.

reticular formation Of the vertebrate brainstem, a major network of interneurons that helps govern activity of the whole nervous system.

reverse transcriptase Viral enzyme required for reverse transcription of mRNA into DNA; also used in recombinant DNA technology.

reverse transcription Assembly of DNA on a single-stranded mRNA molecule by viral enzymes.

RFLPs (restriction fragment length polymorphisms) Slight but unique differences in the banding pattern of DNA fragments from different individuals of a species; result from individual differences in the number and location of DNA sites that restriction enzymes can recognize and cut.

ribosomal RNA (rRNA) Type of RNA molecule that combines with proteins to form ribosomes, on which polypeptide chains are assembled.

ribosome Of cells, a structure having two subunits, each composed of RNA and protein molecules; the site of protein synthesis.

RNA Ribonucleic acid; a category of nucleotides used in translating the genetic message of DNA into protein.

rod cell A vertebrate photoreceptor sensitive to very dim light and that contributes to coarse perception of movement.

root hair Of vascular plants, an extension of a specialized root epidermal cell; root hairs collectively enhance the surface area available for absorbing water and solutes.

RuBP Ribulose bisphosphate, a five-carbon compound required for carbon fixation in the Calvin-Benson cycle of photosynthesis.

salivary gland Any of the glands that secrete saliva, a fluid that initially mixes with food in the mouth and starts the breakdown of starch.

salt An ionic compound formed when an acid reacts with a base.

saltatory conduction In myelinated neurons, rapid, node-to-node hopping of action potentials.

saprobe Heterotroph that obtains its nutrients from nonliving organic matter. Most fungi are saprobes.

sarcomere (SAR-koe-meer) Of skeletal and cardiac muscles, the basic unit of contraction; a region of organized myosin and actin filaments between two Z lines of a myofibril inside a muscle cell.

sarcoplasmic reticulum (sar-koe-PLAZ-mik reh-TIK-you-lum) A calcium-storing membrane system surrounding myofibrils of a muscle cell.

Schwann cells Specialized neuroglial cells that grow around neuron axons, forming a *myelin sheath*.

sclerenchyma Of vascular plants, a ground tissue that provides mechanical support and protection in mature plant parts.

second law of thermodynamics The spontaneous direction of energy flow is from high-quality to low-quality forms. With each conversion, some energy is randomly dispersed in a form that is not as readily available to do work.

second messenger A molecule inside a cell that mediates and generally triggers amplified response to a hormone.

secondary immune response Rapid, prolonged immune response by white blood cells, memory cells especially, to a previously encountered antigen.

secretion Generally, the release of a substance for use by the organism producing it. (Not the same as *excretion*, the expulsion of excess or waste material.) Of kidneys, a regulated stage in urine formation, in which ions and other substances move from capillaries into nephrons.

sedimentary cycle A biogeochemical cycle without a gaseous phase; the element moves from land to the seafloor, then returns only through long-term geological uplifting.

seed Of gymnosperms and flowering plants, a fully mature ovule (contains the plant embryo), with its integuments forming the seed coat.

segmentation Of earthworms and many other animals, a series of body units that may be externally similar to or quite different from one another.

segregation (Mendelian principle of) [L. *se-*, apart, + *grex*, herd] The principle that diploid organisms inherit a pair of genes for each trait (on a pair of homologous chromosomes) and that the two genes segregate during meiosis and end up in separate gametes.

selective gene expression Of multicelled organisms, activation or suppression of a fraction of the genes in unique ways in different cells, leading to pronounced differences in structure and function among different cell lineages.

selfish behavior Form of behavior by which an individual protects or increases its own chance of producing offspring, regardless of the consequences for the group to which it belongs.

semen (SEE-mun) [L. *serere*, to sow] Sperm-bearing fluid expelled from a penis during male orgasm.

semiconservative replication [Gk. *hēmi*, half, + L. *conservare*, to keep] Reproduction of a DNA molecule when a complementary strand forms on each of the unzipping strands of an existing DNA double helix, the outcome being two "half-old, half-new" molecules.

senescence (sen-ESS-cents) [L. *senescere*, to grow old] Sum total of processes leading to the natural death of an organism or some of its parts.

sensory neuron Any of the nerve cells that act as sensory receptors, detecting specific stimuli (such as light energy) and relaying signals to the brain and spinal cord.

sessile animal Animal that remains attached to a substrate during some stage (often the adult) of its life cycle.

sex chromosomes Of most animals and some plants, chromosomes that differ in number or kind between males and females but that still function as homologues during meiosis. Compare *autosomes*.

sexual dimorphism Phenotypic differences between males and females of a species.

sexual reproduction Production of offspring by way of meiosis, gamete formation, and fertilization.

sexual selection Natural selection based on a trait that provides a competitive edge in mating and producing offspring.

shoot system Stems and leaves of vascular plants.

sieve tube member Of flowering plants, a cellular component of the interconnecting conducting tubes in phloem.

sink region In a vascular plant, any region using or stockpiling organic compounds for growth and development.

sliding filament model Model of muscle contraction, in which actin filaments physically slide over myosin filaments toward the center of the sarcomere. The sliding requires ATP energy and the formation of cross-bridges between actin and myosin filaments.

small intestine Of vertebrates, the portion of the digestive system where digestion is completed and most nutrients absorbed.

smog, industrial Gray-colored air pollution that predominates in industrialized cities with cold, wet winters.

smog, photochemical Form of brown, smelly air pollution occurring in large cities in warm climates.

social behavior Tendency of individual animals to enter into cooperative, interdependent relationships with others of their kind; based on the ability to use communication signals.

social parasite Animal that depends on the social behavior of another species to gain food, care for young, or some other factor to complete its life cycle.

sodium-potassium pump A transport protein spanning the lipid bilayer of the plasma membrane. When activated by ATP, its shape changes and it selectively transports sodium ions out of the cell and potassium ions in.

solute (SOL-yoot) [L. *solvere*, to loosen] Any substance dissolved in a solution. In water, this means its individual molecules are surrounded by spheres of hydration that keep their charged parts from interacting, so the molecules remain dispersed.

solvent Fluid in which one or more substances are dissolved.

somatic cell (so-MAT-ik) [Gk. *sōmā*, body] Of animals, any cell that is not a germ cell (which develops by meiosis into sperm or eggs).

somatic nervous system Those nerves leading from the central nervous system to skeletal muscles.

source region Of vascular plants, any of the sites of photosynthesis.

speciation (spee-cee-AY-shun) The time at which a new species emerges, as by divergence or polyploidy.

species (SPEE-sheez) [L. *species*, a kind] Of sexually reproducing species in nature, one or more populations composed of individuals that are interbreeding and producing fertile offspring, and that are reproductively isolated from other such groups.

sperm [Gk. *sperma*, seed] Mature male gamete.

spermatogenesis (sperm-AT-oh-JEN-ih-sis) Formation of mature sperm following meiosis in a germ cell.

sphere of hydration Through positive or negative interactions, a clustering of water molecules around the individual molecules of a substance placed in water. Compare *solute*.

sphincter (SFINK-tur) Ring of muscle between regions of a tubelike system (as between the stomach and small intestine).

spinal cord Of central nervous systems, the portion threading through a canal inside the vertebral column and providing direct reflex connections between sensory and motor neurons as well as communication lines to and from the brain.

spindle apparatus A bipolar structure composed of microtubules that forms during mitosis or meiosis and that moves the chromosomes.

sporangium, plural **sporangia** (spore-AN-gee-um) [Gk. *spora*, seed] The protective tissue layer that surrounds haploid spores in a sporophyte.

spore Of fungi, a walled, resistant cell or multicelled structure, produced by mitosis or meiosis, that can germinate and give rise to a new mycelium. Of land plants, a reproductive cell formed by meiosis that can develop into a gametophyte (gamete-producing body).

sporophyte [Gk. *phyton*, plant] Diploid, spore-producing stage of plant life cycles.

stabilizing selection Mode of natural selection in which the most common phenotypes in a population are favored, and the underlying allele frequencies persist over time.

stamen (STAY-mun) Of flowering plants, a male reproductive structure; commonly consists of pollen-bearing structures (anthers) on single stalks (filaments).

start codon Of protein synthesis, a base triplet in a strand of mRNA that serves as the start signal for mRNA translation.

steroid (STAIR-oid) A lipid with a backbone of four carbon rings. Steroids differ in the number and location of double bonds in the backbone and in the number, position, and type of functional groups.

stimulus [L. *stimulus*, goad] A specific form of energy, such as light, heat, and mechanical pressure, that the body can detect through sensory receptors.

stoma (STOW-muh), plural **stomata** [Gk. *stoma*, mouth] A controllable gap between two guard cells in stems and leaves; any of the small passageways across the epidermis through which carbon dioxide moves into the plant and water vapor moves out.

stomach A muscular, stretchable sac that receives ingested food; of vertebrates, an organ between the esophagus and intestine in which considerable protein digestion occurs.

stop codon Of protein synthesis, a base triplet in a strand of mRNA that serves as the stop signal for translation, so that no more amino acids are added to the polypeptide chain.

stroma [Gk. *strōma*, bed] Of chloroplasts, the semifluid matrix surrounding the thylakoid membrane system; the zone where sucrose, starch, and other end products of photosynthesis are assembled.

substrate Specific molecule or molecules that an enzyme can chemically recognize, bind briefly to itself, and modify in a specific way.

substrate-level phosphorylation Enzyme-mediated reaction in which a substrate gives up a phosphate group to another molecule, as when an intermediate of glycolysis donates phosphate to ADP, producing ATP.

succession, primary (suk-SESH-un) [L. *succedere*, to follow] Orderly changes from the time pioneer species colonize a barren habitat through replacements after replacements by various species; the changes lead to a *climax community*, when the composition of species remains steady under prevailing conditions.

succession, secondary Orderly changes in a community or patch of habitat toward the climax state after having been disturbed, as by fire.

surface-to-volume ratio A mathematical relationship in which volume increases with the cube of the diameter, but surface area increases only with the square. Of growing cells, the volume of cytoplasm increases more rapidly than the surface area of the plasma membrane that must service the cytoplasm. Because of this constraint, cells generally remain small or elongated, or have elaborate membrane foldings.

symbiosis (sim-by-OH-sis) [Gk. *sym*, together, + *bios*, life, mode of life] A form of mutualism in which interacting species have become intimately and permanently dependent on each other for survival and reproduction.

sympathetic nerve Of the autonomic nervous system, any of the nerves generally concerned with increasing overall body activities during times of heightened awareness, excitement, or danger; such nerves also work continually in opposition with *parasympathetic nerves* to bring about minor adjustments in internal organs.

sympatric speciation [Gk. *syn*, together, + *patria*, native land] Speciation that follows after ecological, behavioral, or genetic barriers arise within the geographical boundaries of a single population.

synaptic integration (sin-AP-tik) Moment-by-moment combining of excitatory and inhibitory signals arriving at a trigger zone of a neuron.

systematics Branch of biology that deals with patterns of diversity among organisms in an evolutionary context; its three approaches include taxonomy, phylogenetic reconstruction, and classification.

systemic circuit (sis-TEM-ik) Circulation route in which oxygen-enriched blood flows from the lungs to the left half of the heart, through the rest of the body (where it gives up oxygen and picks up carbon dioxide), then back to the right side of the heart.

taproot system A primary root and its lateral branchings.

taxonomy (tax-ON-uh-mee) Approach in biological systematics that involves identifying organisms and assigning names to them.

telophase (TEE-low-faze) Of mitosis, final stage when chromosomes decondense into threadlike structures and two daughter nuclei form.

telophase I Of meiosis, stage when one of each type of duplicated chromosome has arrived at one or the other end of the spindle pole.

telophase II Of meiosis, final stage when four daughter nuclei form.

temperate pathway Mode of viral replication in which the virus enters latency instead of killing the host cell outright; viral genes remain inactive and, if integrated into the bacterial chromosome, may be passed on to any of the cell's descendants—which will be destroyed when the viral genes do become activated.

testcross Experimental cross in which hybrids of the first generation of offspring (F_1) are crossed with an individual known to be true-breeding for the same recessive trait as the recessive parent.

testis, plural **testes** Male gonad; primary reproductive organ in which male gametes and sex hormones are produced.

testosterone (tess-TOSS-tuh-rown) In male mammals, a major sex hormone that helps control male reproductive functions.

theory A related set of hypotheses that, taken together, form a broad-ranging explanation about some aspect of the natural world; differs from a scientific hypothesis in its breadth of application. In modern science, only explanations that have been extensively tested and can be relied upon with a very high degree of confidence are accorded the status of theory.

thermal inversion Situation in which a layer of dense, cool air becomes trapped beneath a layer of warm air; can cause air pollutants to accumulate to dangerous levels close to the ground.

thermoreceptor Sensory cell that can detect radiant energy associated with temperature.

thigmotropism (thig-MOTE-ruh-pizm) [Gk. *thigm*, touch] Of vascular plants, growth oriented in response to physical contact with a solid object, as when a vine curls around a fencepost.

threshold Of neurons and other excitable cells, a certain minimum amount by which the steady voltage difference across the plasma membrane must change to produce an action potential.

thylakoid membrane Of chloroplasts, an internal membrane commonly folded into flattened channels and stacked disks (*grana*); contains light-absorbing pigments and enzymes used in the formation of ATP, NADPH, or both during photosynthesis.

thymine Nitrogen-containing base in some nucleotides.

thymus gland Of endocrine systems, a gland in which certain white blood cells multiply, differentiate, and mature, and which secretes hormones that affect their functions.

thyroid gland Of endocrine systems, a gland that produces hormones that affect overall metabolic rates, growth, and development.

tissue Of multicelled organisms, a group of cells and intercellular substances that function together in one or more specialized tasks.

tonicity The relative concentrations of solutes in two fluids, such as inside and outside a cell. When solute concentrations are *isotonic* (equal in both fluids), water shows no net osmotic movement in either direction. When one fluid is *hypotonic* (has less solutes than the other), the other is *hypertonic* (has more solutes) and is the direction in which water tends to move.

trachea (TRAY-kee-uh), plural **tracheae** Of insects, spiders, and some other animals, a finely branching air-conducting tube that functions in respiration; of land vertebrates, the windpipe that carries air between the larynx and bronchi.

tracheid (TRAY-kid) Of flowering plants, one of two types of cells in xylem that conduct water and dissolved minerals.

transcript-processing controls Of eukaryotic cells, controls that govern modification of new mRNA molecules into mature transcripts before shipment from the nucleus.

transcription [L. *trans*, across, + *scribere*, to write] Of protein synthesis, the assembly of an RNA strand on one of the two strands of a DNA double helix; the base sequence of the resulting transcript is complementary to the DNA region on which it is assembled.

transcriptional controls Controls influencing when and to what degree a particular gene will be transcribed.

transfer RNA (tRNA) Of protein synthesis, any of the type of RNA molecules that bind and deliver specific amino acids to ribosomes *and* pair with mRNA code words for those amino acids.

translation Of protein synthesis, the conversion of the coded sequence of information in mRNA into a particular sequence of amino acids to form a polypeptide chain; depends on interactions of rRNA, tRNA, and mRNA.

translational controls Of eukaryotic cells, controls governing the rates at which mRNA transcripts that reach the cytoplasm will be translated into polypeptide chains at ribosomes.

translocation Of cells, the transfer of part of one chromosome to a nonhomologous chromosome. Of vascular plants, the conduction of organic compounds through the plant body by way of the phloem.

transmitter substance Any of the class of signaling molecules that are secreted from neurons, act on immediately adjacent cells, and are then rapidly degraded or recycled.

transpiration Evaporative water loss from stems and leaves.

transport control Of eukaryotic cells, controls governing when mature mRNA transcripts are shipped from the nucleus into the cytoplasm.

transposable element DNA element that can spontaneously "jump" to new locations in the same DNA molecule or a different one. Such elements often inactivate the genes into which they become inserted and give rise to observable changes in phenotype.

trisomy (TRY-so-mee) Of diploid cells, the abnormal presence of three of one type of chromosome.

trophic level (TROE-fik) [Gk. *trophos*, feeder] All organisms in an ecosystem that are the same number of transfer steps away from the energy input into the system.

tropism (TROE-pizm) Of vascular plants, a growth response to an environmental factor, such as growth toward light.

tumor A tissue mass composed of cells that are dividing at an abnormally high rate.

turgor pressure (TUR-gore) [L. *turgere*, to swell] Internal pressure applied to a cell wall when water moves by osmosis into the cell.

upwelling An upward movement of deep, nutrient-rich water along coasts to replace surface waters that winds move away from shore.

uracil (YUR-uh-sill) Nitrogen-containing base found in RNA molecules; can base-pair with adenine.

urinary system Of vertebrates, an organ system that regulates water and solute levels.

urine Fluid formed by filtration, reabsorption, and secretion in kidneys; consists of wastes and excess water and solutes.

uterus (YOU-tur-us) [L. *uterus*, womb] Chamber in which the developing embryo is contained and nurtured during pregnancy.

vagina Part of a female reproductive system that receives sperm, forms part of the birth canal, and channels menstrual flow to the exterior.

variable Of the factors characterizing or influencing an experimental group under study, the only one (ideally) that is *not* identical to those of a control group.

vascular bundle One of several to many strandlike arrangements of primary xylem and phloem embedded in the ground tissue of roots, stems, and leaves.

vascular cambium A lateral meristem that increases stem or root diameter of vascular plants showing secondary growth.

vascular cylinder Of plant roots, the arrangement of vascular tissues as a central cylinder.

vascular plant Plant having tissues that transport water and solutes through well-developed roots, stems, and leaves.

vein Of the circulatory system, any of the large-diameter vessels that lead back to the heart; of leaves, one of the vascular bundles that thread lacily through photosynthetic tissues.

vernalization Of flowering plants, stimulation of flowering by exposure to low temperatures.

vertebra, plural **vertebrae** Of vertebrate animals, one of a series of hard bones, arranged with intervertebral disks, into a backbone.

vertebrate Animal having a backbone of bony segments, the *vertebrae*.

vesicle (VESS-ih-kul) [L. *vesicula*, little bladder] Of cells, a small membranous sac that transports or stores substances in the cytoplasm.

villus (VIL-us), plural **villi** Any of several types of absorptive structures projecting from the free surface of an epithelium.

viroid An infectious nucleic acid that has no protein coat; a tiny rod or circle of single-stranded RNA.

virus A noncellular infectious agent, consisting of DNA or RNA and a protein coat; can replicate only after its genetic material enters a host cell and subverts that cell's metabolic machinery.

vision Precise light focusing onto a layer of photoreceptive cells that is dense enough to sample details concerning a given light stimulus, followed by image formation in the brain.

vitamin Any of more than a dozen organic substances that animals require in small amounts for normal cell metabolism but generally cannot synthesize for themselves.

water potential The sum of two opposing forces (osmosis and turgor pressure) that can cause the directional movement of water into or out of a walled cell.

watershed A region where all precipitation becomes funneled into a single stream or river.

wax A type of lipid with long-chain fatty acid tails; waxes help form protective, lubricating, or water-repellent coatings.

white blood cell Leukocyte; of vertebrates, any of the macrophages, eosinophils, neutrophils, and other cells which, together with their products, comprise the immune system.

white matter Of spinal cords, major nerve tracts so named because of the glistening myelin sheaths of their axons.

wild-type allele Of a population, the allele that occurs normally or with greatest frequency at a given gene locus.

wing Of birds, a forelimb of feathers, powerful muscles, and lightweight bones that functions in flight. Of insects, a structure that develops as a lateral fold of the exoskeleton and functions in flight.

X-linked gene Any gene on an X chromosome.

X-linked recessive inheritance Recessive condition in which the responsible, mutated gene occurs on the X chromosome.

Y-linked gene Any gene on a Y chromosome.

xylem (ZYE-lum) [Gk. *xylon*, wood] Of vascular plants, a tissue that transports water and solutes through the plant body.

yolk sac Of many vertebrates, an extraembryonic membrane that provides nourishment (from yolk) to the developing embryo; of humans, the sac does not include yolk but helps give rise to a digestive tube.

zooplankton A freshwater or marine community of floating or weakly swimming heterotrophs, mostly microscopic, such as rotifers and copepods.

zygote (ZYE-goat) The first diploid cell formed after fertilization (fusion of nuclei from a male and a female gamete).

CREDITS AND ACKNOWLEDGMENTS

Page 531 © Kevin Schafer

Chapter 31

31.1 David Macdonald / 31.2 Photographs (a) Lennart Nilsson from *Behold Man*, © 1974 Albert Bonniers Forlag and Little, Brown and Company, Boston; (b) Manfred Kage/Bruce Coleman Ltd.; (c) Ed Reschke/Peter Arnold Inc. / 31.3 (a) Art by Palay/Beaubois; (b) Focus on Sports; (inset) Manfred Kage/Bruce Coleman Ltd. / 31.4 Art by Palay/Beaubois / 31.5 Photographs Ed Reschke / 31.6 (left) Art by L. Calver / 31.7 Photographs Ed Reschke / 31.8 Lennart Nilsson from *Behold Man*, © 1974 Albert Bonniers Forlag and Little, Brown and Company, Boston / 31.9 Art by L. Calver / 31.10 Art by Palay/Beaubois / **Page 544** (a) Manfred Kage/Bruce Coleman Ltd.; (b–d) Ed Reschke

Chapter 32

32.1 Adrian Warren/Ardea, London / 32.2 Manfred Kage/Peter Arnold, Inc. / 32.3 (left) Art by Kevin Somerville; (right) art by L. Calver / 32.4, 32.6 Art by Leonard Morgan / 32.7 (b) A.L. Hodgkin, *Journal of Physiology*, vol. 131, 1956 / 32.8 (top) Art by Leonard Morgan; (bottom) art by Jeanne Schreiber / 32.9 (a) Carolina Biological Supply Company; art by Leonard Morgan / 32.10 Art by D. & V. Hennings; (c) J.E. Heuser and T.S. Reese / **Page 558** Painting by Sir Charles Bell, 1809, courtesy of Royal College of Surgeons, Edinburgh / 32.12 (a) Art by Robert Demarest; (b) from *Tissues and Organs: A Text-Atlas of Scanning Electron Microscopy*, by R.G. Kessel and R.H. Kardon. Copyright © 1979 by W.H. Freeman and Company. Reprinted with permission / 32.13 Art by K. Kasnot

Chapter 33

33.1 Comstock/Comstock Inc. / **Page 564** Art by D. & V. Hennings / 33.2 Photograph Francois Gohier/Photo Researchers; art by Raychel Ciemma / 33.3 Art by Palay/Beaubois / 33.4–33.5 Art by Kevin Somerville / 33.6 (b) Art by Kevin Somerville / 33.8 Art by Robert Demarest / 33.9 (a) Art by Kevin Somerville; (b) Manfred Kage/Peter Arnold, Inc. / 33.11 Art by Kevin Somerville / 33.12 C. Yokochi and J. Rohen, *Photographic Anatomy of the Human Body*, second edition, Igaku-Shoin Ltd., 1979 / **Page 573** Art by Palay/Beaubois / 33.13 Art by Joel Ito / 33.14 Art by Palay/Beaubois after Penfield and Rasmussen, *The Cerebral Cortex of Man*, copyright © 1950 Macmillan Publishing Company, Inc. Renewed 1978 by Theodore Rasmussen / 33.15 Art by Robert Demarest / 33.16 After H. Jasper, 1941

Chapter 34

34.1 Hugo van Lawick / 34.2–34.3 Art by Kevin Somerville / 34.4–34.7 Art by Robert Demarest / 34.8 (a) Mitchell Layton; (b) Syndication International (1986) Ltd. / 34.9 Photographs courtesy of Dr. William H. Daughaday, Washington University School of Medicine. From A.I. Mendelhoff and D.E. Smith, eds., *American Journal of Medicine*, 20:133 (1956) / 34.12 The Bettmann Archive / 34.13 Biophoto Associates /SPL/Photo Researchers / 34.14 Art by Leonard Morgan / **Page 592** Evan Cerasoli

Chapter 35

35.1 Merlin D. Tuttle, Bat Conservation International / 35.2 Eric A. Newman / 35.3 Art by Kevin Somerville / 35.4 Art by Palay/Beaubois after Penfield and Rasmussen, *The Cerebral Cortex of Man*, copyright © 1950 Macmillan Publishing Company, Inc. Renewed 1978 by Theodore Rasmussen / 35.5 From Hensel and Bowman, *Journal of Physiology*, 23:564–568, 1960 / 35.6 Art by Ron Ervin; photograph Ed Reschke / 35.7 Art by D. & V. Hennings / 35.8 Art by Robert Demarest; micrograph Omikron/SPL/Photo Researchers / 35.9 Art by Kevin Somerville / 35.10 Art by Robert Demarest / 35.11 (a), (b) Robert E. Preston, courtesy Joseph E. Hawkins, Kresge Hearing Research Institute, University of Michigan Medical School / 35.12 Photograph Edward W. Bower/ © 1991 TIB/West; art by Kevin Somerville / 35.13 (a) Keith Gillett/Tom Stack & Associates; (b–d) after M. Gardiner, *The Biology of Vertebrates*, McGraw-Hill, 1972 / 35.14 (a) E.R. Degginger / 35.15 G.A. Mazohkin-Porshnykov (1958). Reprinted with permission from *Insect Vision*, © 1969 Plenum Press / 35.16 Art by Robert Demarest / 35.17–35.18 Art by Kevin Somerville / 35.19 Micrograph Lennart Nilsson © Boehringer Ingelheim International GmbH / **Page 613** Photographs Gerry Ellis/The Wildlife Collection; art by Kevin Somerville / 35.20 Art by Robert Demarest / 35.21 Art by Palay/Beaubois after S. Kuffler and J. Nicholls, *From Neuron to Brain*, Sinauer, 1977 / **Page 616** Art by Robert Demarest

Chapter 36

36.1 Jeff Schultz/AlaskaStock Images / 36.2 Art by L. Calver / 36.3 (a) Robert & Linda Mitchell; (b) Jane Burton/Bruce Coleman Ltd. / 36.4 Chaumeton-Lanceau/Agence Nature / 36.5 Art by Robert Demarest / 36.6 Ed Reschke / 36.7 Michael Keller/FPG / 36.8 CNRI/SPL/Photo Researchers / 36.9 Linda Pitkin/Planet Earth Pictures / 36.10 Photograph Stephen Dalton/Photo Researchers; art by Raychel Ciemma / 36.11 D.A. Parry, *Journal of Experimental Biology*, 36:654, 1959 / 36.12 Art by D. & V. Hennings / 36.13 Art by Joel Ito; micrograph Ed Reschke / 36.14 Art by K. Kasnot / 36.15 National Osteoporosis Foundation / 36.16 Art by Ron Ervin / **Page 629** Photograph C. Yokochi and J. Rohen, *Photographic Anatomy of the Human Body*, second edition, Igaku-Shoin Ltd., 1979 / 36.17 (b), (c) Art by L. Calver / 36.18 (a) Ed Reschke; (b) D. Fawcett, *The Cell*, Philadelphia: W.B. Saunders Co., 1966 / 36.21 Art by R.M. Jensen / 36.22 Adapted from R. Eckert and D. Randall, *Animal Physiology: Mechanisms and Adaptations*, second edition, W.H. Freeman and Co., 1983 / 36.23 (a) Art by Kevin Somerville; (b) Ed Reschke / **Page 637** Photograph Michael Neveux / 36.25 N.H.P.A./A.N.T. Photo Library

Chapter 37

37.1 (a) D. Robert Franz/Planet Earth Pictures; (b) art by D. & V. Hennings adapted from *Mammalogy*, third edition, by Terry Vaughan, copyright © 1986 by Saunders College Publishing. Used by permission of the publisher / 37.2 (a) Kim Taylor/Bruce Coleman Ltd.; (b) Wardene Weisser/Ardea, London / 37.4 Art by Robert Demarest / 37.6 Art by Raychel Ciemma; (b) after A. Vander et al., *Human Physiology: Mechanisms of Body Function*, fifth edition, McGraw-Hill, 1990. Used by permission / 37.8 (a), (c) Lennart Nilsson © Boehringer Ingelheim International GmbH; (b) Biophoto Associates/SPL/Photo Researchers; art by Victor Royer / 37.9 Art by Robert Demarest / 37.10 Art by L. Calver / 37.11 (b) Steven Jones/FPG / **Page 650** Photograph CNRI/Phototake / 37.12 Photograph Ralph Pleasant/FPG / 37.13 Modified after A. Vander et al. *Human Physiology*, fourth edition, McGraw-Hill, 1985 / **Page 655** Photograph courtesy of David Steinberg

Chapter 38

38.1 (a) From A.D. Waller, *Physiology, The Servant of Medicine*, Hitchcock Lectures, University of London Press, 1910; (b) photograph courtesy of The New York Academy of Medicine Library / 38.3 (b) (below) After M. Labarbera and S. Vogel, *American Scientist*, 70:54–60, 1982 / **Page 662** Art by Palay/Beaubois / 38.4 (a) CNRI/SPL/Photo Researchers; (b) Lennart Nilsson from *Behold Man*, © 1974 by Albert Bonniers Forlag and Little, Brown and Company, Boston / 38.5 (left) Art by L. Calver and Victor Royer; (right) art by Victor Royer / 38.6 (a) Art by Leonard Morgan; (b) art by Kevin Somerville / 38.7 (a) Art by Joel Ito; (b) C. Yokochi and J. Rohen, *Photographic Anatomy of the Human Body*, second edition, Igaku-Shoin Ltd., 1979 / 38.11 Art by Robert Demarest based on A. Spence, *Basic Human Anatomy*, Benjamin-Cummings, 1982 / 38.14 Art by Raychel Ciemma / 38.15 After J. A. Gosling et al., *Atlas of Human Anatomy with Integrated Text*, copyright © 1985 by Gower Medical Publishers. **Page 671** (a) (above) Ed Reschke; (below) F. Sloop and W. Ober/Visuals Unlimited / 38.16 Photograph Lennart Nilsson © Boehringer Ingelheim International GmbH / 38.17 (a) After F. Ayala and J. Kiger, *Modern Genetics*, © 1980 Benjamin-Cummings; (b) Lester V. Bergman & Associates, Inc. / 38.18 After Gerard J. Tortora and Nicholas P. Anagnostakos, *Principles of Anatomy and Physiology*, sixth edition, copyright © 1990 by Biological Sciences Textbooks, Inc., A & P Textbooks, Inc. and Elia-Sparta, Inc. Reprinted by permission of Harper Collins Publishers / 38.19 Art by Kevin Somerville

Chapter 39

39.1 (a) The Granger Collection, New York; (b) Lennart Nilsson © Boehringer Ingelheim International GmbH / 39.2 Lennart Nilsson © Boehringer Ingelheim International GmbH / 39.5 Art by Palay/Beaubois / 39.6 Art by L. Calver and Victor Royer / 39.7 Art by Palay/Beaubois after S. Tonegawa, *Scientific Ameican*, October 1965 / **Page 689** Photographs Dr. Gilla Kaplan / 39.8 Art by Palay/Beaubois / 39.9 Art by Palay/Beaubois after B. Alberts et al., *Molecualr Biology of the Cell*, Garland Publishing Company, 1983 / **Page 693** (a) Art by L. Calver / **Page 694** (b), (c) Micrographs Z. Salahuddin, National Institutes of Health

Chapter 40

40.1 Galen Rowell/Peter Arnold, Inc. / **40.2** (b) Steve Lissau/Rainbow; (c) Peter Parks/Oxford Scientific Films / **40.4** Ed Reschke / **40.5** Art by D. & V. Hennings after C. P. Hickman et al., *Integrated Principles of Zoology*, sixth edition, St. Louis: C. V. Mosby Co., 1979 / **40.6** After C. P. Hickman et al., *Integrated Principles of Zoology*, sixth edition, St. Louis: C. V. Mosby Co., 1979 / **40.7** Art by Palay/ Beaubois adapted from H. Scharnke, *Z. vergl. Physiol.*, 25:548–583 (1938) in *Form and Function in Birds*, Vol. 4, A. King and J. McLelland, Eds., Academic Press, 1989; micrograph H.R. Duncker, Justus-Liebig University, Giessen, Germany / **40.8** Art by L. Calver / **40.9** Art by Kevin Somerville / **40.11** CNRI/SPL/Photo Researchers / **40.12** After A. Vander et al., *Human Physiology*, third edition, McGraw-Hill, 1980 / **40.13** Art by K. Kasnot / **40.14** From L.G. Mitchell, J.A. Mutchmor, and W.D. Dolphin, *Zoology*, © 1988 Benjamin-Cummings Publishing Company / **40.16** Art by Leonard Morgan / **Page 710** (a) Gerard D. McLane / **Page 711** (b) Lennart Nilsson from *Behold Man*, © 1974 by Albert Bonniers Forlag and Little, Brown and Company, Boston / **Page 713** Christian Zuber/Bruce Coleman Ltd. / **40.17** (b) Giorgio Gualco/Bruce Coleman Ltd.

Chapter 41

41.1 Claude Steelman/Tom Stack & Associates / **41.3** Art by Kevin Somerville / **41.4** Art by Robert Demarest / **41.8** Art by Joel Ito / **41.9** Thomas D. Mangelsen/Images of Nature / **41.10** (left) David Jennings/Image Works; (right) Evan Cerasoli / **41.11** Art by Kevin Somerville / **41.12** The Bettmann Archive / **41.13** Terry Vaughan / **41.14** Fred Bruemmer

Chapter 42

42.1 (a) Hans Pfletschinger; (b) Carolina Biological Supply Company; (c–e) John H. Gerard / **42.2** (a) Frieder Sauer/Bruce Coleman Ltd.; (b) Evan Cerasoli; (c) Fred McKinney/FPG; (d) Carolina Biological Supply Company; (e) Leonard Lee Rue III / **42.4** Art by Palay/Beaubois adapted from R.G. Ham and M.J. Veomett, *Mechanisms of Development*, St. Louis: C.V. Mosby Co., 1980 / **42.5** Photographs Carolina Biological Supply Company; sketch after M.B. Patten, *Early Embryology of the Chick*, fifth edition, McGraw-Hill, 1971 / **42.6** J.R. Whittaker / **42.8** Photographs Carolina Biological Supply Company / **42.9** (a), (b) Micrographs J.B. Morrill and N. Ruediger; (c), (d) Micrographs J.B. Morrill; art by Raychel Ciemma after V.E. Foe and B.M. Alberts, *Journal of Cell Science*, 61:32, © The Company of Biologists 1983 / **42.10** Micrographs F.R. Turner; art by Raychel Ciemma / **42.11** Sketches after B. Burnside, *Developmental Biology*, 26:416–441, 1971; micrograph K.W. Tosney / **42.12** (a–c) Adapted by permission of Macmillan Publishing Company from *Developmental Biology: Patterns, Problems, Principles* by John W. Saunders, Jr., Copyright © 1982 by John W. Saunders, Jr.; (d) S.R. Hilfer and J.W. Yang, *The Anatomical Record*, 197:423–433, 1980 / **42.13** (a) K.W. Tosney / **42.14** Art by Palay/Beaubois after Robert F. Weaver and Philip W. Hedrick, *Genetics*. Copyright © 1989 Wm. C. Brown Publishers, Dubuque, Iowa. All rights reserved. Reprinted by permission / **42.15** After J.W. Fristrom et al., in E.W. Hanly, ed., *Problems in Biology: RNA Development*, University of Utah Press / **42.16** Carolina Biological Supply Company / **42.17** Sketches after Willier, Weiss, and Hamburger, *Analysis of Development*, Philadelphia: W.B. Saunders Co., 1955; photograph Roger K. Burnard / **42.18** Art by Raychel Ciemma adapted from L.B. Arey, *Developmental Anatomy*, Philadelphia: W.B. Saunders Co., 1965 / **Page 755** Dennis Green/Bruce Coleman Ltd.

Chapter 43

43.1 Lennart Nilsson from *A Child Is Born*, © 1966, 1977 Dell Publishing Company, Inc. / **43.2** (left) Art by Ron Ervin; (right) art by L. Calver / **43.3** Art by L. Calver; (c) R.G. Kessel and R.H. Kardon, *Tissues and Organs: A Text-Atlas of Scanning Electron Microscopy*, W.H. Freeman and Co., copyright © 1979 / **43.4–43.5** Art by Ron Ervin / **43.7** (left) Art by Ron Ervin; (right) art by L. Calver / **43.8** (top) Art by Robert Demarest; photograph Lennart Nilsson from *A Child Is Born*, © 1966, 1977 Dell Publishing Company, Inc. / **43.10** Art by Robert Demarest / **43.11** Art by Robert Demarest; (left) micrograph from Lennart Nilsson, *A Child Is Born*, © 1966, 1977 Dell Publishing Company, Inc.; (right) from Lennart Nilsson, *Behold Man*, © 1974 by Albert Bonniers Forlag and Little, Brown and Co., Boston / **43.12** Art by Robert Demarest / **43.13** Art by L. Calver; (c) after A.S. Romer and T.S. Parsons, *The Vertebrate Body*, sixth edition, Saunders College Publishing, © CBS College Publishing / **43.14** Art by L. Calver after Bruce Carlson, *Patten's Foundations of Embryology*, fourth edition, McGraw-Hill, 1981 / **43.15** From Lennart Nilsson, *A Child Is Born*, © 1966, 1977 Dell Publishing Company, Inc. / **Page 772** Modified from Keith L. Moore, *The Developing Human: Clinically Oriented Embryology*, fourth edition, Philadelphia: W.B. Saunders Co., 1988 / **Page 773** James W. Hanson, M.D. / **43.16–43.17** From Lennart Nilsson, *A Child Is Born*, © 1966, 1977 Dell Publishing Company, Inc. / **43.18** Art by Robert Demarest / **Page 776** Mills-Peninsula Hospitals / **43.19** Art by Ron Ervin / **Page 781** (a) Cheun-mo To and C.C. Brinton / **Page 782** (b) Joel B. Baseman

INDEX